INSECT FLIGHT MUSCLE

INSECT FLIGHT MUSCLE

Proceedings of the Oxford Symposium.
3rd - 5th April, 1977.

Editor

Richard T. Tregear

Agricultural Research Council Unit of Muscular Mechanisms, Department of Zoology, South Parks Road, Oxford.

1977

NORTH-HOLLAND PUBLISHING COMPANY
AMSTERDAM · NEW YORK · OXFORD

ISBN 0-444-41612-9

Published by:
Elsevier/North-Holland Biomedical Press
335 Jan van Galenstraat, P.O. Box 211
Amsterdam, The Netherlands.

Sole distributors for the USA and Canada:
Elsevier/North-Holland Inc.
52 Vanderbilt Avenue
New York, N.Y. 10017

Library of Congress Cataloging in Publication Data

Main entry under title:

Insect flight muscle.

Bibliography: p.
Includes index.
1. Insects--Flight--Congresses. 2. Insects--Physiology--Congresses. 3. Muscle--Congresses.
I. Tregear, R. T.
QL496.7.I59 595.7'01'852 77-8986
ISBN 0-444-41612-9

Printed in the Netherlands

To J. Pringle and L. Cordofanus

Deus ex machina et machina ex deo

PREFACE

Here is what has been done with insect flight muscle to show how muscle works. The authors took part in the first symposium on the subject, held in Oxford whence all the work originally sprang. Their contributions are diverse, from how to hunt a water bug to the theoretical interpretation of its X-ray diffraction pattern, but behind all of them lies the belief that this is a good preparation which often gives answers when vertebrate muscle does not.

ACKNOWLEDGEMENTS

I thank the contributors for their prompt supply of manuscripts, and the editorial staff, Tony Allen, Maxine Clarke, Neela Connolly, Pat Messenger and Margaret Miles for their equally prompt handling of the same.

I also gratefully acknowledge the financial support of the Agricultural Research Council Underwood Fund, which made it possible for the meeting to take place.

CONTENTS

INTRODUCTION

Biophysical work on the flight muscle of the belostomid water bug *Lethocerus spp* began in 1963, when a means of handling the glycerol extracted tissue was devised. Since then a body of knowledge has been created which now allows us the same advantages in using insect flight muscle as are enjoyed by those who work on frog sartorius. We can tell from previous experience whether the preparation is performing sensibly in a number of ways before proceeding to make new observations; *Lethocerus* flight muscle has become a standard preparation.

The work which has been done can be roughly divided into four parts: on filament structure and content, on crossbridge conformation, on mechanical response and on biochemical kinetics. These are much the same divisions as might be applied to the vertebrate striated work but the emphasis is different. The beautiful regularity of the flight muscle's structure allows elegant elucidation of its geometry and particularly the geometry of its crossbridge arrangements. In this field the results far exceed those that have been obtained from vertebrates. On the other hand *Lethocerus* flight muscle cannot be reversibly extended by a large fraction of its rest length, so the classical tension-length and force-velocity measurements cannot be applied. However, there is another side to the coin; the self-same inextensibility provides a stable and structurally homogeneous glycerol-extracted preparation which allows ready measurement of small-amplitude kinetics. The interpretation of such measurements is at the moment a matter of hot debate. The biochemistry of insect muscle has been less investigated than that of rabbit psoas, primarily because insects are much smaller than rabbits, but it has produced one definitive result which may yet apply in the vertebrate case: insect flight muscle is dually calcium-activated. In addition it is, of course, stretch activated - and so is vertebrate muscle. Thus in the end apparent specialisations may prove to be extreme cases of the general rules.

The articles in the present volume fall naturally within the four headings above apart from three on how to catch and keep tropical water bugs. The sections overlap somewhat because authors consider more than one aspect of subject, but in all cases their main interest is reasonably closely defined. In the rest of the introduction I have attempted to outline the broad scope of their contributions.

FILAMENT STRUCTURE AND CONTENT

The great regularity of the filament array of insect flight muscle became apparent as soon as electron micrographs were made of the tissue and has intrigued microscopists ever since. The classical microscopy is summarised by

Ashhurst and Cullen in the first paper of the volume.

Further development has demanded a combination of protein chemistry with microscopy by means of antibody staining. In this way the locations of various 'minor' proteins have been determined, in particular α-actinin, paramyosin and 'connecting protein' which lies in the region between the thick filaments and the Z-line. Goll, Stromer, Robson, Luke and Hammond describe the search for, and purification of, α-actinin. They show how the difficulties inherent in biochemical separation of minor components from small quantities of tissue have been and can be overcome. Bullard, Bell and Luke demonstrate the location of paramyosin and connecting protein, as determined with antibodies. The separation of the two antibody components was only achieved after a painstaking re-investigation and illustrates the pitfalls as well as the potential of the method. This point is also brought home by the short paper from Trombitás, Tigyi-Sebes and Palloy.

Whatever its protein content, there have been persistent reports of structural connections between the thick filaments and the Z-line; equally, over the years there have been repeated denials of the existence of the so-called 'C-filaments'. The dialogue is taken up here by Ashhurst, who suggests that the controversy arises from the fact that "On stretching, the molecules of this material become more highly organised and orientated to form filaments", (Ashhurst, this volume). The existence and mechanical strength of this 'connecting material' - probably a better name than C-filament - is strongly documented by Trombitás and Tigyi-Sebes, who also show that its strength can actually elongate the thick filaments themselves.

The great regularity of the insect flight muscle filament array has led towards purely geometrical studies. The Z-line is an exceedingly ordered structure. The general though not universal consensus of opinion is that its regularity arises from the interdigitation of the hollow-hexagonal arrays of thin filaments from neighbouring sarcomeres. The evidence for this concept is reviewed by Ashhurst, while the short paper by Sainsbury and Hulmes provides an additional plank for Ashhurst's thesis.

The thick filament assembly requires a much more complex geometrical study, involving both electron microscopic and X-ray diffraction evidence. Squire comprehensively reviews the evidence available and concludes that continuous layers of myosin molecules probably pack to form a six-stranded helix. This study leads naturally into the subject of the second section.

CROSSBRIDGE CONFORMATION

Crossbridges are the electron micrographic appearance of attachment between myosin heads and actin molecules in the filament array of muscle. As soon as good pictures were taken of insect flight muscle in rigor it was obvious that this tissue gave a clearer representation of their position and orientation than could be obtained from vertebrate muscle. Reedy and Garrett consider the geometrical situation in great detail, using many different orientations of section and view to arrive at their conclusions. They show that the myosin heads are orientated on the actin monomers at a similar angle to that shown by vertebrate subfragment-1 on an isolated actin filament, and that they occur preferentially in two groups related to the orientation of the actin helix relative to the adjacent myosin filament. The extreme regularity of the images seen is explained by the exact, staggered, axial register of the actin filaments within the array and the apparently total relation of the crossbridges to their actin attachment rather than to their myosin origin.

These points are taken up by Barrington-Leigh, Goody, Hofman, Holmes, Mannherz, Rosenbaum and Tregear who show in outline how the X-ray diffraction pattern of the rigor flight muscle arises from the particular geometry described in the previous paper, and how it is simplified when subfragment-1 monomers are attached to the vacant actins within the lattice. They further consider the effects of AMP.PNP and ATP on the diffraction pattern in terms of an increased geometrical dependence of the crossbridge conformation on its origin in the myosin filament and a concomitant decrease in the dependence on the actin attachment geometry.

If such a process occurs it should be possible to find corroborative evidence from electron microscopy and indeed such evidence has been presented. However, Beinbrech's detailed study of thin sections from *Lethocerus* flight muscle in the presence of ADP, AMP.PNP and pyrophosphate shows no evidence of thick filament modulation, but rather evidence of a rotation of the myosin head about its actin attachment as in the now classical actomyosin enzymatic cycle.

Neutron diffraction is a recently developed technique which, in principle, might allow the direct differentiation of order from different parts of the structure. Rodger, Worcester and Miller have applied the method to *Lethocerus* flight muscle and obtained comparable results to those from X-rays; so far, unfortunately, the resolution available has not exposed the anticipated structural differentiation.

MECHANICAL KINETICS

Asynchronous insect flight muscle, of which *Lethocerus* is an example, was first recognised by the asynchrony of its nerve impulses to its muscular twitches, and in the 1950's its mechanism was determined as a form of stretch activation. Asynchronous flight muscles also have a high passive elasticity even when relaxed, and this has commonly been supposed to be related to stretch activation. In his article Pringle sets out the arguments for various structural causes of the passive elasticity itself. He concludes that slowly cycling crossbridges and the aforementioned connecting material are both responsible, the former in high and the latter in low frequency responses. White, Donaldson, Pearce and Wilson provide fresh evidence that even in activated muscle there is indeed a mechanical structure in parallel with the cycling crossbridges.

The manner in which passive elasticity causes stretch activation remains unknown, although Pringle points out that if the connecting material is responsible the sensor must lie within the thick filaments themselves. Whatever its cause stretch activation is a useful tool, since it may be employed in much the same way as a temperature-jump experiment in solution chemistry. Both White *et al.* and Herzig present experiments performed in this manner. They obtained different results: White *et al.* find that the muscle's stiffness increases linearly as the tension increases, while Herzig finds that with larger extensions stiffness reaches a plateau value while tension is still rising. It is curious that both authors invoke the Huxley-Simmons bridge rotation as a probable explanation of their results.

Stretch activation is not confined to insect flight muscle. Over the years there have been many reports of 'delayed tension' changes in various muscle preparations after a quick stretch. Steiger presents the full experimental evidence concerning the phenomenon in mammalian cardiac muscle and suggests that it may aid the operation of the heart by providing " ... an optimal tuning between the biochemical and mechanical cycles of the crossbridges". (Steiger, this volume). He contrasts the large effects in mammalian cardiac and insect flight muscle to the small delayed tension response of rabbit psoas and relates this to their passive elasticities.

Stretch activation is a relatively slow mechanical response. It is preceded by at least two phases of faster response, presumably analogous to the phenomena observed in tetanically-activated intact frog muscle by Huxley and Simmons. Steiger provides the experimental data on these phases in cardiac muscle and compares them with those already published from insect flight and rabbit psoas muscle. Such data have been extremely controversial and Steiger is cautious in his interpretation, but it is evident that sufficient experimental evidence is now becoming available for general conclusions to be drawn.

Mechanical kinetics work requires models for its interpretation. Such models have frequently been highly individualistic and difficult to correlate with one another. Abbott sets up a model which is closely constrained by both the mechanical and biochemical kinetic evidence. He deduces that "The same states may have indeed been identified in both types of experiments". (Abbott, this volume). It is certainly to be hoped that this short essay heralds a new generation of generalised, fully constrained modelling.

BIOCHEMICAL KINETICS

Arthropod muscle was the first and still remains the best source of myosin for the demonstration of dual calcium activation; both troponin-linked and myosin-linked systems exist and their combined inhibitory effect on the acto-myosin ATPase is the product of each in isolation. Lehman sets out the evidence for this from insect flight and lobster muscle. He further considers whether myosin linkage has been missed in some arthropod muscles, and perhaps even in vertebrate striated muscle, because of the low ionic strength experimentally employed. Winkelman and Bullard show that myosin-linked calcium control in locust flight muscle is probably dependent upon the phosphorylation of a particular light chain which can be accomplished by an endogenous kinase. Beinbrech has also noted the phosphorylation of the same light chain.

Because of both the high concentration of myosin enzymatic sites in the myofibrillar matrix and their high affinity for nucleotide it is possible to measure nucleotide binding directly in glycerol-extracted muscle fibres. Marston has made such measurements on *Lethocerus* flight muscle. He concludes that the myosin-product complex dominates in relaxation, as it does in vertebrate muscle. He has also measured the direct binding of the products and of the unhydrolysable nucleotide, AMP.PNP, and shows that these bind to the same site as ATP but without causing actomyosin detachment.

Such a ternary complex between actin, myosin and an unhydrolysable ATP analogue is a natural choice for attempted simulation of a 'frozen' state within the contractile cycle; the structural evidence for this has been reviewed in the section on crossbridge conformation. Kuhn states the mechanical case for the simulation. He has further tested the concept by measuring the effect of extension on the binding AMP.PNP. He finds that stretch increases AMP.PNP binding, as theoretically predicted; the hypothesis that the ternary complex represents a frozen contractile state has passed a further experimental test.

Any enzymatic cycle is in principle reversible and this has been demonstrated for the myosin ATPase cycle. Ulbrich and Rüegg have used insect flight muscle to show that the incorporation of phosphate into ATP is closely related to stretch activation, and speculate that both may " ... reflect the formation of a high-energy intermediate". (Ulbrich and Rüegg, this volume).

CAPTURE AND CULTURE

So much work has now been done on *Lethocerus spp.* that there is great advantage in continuing its use. The difficulty is its availability. There are temperate species of water bug but they are small and therefore less suitable for much work. The large species are all tropical and only effectively available when they fly. They can then be caught at light, packaged and sent direct to the research laboratory, where they can be maintained in stable condition for several months before use. This does not sound a difficult process, but in practise it has often been very frustrating.

Pringle considers the alternative sources of asynchronous flight muscle, some of which are certainly preferable for large-scale biochemistry, and then annotates the way in which a routine source of *Lethocerus cordofanus* has been obtained from Africa. Reedy describes his experience in encouraging collection from Central and North America and his hopes for future success.

Finally Cullen describes his attempts to breed water bugs in the laboratory leading to a typical experimental conundrum; " ... those species with the most suitable muscles have not bred in the laboratory, while those species which do breed do not contain normal mature flight muscles". (Cullen, this volume). Thus biophysical work on *Lethocerus* flight muscle is still dependent on a continued tropical supply.

Richard T. Tregear
MAY 1977
OXFORD.

FILAMENT STRUCTURE AND CONTENT

THE STRUCTURE OF FIBRILLAR FLIGHT MUSCLE

Doreen E. Ashhurst[1]
M.J. Cullen[2]

1. *Department of Anatomy, Medical School, University of Birmingham, Birmingham, B15 2TJ, England.*

2. *Muscular Dystrophy Group Research Laboratories, Newcastle General Hospital, Westgate Road, Newcastle-upon-Tyne NE4 6BE, England.*

The papers in this volume are concerned primarily with the peculiar properties of the myofibrils of fibrillar flight muscle, in particular their biochemical and biophysical properties. These peculiarities can be correlated with structural features, which are again found only in this type of muscle. This brief paper attempts to illustrate those features of fibrillar muscle which are not included in the other papers in this volume. No attempt is made to review the literature on the fine structure of fibrillar muscle; descriptions of this muscle have been published by Ashhurst (1967), Cullen (1971, 1974), Elder (1975), Shafiq (1964) and Smith (1966).

Fibrillar muscle fibres are unbranched and are in the order of 50 to 100 μm or more in diameter. The fibres are arranged in blocks separated by tracheae. They are multinucleate; often the nuclei are situated peripherally (Fig. 3) immediately under the sarcolemma.

The myofibrils are very characteristic of this type of muscle. They are cylindrical (Fig.2) and remain discrete throughout the length of the muscle fibre. They vary between 1 and 5 μm in diameter. The sarcomeres are between 1.7 and 2.5 μm long and the A-band occupies about 90% of their length, leaving only very narrow I-bands (Fig.1). The Z-lines are between 90 and 140 nm wide. The filament lattice in the A-band differs from that in other striated muscles (inset, Fig.2); each A-filament is surrounded by 6 I-filaments and the arrangement is such that the ratio of A:I filaments is 1:3. One striking feature of the lattice is its precise regularity; this regularity is much greater than that seen in other muscles, including vertebrate ones. This regularity is also reflected in the moiré patterns seen in longitudinal sections of sarcomeres (Fig. 1). It is rare to get a section which is cut exactly along one lattice plane throughout a sarcomere, most cut through several lattice layers. When such a thin section is cut through a very regular lattice moiré patterns are generated. This phenomenon is not seen in other muscles because their lattices are not sufficiently regular

The membrane systems, especially the sarcoplasmic reticulum, are sparse in all fibrillar muscles when compared with sychronous muscles. The T-tubules penetrate the fibres and ramify among the myofibrils, usually at either the level of the Z-line or M-line (Figs. 4 & 5). They associate with small cisternae of the sarcoplasmic reticulum (dense bodies) to form dyads. In the *Lethocerus* muscle illustrated here the dyads occur at the M-line and a further series of small vesicles of sarcoplasmic reticulum encircle the myofibrils at the level of the Z-line (Fig.5; Ashhurst, 1967).

The space between the myofibrils is filled with large mitochondria which appear dense due to the presence of many closely packed cisternae (Figs. 1 & 2). They occupy about 30% of the volume of the fibre (Tribe and Ashhurst, 1972), which reflects the energy requirements of this muscle. In many species, such as *Calliphora* and *Leptopterna*, masses of glycogen granules occur around the myofibrils (Fig.3) while in *Lethocerus* lipid droplets occur (Figs. 1 & 2). The nature of the fuel stored reflects the feeding habits of the insect; phytophagous species store glycogen, sarcophagous species store lipid (Cullen 1971, 1974).

The structure of the fibrillar muscle fibres of *Lethocerus* spp. is summarized in Figure 6.

REFERENCES

1 Ashhurst, D.E. *J. Cell Sci.* 2 (1967) 435.

2 Cullen M.J. *D. Phil. thesis, University of Oxford* (1971).

3 Cullen M.J. *J. Ent. (A)* 49 (1974) 17.

4 Elder H.Y. *In : Insect Muscle* Ed., P.N.R. Usherwood, Academic Press, London & New York.

5 Shafiq S.A. *Q.J. microsc. Sci* 105 (1964) 1.

6 Smith D.S. *Prog. Biophys. molec. Biol.* 16 (1966) 107.

7 Tribe M.A. & Ashhurst D.E. *J. Cell Sci.* 10 (1972) 443.

Fig. 1. Longitudinal section of part of a muscle fibre of *L. maximus*. The sarcomeres of the myofibrils (MF) are delimited by the Z-lines (Z) and the A-band (A) can be seen to fill almost their entire length. An H-zone is present in the centre of each sarcomere. Moire patterns can be seen in several sarcomeres. Mitochondria (M) and lipid droplets (L) are present between the myofibrils. Sarcoplasmic reticulum (SR) and dyads (D) are also present.

Fig. 2. Transverse section of a muscle fibre of *L. maximus*. The circular, discrete myofibrils (MF) are most obvious. The lattice pattern can be seen more clearly in the inset; note its precise regularity. The myofibrils are separated by mitochondria (M) and lipid droplets (L). Dyads (D) can also be seen.

Fig. 3. Part of the peripheral region of a muscle fibre of *Leptoterna dolabrata* showing two nuclei (N) immediately beneath the sarcolemma (S). A mass of glycogen (G) fills the space between the nuclei and the myofibrils (MF).

Fig. 4. Longitudinal section of a myofibril of *Lethocerus* sp. which has been cut along its periphery so that the vesicles of sarcoplasmic reticulum (ZSR) surrounding the Z-line are seen. T-tubules (T) are also present and a dense body (DB) of the sarcoplasmic reticulum forms a dyad with one T-tubule.

Fig. 5. Longitudinal section of a myofibril of *L. maximus* cut peripherally. Sarcoplasmic reticulum (ZSR) can be seen surrounding one Z-line, while at the other (Z), the vesicles are present at the edges. A dyad (D) is present opposite the centre of one sarcomere, while a T-tubule (T) only occurs opposite the other two sarcomeres.

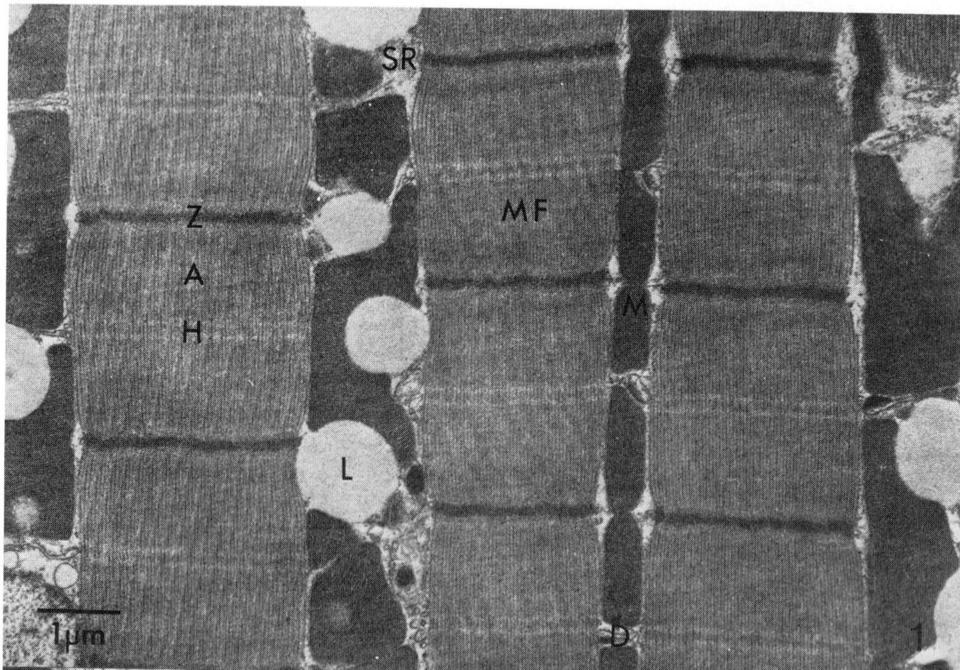
SR
Z
A
H
MF
M
L
D
1µm
1

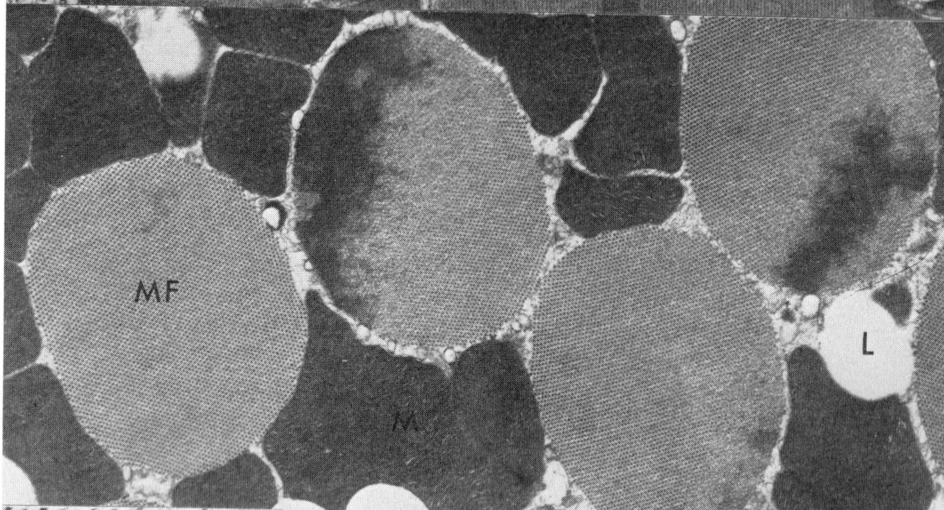
MF
M
L
D
100nm
1µm
2

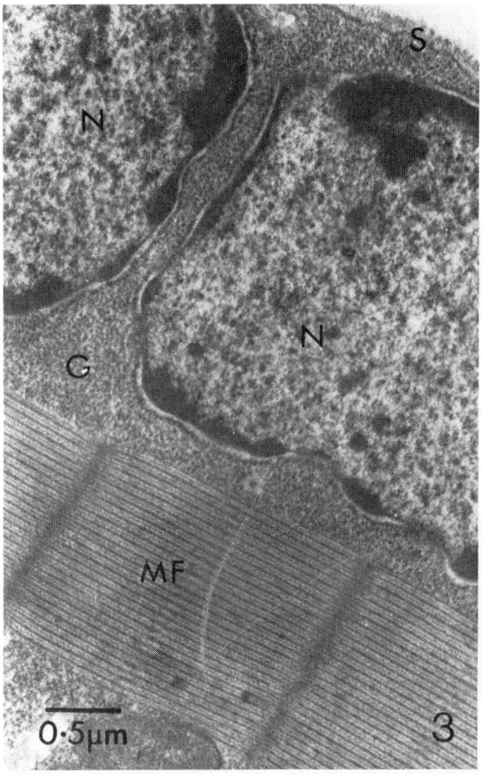
S
N
N
G
MF
0·5μm
3

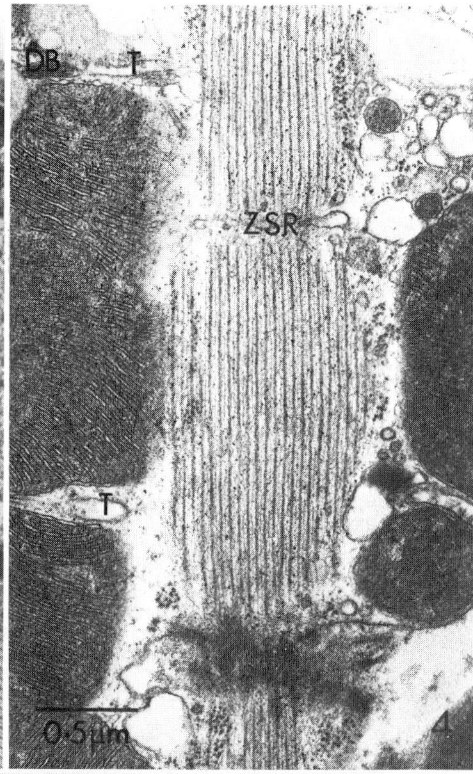
DB
T
ZSR
T
0·5μm
4

D
T
T
ZSR
Z
0·5μm
5

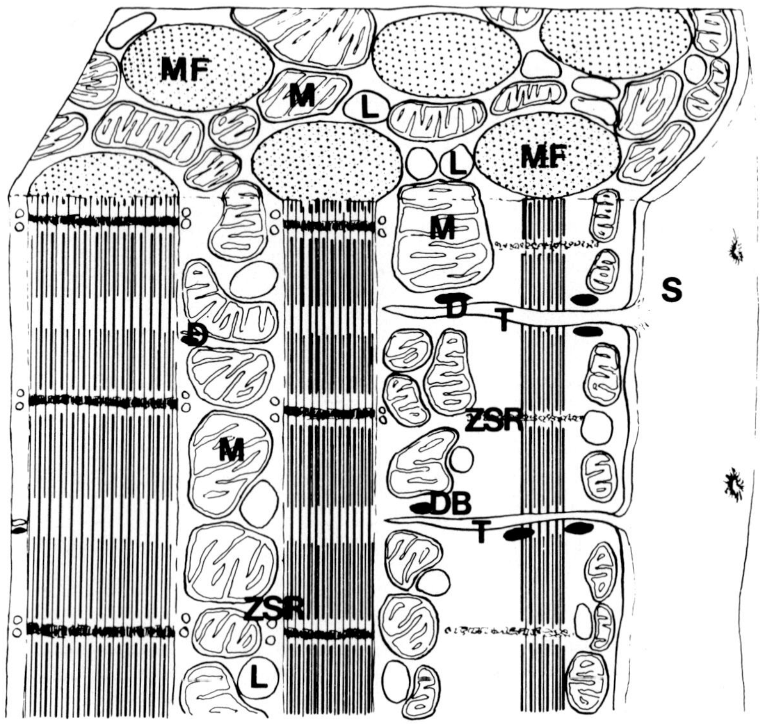

Fig. 6. A diagrammatic representation of the structure of the fibrillar flight muscle fibres of *Lethocerus* spp.;

MF = myofibril
D = dyad
DB = dense body of dyad
T = T-tubule
ZSR = vesicles of sarcoplasmic reticulum at Z-line
S = sarcolemma
M = mitochondrion
L = lipid droplet

EXTRACTION, PURIFICATION, AND LOCALIZATION OF α-ACTININ FROM ASYNCHRONOUS INSECT FLIGHT MUSCLE

Darrel E. Goll, Marvin H. Stromer*, Richard M. Robson,* Barbara M. Luke** and Kathleen S. Hammond+

Nutrition and Food Science, University of Arizona, Tucson, Arizona 85721

**Present address*: *Muscle Biology Group, Iowa State University, Ames, Iowa 50011*

***Present address*: *A.R.C. Unit, South Parks Road, University of Oxford, Oxford, England OX1 3PS.*

+ *Present address*: *Department of Pathology, Yale University, New Haven, Connecticut 06510.*

INTRODUCTION

Insect fibrillar flight muscle possesses several unique characteristics that make it especially useful for the study of striated muscle contraction (Pringle, 1967a; 1967b; 1975). First, the highly organized and very regular three-dimensional structure of insect fibrillar flight muscle has substantially aided electron microscope and X-ray diffraction studies of structural changes (Armitage, Miller, Rodger and Tregear, 1973; Reedy, 1968; Tregear and Miller, 1969) and has greatly facilitated the study of muscle mechanics during activation and contraction. For example, the differing orientation of myosin cross-bridges in rigor and relaxed muscle was first observed in electron micrographs of insect fibrillar flight muscle (Reedy, Holmes and Tregear, 1965). Electron micrographs of insect fibrillar flight muscle have also provided important information about the arrangement of myosin cross-bridges on thick filaments (Reedy, 1968). More recently, insect fibrillar muscle has been used to investigate the effects of ATP analogues on orientation of myosin cross-bridges (Goody, Barrington-Leigh, Mannherz, Tregear, and Rosenbaum, 1976; Marston, Rodger and Tregear, 1976) and to study Z-disk structure (Bullard and Sainsbury, 1977) because Z-disks of insect fibrillar muscle seem unusually resistant to extraction. Second, the phenomenon of "stretch activation" (Jewell and Rüegg, 1966), although it seems to exist in other muscles as well (Rüegg, Steiger and Schadler, 1970; Steiger, 1971), is unusually pronounced in insect fibrillar flight muscle; it may be hoped that the molecular features responsible for this phenomenon can be deduced from study of the biochemical composition and ultrastructural properties of insect fibrillar flight muscle.

Although the several ultrastructural (Ashhurst, 1967; Reedy, 1968; Reedy *et al.*, 1965; Trombitas and Tigyi-Sebes, 1974), X-ray diffraction (Armitage *et al.*, 1973; Goody *et al.*, 1976; Marston *et al.*, 1976; Tregear and Miller, 1969), and mechanical (Pybus and Tregear, 1973; White, 1970; White and Thorson, 1972; 1973) studies of insect fibrillar flight muscle have produced valuable insights into the mechanism of muscle contraction, the very small amounts of fibrillar muscle usually available have made biochemical studies of contractile proteins in insect fibrillar muscle very difficult. Improvements in handling or breeding large water bugs or beetles having asynchronous flight muscle obviously would facilitate biochemical studies of fibrillar muscle by increasing the number of specimens available. Because only one gram or less of fibrillar muscle can be obtained from even the largest water beetles, biochemical studies of asynchronous flight muscle would continue to be hampered by the time required to dissect this muscle from large numbers of beetles, even if numbers of insects were unlimited. For this reason, it is very important for future biochemical studies of the contractile proteins in insect fibrillar flight muscle to devise procedures that will maximize yield of the different contractile proteins and that will also permit preparation of several different contractile proteins from a single muscle sample rather than use a different muscle sample for preparation of each contractile protein as is normally done with vertebrate skeletal muscle. Availability of such procedures is especially critical when attempting to study contractile proteins such as α-actinin, which make up less than 3% of total contractile protein in muscle cells. Our initial efforts in determining whether insect fibrillar flight muscle possesses α-actinin that is qualitatively similar to α-actinin from mammalian skeletal muscle were directed toward development of a procedure that would permit preparation of as many of the other contractile proteins as possible from the same muscle sample used to prepare α-actinin.

CONTRACTILE PROTEIN PREPARATION FROM SMALL MUSCLE SAMPLES

The procedure that was ultimately developed for extraction of α-actinin from insect fibrillar flight muscle permitted extraction of myosin, paramyosin, tropomyosin, and troponin before extraction of α-actinin. It was also possible to extract M-lines selectively before any other contractile protein was extracted, but no attempt was made to purify M-protein from insect fibrillar flight muscle. The residue remaining after α-actinin extraction contained much of the actin present in the original muscle sample, but again, no attempt was made to actually prepare actin from this residue. Hence, the procedure permitted extraction of five of the eight or nine known contractile proteins, and it may be applicable for preparation of as many as seven or eight different contractile proteins from a single muscle sample.

The first step in the procedure involves preparation of myofibrils. Myofibrils were originally made as described by Bullard *et al.*(1973b), but we have subsequently used a procedure developed in our laboratory for large scale preparation of myofibrils from mammalian skeletal muscle (Table 1). Both procedures were satisfactory and no distinction will be made in this paper between myofibrils made by either of these two procedures. Myofibrils prepared by these procedures had all the contractile proteins that have been described for insect fibrillar flight muscle (Figure 1). Paramyosin (110,000 daltons) and α-actinin (100,000 daltons) are separated in the 7.5% polyacrylamide gels used for the experiment in Figure 1. To establish identity of the lower of these two bands as α-actinin different amounts of purified chicken gizzard α-actinin were added to the myofibrils before applying them to the polyacrylamide gels. In all instances density of the lower of the two bands was increased by addition of purified α-actinin (Figure 1). The experiment also demonstrates that the subunit molecular weight of α-actinin from insect fibrillar flight muscle is very similar to that of α-actinin from chicken gizzard smooth muscle; we have previously shown that the subunit molecular weights of rabbit skeletal muscle and insect fibrillar flight muscle α-actinins is very similar (Hammond and Goll, 1975). The nature of the protein band at 85,000 to 90,000 daltons in the SDS-polyacrylamide gels of insect fibrillar flight muscle is unknown, although Saide and Ullrich (1974) have indicated that Z-disks prepared by 0.43% lactic acid extraction of bee fibrillar flight muscle contain a protein with a subunit molecular weight near 87,000. The SDS-polyacrylamide gels shown in Figure 1 also contain two protein bands just above the actin band; these two bands migrate at rates corresponding to polypeptide molecular weights of approximately 57,000 and 54,000. It seems likely that the upper of these two protein bands originates from troponin-T, which has a polypeptide molecular weight of approximately 57,000 in insect fibrillar flight muscle, and that the lower of these two bands is from the 55,000 dalton protein that has been observed in cardiac and smooth muscle fibrils (Huiatt, Robson, Arakawa, Goll and Stromer, unpublished results). No tests were done to establish the identity of these two protein bands conclusively, however. The 7.5% polyacrylamide gels used for the experiment shown in Figure 1 are not as useful for separation and identification of proteins having subunit molecular weights below 40,000 as they are for separation of polypeptides larger than 40,000 daltons.

TABLE 1 : PREPARATION OF PURIFIED MYOFIBRILS

Cround or minced muscle

1) Suspend in 10 volumes (v/w) 100 mM KCl, 20 mM K phosphate, pH 6.8, 2 mM $MgCl_2$, 2 mM EGTA, 1 mM NaN_3 (standard salt solution) by homogenizing for 10 sec in Waring Blendor.
2) Centrifuge at 1000 x g for 10 min.

II. Sediment

1) Suspend in 6 volumes (v/w) of standard salt solution by homogenizing for 10 sec in Waring Blendor.
2) Centrifuge at 1000 x g for 10 min.

III. Sediment

1) Suspend in 8 volumes (v/w) of standard salt solution by homogenising for 10 sec in Waring Blendor.
2) Pass suspension through household nylon net strainer.
3) Centrifuge at 1000 x g for 10 min.

IV. Sediment

1) Suspend in 8 volumes (v/w) of standard salt solution by homogenizing for 10 sec in Waring Blendor.
2) Pass suspension through household nylon net strainer.
3) Centrifuge at 1000 x g for 10 min.

V. Sediment

1) Suspend in 6 volumes (v/w) of standard salt solution plus 1% (v/w) Triton X-100 by homogenizing for 10 sec in Waring Blendor.
2) Centrifuge at 1500 x g for 10 min.

VI. Sediment

1) Suspend in 6 volumes (v/w) of standard salt solution plus 1% (v/w) Triton X-100 by homogenizing for 10 sec in Waring Blendor.
2) Centrifuge at 1500 x g for 10 mon.

VII. Sediment

1) Suspend in 8 volumes (v/w) of standard salt solution by stirring vigorously with polyethylene stirring rod.
2) Centrifuge at 1500 x g for 10 min.

VIII to XI. Sediment

1) Repeat Step VII four times, but suspending in 8 volumes (v/w) of 100 mM KCl instead of standard salt solution; centrifuge at 1500 x g after each suspension.

XII. Sediment

1) Suspend in 8 volumes (v/w) of 100 mM KCl by homogenizing for 3 sec. in Waring Blendor.
2) Centrifuge at 1500 x g for 10 min.

XIII. Sediment

1) Suspend in 8 volumes (v/w) of 100 mM KCl by homogenizing for 3 sec in Waring Blendor.
2) Centrifuge at 1500 x g for 10 min.

XIV. Sediment

1) Suspend in 2 volumes (v/w) of 100 mM KCl by homogenizing for 3 sec in Waring Blendor.

   Purified Myofibrils free of membranes.

   N.B. The supernatants were discarded throughout.

Although protein bands that likely correspond to tropomyosin (about 36,000 daltons), the two light chains of myosin (30,000 and 17,000 daltons), troponin-I (about 27,000 daltons), and troponin-C (19,000 daltons; possibly overlapped with the 17,000 dalton light chain of myosin in Figure 1) can be detected in the polyacrylamide gels shown in Figure 1, the identification of these bands is very tenuous. It is interesting to note that no protein bands clearly corresponding to the 170,000-dalton M-protein and 140,000 dalton C-protein bands seen in SDS-polyacrylamide gels of mammalian skeletal muscle (See for example Goll, Robson and Stromer, 1977) can be detected in the SDS-polyacrylamide gels of insect fibrillar flight muscle (Figure 1), even though insect fibrillar flight muscle has distinct M-lines (Figure 2).

Effects of the extraction procedure developed in this study were monitored by using both electron microscopy and SDS-polyacrylamide gel electrophoresis. Purified myofibrils made with either of the two procedures used in this study had all the structural features that have already been described for insect fibrillar flight muscle (Figure 2), although the un-orientated nature of myofibril preparations makes it extremely difficult to obtain the precisely oriented sections necessary for careful structural analysis. Most myofibrils were relaxed with sarcomere lengths of 2.5 to 2.6 μm (Figure 2), although stretched sarcomeres with lengths up to 3.0 μm and contracted sarcomeres with lengths down to 1.9 μm were also observed occasionally (Figure 2). As described earlier for insect fibrillar flight muscle, Z-disks are wider (90 to 120 nm) and I-bands are narrower (0.1 to 0.3 μm) than in mammalian skeletal muscle. Although these measurements are all made on myofibril suspensions and are therefore not necessarily related to actual dimensions *in vivo* of insect fibrillar flight muscle myofibrils, they are similar to the 2.5 μm sarcomere lengths (Ashhurst,1976b) and 100 to 140 nm Z-disk widths (Ashhurst,1976a) that have been reported. The narrow I-band in sarcomeres 2.5 to 2.6 μm long suggests that thick filaments in insect fibrillar flight muscle are longer than the 1.5 to 1.6 μm lengths observed for thick filaments from vertebrate skeletal muscle. Long thick filaments in insect fibrillar flight muscle would be consistent with the reported presence of paramyosin in this muscle (Bullard *et al.*, 1973a; 1973b) because paramyosin is associated with increased length of myosin filaments (Ikemoto and Kawaguti, 1967).

After purified myofibrils were available, the next step in the procedure in this study for preparation of several different contractile proteins from a single muscle sample was to extract the purified myofibrils with a 0.25 M sucrose solution at pH 8.20 (Table 2). We have found that this solution causes a remarkably specific extraction of M-lines from mammalian skeletal muscle (Stromer, Goll, Robson and Richardson, in preparation). As shown in Figure 3,

TABLE 2. PREPARATION OF SEVERAL DIFFERENT MYOFIBRILLAR PROTEINS FROM ONE MUSCLE SAMPLE.

I. Prepare Myofibrils

1) Wash twice with 0.25 M sucrose to remove KCl
2) Extract overnight with 2 vol (v/w) 0.25 M sucrose, 50 mM Tris·acetate, 1 mM EDTA, 1 mM NaN_3, pH 8.20.
3) Centrifuge at 2000 x g for 20 min.

Supernatant (Contains M-protein)

II. Sediment - M-line-Free Myofibrils

1) Extract successively for 2 hr, 2 hr, and 16 hr with 3 vol (v/w) 0.1 M K pyrophosphate, 10 mM $MgCl_2$, pH 7.2.
2) Centrifuge at 2000 x g for 30 min after each extraction.

Combined Supernatants (contain myosin, paramyosin, and C-protein)

III. Sediment - IZI Brushes

1) Wash twice with 4 vol (v/w) each time 100 mM KCl, 5 mM EDTA, pH 7.2.
2) Wash four times with 3 vol (v/w) each time deionized water, centrifuging at 2000 x g for 30 min after each wash.

Combined Water Washes (Extract A)

IV. Sediment

1) Extract for 16 hr at $2^{\circ}C$ with 1 vol (v/w) 5 mM MCE, 1 mM Tris·acetate, pH 8.5.
2) Centrifuge at 14,000 x g for 10 min.

Supernatant (Extract B)

V. Sediment

1) Suspend in 1 vol (v/w) 5 mM MCE, 1 mM Tris·acetate, pH 8.5.
2) Centrifuge at 14,000 x g for 10 min.

Supernatant (Extract C)

VI. Sediment - 5-6 nm filaments.

this solution also removes M-lines from insect fibrillar flight muscle myofibrils without causing other ultrastructurally noticeable changes. In one or two instances, an unusual gap or split occurred in the center of the A-band at the level of the former M-line in M-line extracted myofibrils (Figure 3b). The cause of this split is unclear, but thick filaments seemed completely severed in this region. The uncertain orientation of myofibrils makes further analysis of this split difficult. Sucrose extraction as described in Table 2 removed M-lines in approximately 90% of the sarcomeres in insect fibrillar flight muscle. M-line extraction seems highly dependent on pH, and the 0.25 M sucrose seems to protect the remainder of the myofibrillar structure from swelling or other damage that usually accompanies low ionic strength extraction of myofibril suspensions.

In the third step of the extraction procedure developed in this study M-line-free myofibrils are extracted three times successively with a 0.1 M potassium pyrophosphate, 0.01 M $MgCl_2$, pH 7.2 solution (Table 2). These three extractions remove almost all myosin and paramyosin from insect fibrillar muscle myofibrils and leave Z-disks with sheaves of thin filaments attached to either side (Figure 4). We have called these structures 1Z1 brushes (Table 2). Z-disk structure as evaluated by electron microscopy seems unaffected by myosin and paramyosin extraction (Figure 4); Z-disks in 1Z1 brushes are 94 ± 4 nm wide (mean plus or minus standard error of measurements on six different preparations), the same width as Z-disks in the original unextracted myofibrils. It is also remarkable that, in many instances, Z-disks and their attached thin filaments remained in almost the same orientation relative to neighboring Z-disks as they were in the original myofibrils before myosin and paramyosin extraction (Figure 4). Thus, it was not unusual to find 1Z1 brushes aligned in rows as though myosin and paramyosin had been plucked from the original myofibril without disturbing its continuity (Figure 4). Z-disk to Z-disk distances in these rows of 1Z1 brushes were 2.44 ± 0.04 µm (mean plus or minus standard error of measurements on seven different preparations); this distance is approximately the same as the sarcomere lengths of 2.5 to 2.6 µm measured on the myofibrils before pyrophosphate extraction and supports the suggestion that the pyrophosphate extraction used in this study removes thick filaments with only minimal disturbance of the remainder of the myofibril. It is unclear whether the ability of 1Z1 brushes to remain aligned after removal of thick filaments indicates that myofibrils contain connections that are independent of the thick and thin filaments and that pass from one Z-disk to the next (*e.g.*, the filamentous structure that Guba *et al.*, 1968, described as being composed of the protein fibrillin), or that thin filaments from opposite Z-disks are connected at their ends in the middle of the sarcomere (*e.g.*, the S-filaments proposed several years ago, Hanson and Huxley, 1955), or that van der Waals and other noncovalent forces between adjacent filament layers in a myofibril are strong enough to maintain 1Z1 brushes in their original position even after removal of thick filaments. It seems unlikely that preservation of the relative orientation of Z-disks in the 1Z1 brushes is due to a small number of unextracted thick filaments remaining in these preparations because SDS-polyacrylamide gels, not shown here, demonstrate that these 1Z1 brush preparations contain very little myosin and almost no paramyosin. It is also unlikely that preservation of the relative orientation of Z-disks in 1Z1 brush preparations is due to the coiled C-filaments that Garamvölgyi and Belágyi (1968a; 1968b) propose to account for the mechanical properties of bee fibrillar muscle because these C-filaments supposedly attach to the thick filaments that are removed by the pyrophosphate extraction. On the other hand, SDS-polyacrylamide gels indicated that the 1Z1 brush preparations contained most of the actin, α-actinin, tropomyosin, and troponin present in the original myofibrils.

Myosin and paramyosin could be obtained in purified form by salting the pyrophosphate extracts out between 0 and 35% (paramyosin) and between 35 and 45% (myosin) ammonium sulfate saturation as described by Hammond and Goll (1975). The P_{35-45} myosin fraction obtained by this ammonium sulfate fractionation contained over 85% of its protein as myosin as evaluated by SDS-polyacrylamide gel electrophoresis (see Hammond and Goll, 1975, for SDS-polyacrylamide gels of these fractions). Between 6 and 20 mg of purified P_{35-45} myosin is obtained from 1 g wet weight of insect fibrillar flight muscle by this pyrophosphate extraction and ammonium sulfate fractionation procedure. This yield is three to ten times greater than the 2 to 2.5 mg of purified myosin obtained from 1 g wet weight of insect fibrillar flight muscle using conventional procedures (Bullard, Luke and Winkelman, 1973a; Bullard *et al.*, 1973b) to extract and purify myosin. Moreover these conventional myosin preparations contain paramyosin and must be subjected to column chromatography or ammonium sulfate fractionation for final purification. Hence the pyrophosphate extraction of myosin described in Table 2 has the advantage of producing a greater yield of myosin in addition to being part of a scheme that permits preparation of other contractile proteins from the same muscle sample. No attempt was made in these studies to obtain C-protein from insect fibrillar flight muscle. In view of the absence of a band that can be definitely attributed to C-protein in SDS-polyacrylamide gels of whole myofibrils (Figure 1) it is unclear whether insect fibrillar flight muscle contains C-protein; if it does, it seems likely that the C-protein would be in the myosin and paramyosin extract (Table 2).

The 1Z1 brushes remaining after extraction of thick filaments were washed several times with 100 mM KCl, 4 mM EDTA, pH 7.2 to remove the pyrophosphate and Mg^{2+} without extracting any contractile proteins. Extraction of the washed 1Z1 brushes with deionized distilled water, and then with a 1 mM Tris-acetate, pH8.5 solution removed the tropomyosin-troponin complex and some of the α-actinin from 1Z1 brushes of insect fibrillar flight muscle just as it does from mammalian skeletal (Arakawa, Robson and Goll, 1970; Goll, Suzuki, Temple and Holmes, 1972; Suzuki, Goll, Singh, Allen, Robson and Stromer, 1976) or cardiac (Singh, Goll, Robson and Stromer, 1977) muscle. Electron microscope examination of the residue left after the water and pH 8.5, low ionic strength extractions (Steps III, IV and V, Table 2) shows that the water washes and low ionic strength extraction greatly decrease the length and the number of thin filaments attached to Z-disks in the 1Z1 brushes (*cf.* Figures 4 and 5). SDS-polyacrylamide gels of the water and pH 8.5, low ionic strength extracts (not shown here) showed that these extracts contained considerable material migrating at the same rate as actin and also contained material migrating at the same rate as the subunit polypeptides of α-actinin, insect fibrillar troponin and tropomyosin. α-actinin has been reported to be located in Z-disks of mammalian skeletal (Schollmeyer, Stromer and Goll,1972; Schollmeyer, Goll, Robson and Stromer, 1973; Schollmeyer, Goll, Stromer, Dayton, Singh and Robson, 1974a) and cardiac (Schollmeyer *et al.*, 1973) muscle and it would therefore also be expected to be located in Z-disks of insect fibrillar flight muscle. It is surprising, therefore, to find that remnants of Z-disks survive the water and pH 8.5, low ionic strength extractions used to remove α-actinin (Figure 5). Indeed, the average width of Z-disklike structures after the water and pH 8.5, low ionic strength extractions was 101 $\pm$ 5 nm (mean plus or minus standard deviation of measurements on seven different samples), which is virtually the same as the 94 nm width of Z-disks in 1Z1 brushes before extraction (*cf.* Figures 4 and 5). A trough of very slightly decreased density can sometimes be observed along the center of the Z-disk remnants left after the water washes and low ionic strength extraction, but it is clear that Z-diaks in insect fibrillar flight muscle are not nearly as easily disrupted by low ionic strength extraction as Z-disks in rabbit skeletal muscle are (*cf.* Figure 5 with Figure 1 in Arakawa *et al.*, 1970). SDS-polyacrylamide gels of the low ionic residues shown in Figure 5 (gels not shown here) also indicated that these residues contained appreciable quantities of unextracted α-actinin. Therefore it seems that α-actinin is much more difficult to extract from insect fibrillar flight muscle than it is from either porcine (Suzuki *et al.*, 1976) or rabbit (Arakawa *et al.*, 1970) skeletal muscle. Insect fibrillar flight muscle α-actinin is probably also more resistant to extraction than cardiac α-actinin, which itself is more difficult to extract than porcine or rabbit skeletal α-actinin

(Singh *et al.*, 1977).

The cause of the increased resistance of Z-disks in insect fibrillar flight muscle to extraction is unclear although it is known that Z-disk structure in insect fibrillar flight muscle differs from that in mammalian skeletal muscle (Ashhurst, 1967a). Future studies of α-actinin in insect fibrillar flight muscle should attempt to use a longer low ionic strength alkaline extraction, a higher pH in this extraction, or very vigorous homogenization of the residue during its suspension in the low ionic strength alkaline solution (Step IV, Table 2) to increase the amount of α-actinin extracted. Alternatively, it may be possible to use the Ca^{2+}-activated proteolytic enzyme that has recently been purified from mammalian skeletal muscle (Dayton, Goll, Zeece, Robson and Reville, 1976a) and that removes Z-disks (Dayton, Reville, Goll and Stromer, 1976b) without degrading α-actinin (Dayton, Goll, Stromer, Reville, Zeece and Robson, 1975) to release α-actinin from 1Z1 brushes of insect fibrillar flight muscle and leave actin filaments intact. Such intact thin filaments could serve as starting material for preparation of actin.

TABLE 3 PREPARATION OF α-ACTININ, TROPOMYOSIN, AND TROPONIN FROM ONE MUSCLE SAMPLE

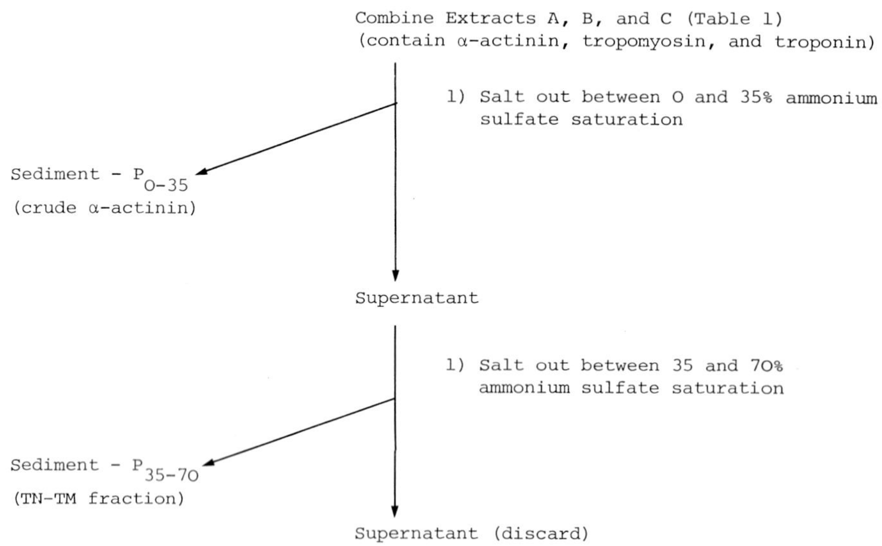

The combined water washes and pH 8.5, low ionic strength extracts (extracts A, B, and C, Table 2) can be salted out between 0 and 35% and between 35 and 70% ammonium sulfate saturation to produce P_{0-35} crude α-actinin and P_{35-75} crude troponin-tropomyosin fractions (Table 3). This is the same general procedure used to prepare similar fractions from mammalian skeletal muscle (Goll, Arakawa, Stromer, Busch and Robson, 1970; Goll *et al.*, 1972) so insect fibrillar flight muscle α-actinin does not differ from mammalian skeletal muscle α-actinin in its salting out properties.

PURIFICATION AND PROPERTIES OF α-ACTININ FROM INSECT ASYNCHRONOUS FLIGHT MUSCLE

α-Actinin can be purified from the P_{0-35} crude α-actinin fraction (Table 3) by using DEAE-cellulose chromatography as first described by Robson, Goll, Arakawa and Stromer (1970). The very small amount of P_{0-35} crude α -actinin obtainable in studies done to date has resulted in only three to four mg of protein being recovered in the α-actinin peak eluted from the first DEAE-cellulose column and has prevented rechromatography of the 1 x DEAE-α-actinin on a second DEAE-cellulose column as recommended by Robson *et al.* (1970) and Goll *et al.* (1972). As evaluated by SDS-polyacrylamide gel electrophoresis, however, approximately 70% of the protein in insect fibrillar flight muscle α-actinin preparations was the 100,000-dalton α-actinin polypeptide, even though those preparations had only been chromatographed once; hence purity of these preparations was fairly high even without rechromatography (see Hammond and Goll, 1975, for examples of SDS-polyacrylamide gels of 1 X DEAE insect fibrillar muscle α-actinin samples). We have recently found that hydroxyapatite chromatography produces a more efficacious purification of 1 X DEAE-purified α-actinin than rechromatography on a second DEAE-cellulose column does (Suzuki *et al*, 1976). It seems likely that hydroxyapatite chromatography would also be a valuable final purification step for insect fibrillar flight muscle α-actinin if it is ever possible to obtain enough 1 X DEAE-purified insect fibrillar muscle α-actinin to permit such rechromatography.

The availability of highly purified myosin and α-actinin prepared by using the procedure developed in this study made it possible to obtain accurate amino acid analyses for these two proteins from insect fibrillar flight muscle (Tables 4 and 5). Insect fibrillar muscle myosin has less lysine and slightly more glycine and glutamic acid than either rabbit skeletal or rabbit cardiac muscle myosins. Insect fibrillar muscle myosin has a slightly lower lysine plus arginine and a slightly higher glutamic acid plus aspartic acid content than either rabbit skeletal or cardiac myosin (Table 4). As a result,

TABLE 4 . AMINO ACID COMPOSITION OF DIFFERENT α-ACTININS

	Source of α-Actinin[a]			
Amino acid	Porcine "fast" skeletal muscle[b]	Porcine cardiac muscle[c]	Dung beetle[d]	*Lethocerus*[d]
Lys	53	58	65	71
His	22	24	26	23
Arg	69	63	66	63
Asp	106	126	99	110
Thr	45	48	58	58
Ser	54	44	53	48
Glu	163	180	170	177
Pro	38	34	37	44
Gly	65	46	60	57
Ala	95	83	79	81
Val	41	36	45	42
Met	28	27	27	27
Ile	38	55	54	47
Leu	101	93	99	102
Tyr	22	26	36	29
Phe	31	30	34	36
Trp	16	14	-	-
Cys	13	11	-	-
Lys + Arg	122	121	131	134
Asp + Glu	269	306	269	287

[a] Figures are residues per 1000 residues

[b] From Suzuki *et al.* (1973)

[c] From Robson and Zeece (1973)

[d] From Hammond and Goll (1975)

TABLE 5. AMINO ACID COMPOSITION OF DIFFERENT MYOSINS

	Source of Myosin[a]				
Amino acid	Rabbit fast skeletal muscle[b]	Rabbit cardiac muscle[c]	Phormia adult[d]	Dung beetle[e]	*Lethocerus*[e]
Lys	109	101	91	92	98
ε-Me-Lys	1.4	0.8	-	-	-
His	18	16	16	16	18
ε-Me-Lys	1	0	-	-	-
Arg	53	50	58	58	56
Asp	98	99	111	100	99
Thr	48	47	39	42	45
Ser	44	46	45	46	47
Glu	191	185	199	202	201
Pro	25	26	24	22	28
Gly	47	48	55	56	51
Ala	88	90	102	97	90
Val	49	49	38	42	46
Met	25	27	19	19	25
Ile	47	44	40	46	40
Leu	94	105	111	109	100
Trp	20	21	21	21	22
Phe	33	34	30	30	30
Cys	10	11	-	-	-
Lys + Arg	163	152	149	150	154
Asp + Glu	289	284	310	302	300

[a] Figures are residues per 1000 residues

[b] From Huszar and Elzinga (1971)

[c] From Huszar and Elzinga (1972)

[d] From Kominz *et al.* (1962)

[e] From Hammond and Goll (1975)

the ratio of negatively to positively charged amino acid side chains at pH 7.0 is 1.95 to 2.01 for insect fibrillar muscle myosin compared with 1.77 to 1.87 for rabbit skeletal or cardiac myosins (Table 4). Except for these small differences, the amino acid composition of these three myosins is very similar (Table 4). The amino acid composition of insect fibrillar muscle myosin is also very similar to that of myosin from nonfibrillar insect muscle (Table 4), although fibrillar muscle myosin may contain slightly more valine than non-fibrillar muscle myosin.

The amino acid composition of insect fibrillar muscle α-actinin is very similar to but not identical with that of porcine skeletal or cardiac muscle α-actinin (Table 5). Insect fibrillar muscle α-actinin contains slightly more lysine, tyrosine, and threonine and slightly less alanine than porcine skeletal or cardiac muscle α-actinin (Table 5). We have previously shown that porcine skeletal and rabbit skeletal α-actinins have very similar amino acid compositions but that mammalian cardiac muscle α-actinin has more aspartic acid than mammalian skeletal muscle α-actinin (Robson and Zeece, 1973; Suzuki, Goll, Stromer, Singh and Temple, 1973). Insect fibrillar muscle α-actinin has an aspartic acid content closer to that of "fast" skeletal muscle than to that of cardiac or "slow" skeletal muscle (Table 5). The ratio of aspartic acid plus glutamic acid content to arginine plus lysine content for insect fibrillar muscle also resembles the corresponding ratio of "fast" skeletal muscle α-actinin more closely than it does that of cardiac muscle α-actinin (Table 5).

The results of this study clearly show that insect fibrillar flight muscle α-actinin is qualitatively very similar to mammalian skeletal or cardiac α-actinins because: 1) all three α-actinins are extracted by using the same general low ionic strength, alkaline solution; 2) all three α-actinins can be precipitated from crude extracts between 0 and 35% ammonium sulfate saturation to produce a P_{0-35} crude α-actinin fraction; 3) all three α-actinins can be purified by using DEAE-cellulose chromatography and all elute similarly from DEAE-cellulose columns, although the exact KCl concentration required for elution may vary slightly (*e.g.*, see Suzuki *et al.*, 1976); 4) all three α-actinins migrate at almost the same rate in SDS-polyacrylamide gel electrophoresis (Figure 1) hence they must have similar subunit molecular weights; 5) these three α-actinins all have similar amino acid compositions (Table 5). The results in the next section will show that insect fibrillar flight muscle α-actinin is also antigenically similar to mammalian α-actinin. These similarities among α-actinin from insect fibrillar flight muscle and different mammalian α-actinins substantiate our earlier conclusion (Singh *et al.* 1977) that α-actinin is a highly conserved contractile protein just as actin seems to be.

LOCALIZATION OF α-ACTININ IN ASYNCHRONOUS INSECT FLIGHT MUSCLE

Studies on localization of binding of antibodies to highly purified α-actinin (Schollmeyer *et al.*, 1972; 1973; 1974a·, Schollmeyer, Furcht, Goll, Robson and Stromer, 1976; Schollmeyer, Goll, Tilney, Mooseker, Robson and Stromer, 1974b) and on very limited trypsin digestion of myofibrils (Goll, Mommaerts, Reedy and Seraydarian, 1969) have shown that α-actinin is located in the Z-disk of skeletal and cardiac muscle myofibrils and in attachment plaques in nonmuscle cells. Because Z-disks of insect fibrillar flight muscle have a different structure than Z-disks in mammalian skeletal muscle (Ashhurst, 1967a; Bullard and Sainsbury, 1977), it was of interest to learn whether Z-disks in insect fibrillar muscle also contained α-actinin.

Anti-α-actinin antibodies were prepared by injecting highly purified α-actinin prepared from chicken gizzard smooth muscle into rabbits. Sheep anti-rabbit IgG antibodies were also prepared and were conjugated with fluorescein isothiocyanate according to procedures used in our earlier studies (Schollmeyer, Goll, Stromer and Robson, in preparation). Purified myofibrils from *Lethocerus* fibrillar flight muscle were incubated with anti-α-actinin antibodies and then with sheep anti-rabbit IgG antibodies conjugated with fluorescein isothiocyanate by using the same procedures employed in our earlier studies (Schollmeyer, Goll, Stromer and Robson, in preparation). Examination of the treated myofibrils with phase and fluorescence light microscopy showed that α-actinin antibodies bound only to Z-disks in insect fibrillar flight muscle fibrils (Figure 6). Occasionally, very faint fluorescence could be observed at the level of the M-line (Figure 6), but this fluorescence is almost certainly due to a very slight contamination of even our highly purified α-actinin with M-protein, which seems to be highly antigenic. We have previously shown that M-protein is a persistent contaminant of α-actinin preparations and that it sometimes is necessary to use preparative polyacrylamide gel electrophoresis to remove all traces of M-protein contamination from α-actinin preparations (Schollmeyer, Goll, Stromer and Robson, in preparation). Consequently, the antibody binding studies shown in Figure 6 indicate that, as in mammalian skeletal and cardiac myofibrils, α-actinin in insect fibrillar flight muscle myofibrils is located only in the Z-disk. Bullard and Sainsbury (1977) have reported that α-actinin is the principal protein component in Z-disks isolated from *Lethocerus* fibrillar flight muscle.

Z-DISK STRUCTURE AND PHYSIOLOGICAL ROLE OF α-ACTININ

Z-disks in vertebrate skeletal muscle have been shown to contain two grossly distinct morphological components (Kelly and Cahill, 1972): 1) filamentous component composed of Z-filaments that in vertebrate skeletal muscle form a rhombohedral lattice structure in cross-section; 2) an amorphous or matrix component composed of densely staining, structureless material that envelops the Z-filaments and fills the rhombohedral lattice openings. Our studies on antibody binding to Z-disks whose amorphous component has been partly removed by low ionic strength extraction and to tadpole Z-disks that contain very little amorphous component have suggested that the filamentous component of Z-disks is composed partly of tropomyosin and that α-actinin is located only in the amorphous component (Schollmeyer *et al.*, 1974a). It seems likely that α-actinin is also located only in the amorphous component that binds overlapping hexagonal arrays of filaments together in Z-disks of insect fibrillar flight muscle (Ashhurst, 1967). That insect fibrillar muscle Z-disks are relatively more resistant to extraction or proteolytic removal than vertebrate skeletal muscle Z-disks must be due either to the hexagonal array of thin filaments entering insect fibrillar muscle Z-disks in contrast to the square lattice array of thin filaments entering vertebrate skeletal muscle Z-disks or to the presence in insect fibrillar muscle Z-disks of one or more proteins that are not present in vertebrate skeletal muscle Z-disks. Although the protein composition of vertebrate skeletal muscle Z-disks is still unclear Bullard and Sainsbury (1977) have indicated that *Lethocerus* fibrillar muscle Z-disks contain, in addition to the α-actinin, tropomyosin, and actin that are presumed components of all Z-disks, a protein having a subunit molecular weight of 175,000 to 180,000 daltons. Saide and Ullrick (1974) found a protein having a subunit molecular weight of 87,000 daltons in honeybee Z-disks. Clearly, more information is needed on both protein composition and the nature of the protein-protein interactions in the Z-disk before the properties of Z-disks can be understood.

Being able to specify the location of α-actinin to the amorphous component of Z-disks does not permit a similar specification of the biological role of α-actinin in contractile systems. α-Actinin has now been found in every motile system that has been examined for its presence. These systems include Z-disks in vertebrate skeletal muscle (Schollmeyer *et al.*, 1972; 1973), vertebrate cardiac muscle (Schollmeyer *et al.*, 1973) and insect fibrillar flight muscle, intercalated disks in cardiac muscle (Schollmeyer *et al.*, 1973), dense bodies and attachment plaques in vertebrate smooth muscle (Schollmeyer *et al.*, 1973; 1976), attachment plaques in epithelial cells and fibroblasts (Schollmeyer *et al.*, 1974b; 1976) and along bundles of microfilaments in the brush borders of intestinal epithelial cells, fibroblasts, and the acrosomal processes of *Limulus* sperm (Schollmeyer *et al.*, 1974b). Many of these motile systems have no Z-disks. Because of its seemingly ubiquitous presence, it seems likely that α-actinin has an important role in biological movement. Suzuki *et al.* (1976) have suggested three possible biological roles for α-actinin: 1) it may nucleate aggregation of actin monomers and thereby assist in regulating the proportion of aggregated to monomeric actin in cells; 2) it may serve as an attachment plaque or organizing center for preformed actin filaments; by exhibiting specificity for binding to actin filaments, α-actinin could control directionality of these filaments; 3) α-actinin may modify the structure of actin monomers while these monomers are in the aggregated filamentous state and this modification may enhance the effectiveness of α-actinin in biological movement. Based on the available evidence, α-actinin could have any one or all of these possible biological roles.

ACKNOWLEDGEMENTS

This paper is dedicated to Professor J.W.S. Pringle who has pioneered in the use and advocacy of insect fibrillar flight muscle as a model system for study of the mechanism of muscle contraction. D.E.G. is also grateful to Professor Pringle for making it possible for him to study α-actinin in insect fibrillar muscle at Oxford while a Special Fellow of the National Institutes of Health (1 FO3 AM52982). The research reported in this paper was supported by the A.R.C. Unit, Oxford, by NIH Grant Nos. AM-12654, AM-19864, and HL-15679, by grants from the Muscular Dystrophy Association, the Iowa Heart Association, and the Arizona Heart Association, and by Arizona Agricultural Experiment Station Project No. 20 and Iowa Agriculture and Home Economics Experiment Station Project No. 2025. We thank Fran Richardson and Mary Bremner for expert technical assistance, Marguerite Santucci for assistance with the manuscript, and Dr. Belinda Bullard and Ms. Lowell Winkelman for helpful discussions.

REFERENCES

Arakawa N., Robson R.M. and Goll D.E. (1970) *Biochim. Biophys. Acta,* 200, 284.

Armitage P., Miller A., Rodger C.D. and Tregear R.T. (1973) *Cold Spring Harbor Symp. Quant. Biol.,* 37, 379.

Ashhurst D.E. (1967a) *J. Mol. Biol.,* 27, 385.

Ashhurst D.E. (1967b) *J. Cell. Sci.,* 2, 435.

Bullard B., Dabrowska R. and Winkelman L., (1973a) *Biochem. J.,* 135,277.

Bullard B., Luke B. and Winkelman L., (1973b) *J. Mol. Biol.,* 75, 359.

Bullard B. and Sainsbury G.M. (1977) *Biochem J.,* 161, 399.

Dayton W.R., Goll D.E., Stromer M.H., Reville W.J. Zeece M.G. and Robson R.M. (1975) *Cold Spring Harbor Conferences on Cell Proliferation, Gol.2 "Proteases and Biological Control"* (Reich E., Rifkin D.B. and Shaw E Ed.) pp.551-577.

Dayton W.R., Goll D.E., Zeece M.G., Robson R.M. and Reville W.J. (1976a) *Biochemistry,* 15, 2150.

Dayton W.R., Reville W.J., Goll D.E. and Stromer M.H. (1976b) *Biochemistry,* 15, 2159.

Garamvölgyi N. and Belágyi J. (1968a) *Acta Biochim. Biophys. Acad. Sci. Hung.* 3, 195.

Garamvölgyi N. and Belágyi J. (1968b) *Acta Biochim. Biophys. Acad. Sci. Hung.* 3, 299.

Goll D.E., Arakawa N., Stromer M.H., Busch W.A. and Robson R.M. (1970) In: *The Physiology and Biochemistry of Muscle as a Food 2* (Briskey E.J. Cassens R.G. and Marsh B.B. ed.) pp. 755-800.

Goll D.E., Mommaerts W.F.H.M., Reedy M.K. and Seraydarian K. (1969) *Biochim. Biophys Acta,* 175, 174.

Goll D.E., Robson R.M. and Stromer M.H. (1977) In: *Funamental Aspects of Proteins Basic to Foods* (Whitaker J. and Tannenbaum S., ed.) (In press).

Goll D.E., Suzuki A., Temple J. and Holmes G.R. (1972) *J. Mol. Biol.,* 67, 469.

Goody R.S., Barrington-Leigh J., Mannherz H.G., Tregear R.T. and Rosenbaum G. (1976) *Nature,* 262, 613.

Guba F., Harsányi V. and Vajda E. (1968) *Acta Biochim. Biophys. Acad. Sci. Hung.* 3, 353.

Hammond K.S. and Goll D.E. (1975) *Biochem. J.,* 151, 189.

Hanson J. and Huxley H.E. (1955) In: *Symposia Soc. Exptl. Biol. 9 Fibrous Proteins and their Biological Significance,* , Cambridge University Press, Cambridge, England. pp.228-264.

Huszar G.and Elzinga M.(1971) *Biochemistry,* 10, 229.

Huszar G. and Elzinga M. (1972) *J. Biol. Chem.,* 247, 745.

Ikemoto N. and Kawaguti S. (1967) *Proc. Japan Acad.,* 43, 974.

Jewell B.R. and Rüegg J.C. (1966) *Proc. Roy. Soc. B,* 164, 428.

Kelly D.E. and Cahill M.A. (1972) *Anat. Rec.,* 172, 623.

Kominz D.R., Maruyama K., Levenbook L. and Lewis M. (1962) *Biochim. Biophys. Acta,* 63, 106.

Lehman W., Bullard B. and Hammond K. (1974) *J. Gen. Physiol.,* 63, 553.

Marston S.B., Rodger C.D. and Tregear R.T. (1976) *J. Mol. Biol.*, 104, 263.

Pringle J.W.S. (1967a) *J. Gen. Physiol.*, 50, 139.

Pringle J.W.S. (1967b) *Prog. Biophys. Mol. Biol.*, 17, 1.

Pringle J.W.S. (1975) In: *Comparative Physiology-Functional Aspects of Structural Materials* (Bolis L., Maddrell H.P. and Schmidt-Nielsen K., ed.) pp. 139-152.

Pybus J. and Tregear R.T. (1973) *Cold Spring Harbor Symp. Quant. Biol.*, 37, 655.

Reedy M.K. (1968) *J. Mol. Biol.*, 31, 155.

Reedy M.K., Holmes K.C. and Tregear R.T. (1965) *Nature*, 207, 1276.

Robson R.M., Goll D.E., Arakawa N. an- Stromer M.H. (1970) *Biochim. Biophys. Acta*, 200, 296.

Robson R.M. and Zeece M.G. (1973) *Biochim. Biophys. Acta*, 295, 208.

Rüegg J.C., Steiger G.J. and Schädler M. (1970) *Pflügers Arch.*, 319, 139.

Saide J.D. and Ullrick W.C. (1974) *J. Mol. Biol.*, 87, 671.

Schollmeyer J.E., Furcht L.T., Goll D.E., Robson R.M. and Stromer M.H. (1976) *Cold Spring Harbor Conferences on Cell Proliferation, Vol.3. "Cell Motility"*. Goldman R., Pollard T. and Rosenbaum J., ed.) pp.361-388.

Schollmeyer J.E., Goll D.E., Robson R.M. and Stromer M.H. (1973) *J. Cell Biol.*, 59, 306a.

Schollmeyer J.V., Goll D.E., Stromer M.H., Dayton W.R., Singh I. and Robson R.M. (1974a) *J. Cell Biol.*, 63, 303a.

Schollmeyer J.V., Goll D.E., Tilney L., Mooseker M., Robson R. and Stromer M., (1974b) *J. Cell Biol.*, 63, 304a.

Schollmeyer J.E., Stromer M.H. and Goll D.E. (1972) *Biophys J.*, 12, 280a.

Singh I., Goll D.E., Robson R.M. and Stromer M.H. (1977) *Biochim. Biophys. Acta*, 491, 29.

Steiger G. (1971) *Pflügers Arch.*, 330, 347.

Suzuki A., Goll D.E., Singh I., Allen R.E., Robson R.M. and Stomer M.H. (1976) *J. Biol. Chem.*, 251, 6860.

Suzuki A., Goll D.E., Stromer M.H., Singh I. and Temple J. (1973) *Biochim. Biophys. Acta*, 295, 188.

Tregear R.T. and Miller A. (1969) *Nature*, 222, 1184.

Trombitas K. and Tigyi-Sebes A. (1974) *Acta Biochim. Biophys. Acad. Sci. Hung.*, 9, 243.

White D.C.S. (1970) *J. Physiol.*, 208, 583.

White D.C.S. and Thorson J. (1972) *J. Gen. Physiol.*, 60, 307.

White D.C.S. and Thorson J. (1973) *Prog. Biophys. Mol. Biol.*, 27, 175.

FIGURE LEGENDS

FIG.1 Seven and one-half per cent SDS-polyacrylamide gels of myofibrils from *Lethocerus* fibrillar flight muscle. A. Myofibrils before treatment with 1% Triton X-100 (Step V, Table I). B. Myofibrils after treatment with Triton X-100 (Step XIV, Table I). Triton X-100 treatment has less effect on SDS-polyacrylamide gel electro-phoresis patterns of myofibrils from insect fibrillar muscle than it does on SDS-electrophoresis patterns of myofibrils from mammalian skeletal muscle (*cf.* Goll *et al.*, 1977). C. Myofibrils after Triton X-100 treatment plus 0.5 μg of purified smooth muscle α-actinin. D. Myofibrils after Triton X-100 treatment plus 1.0 μg purified smooth muscle α-actinin. E. Myofibrils after Triton X-100 treatment plus 2.0 μg purified smooth muscle α-actinin. Addition of purified smooth muscle α-actinin enhances the lower of the two closely spaced bands in the 100,000 to 110,000 dalton region. All gels loaded with 25 μg myofibrillar protein and gels C, D, and E contained the additional amount of purified α-actinin specified.

FIG.2 Electron micrographs of myofibrils from *Lethocerus* fibrillar flight muscle. A. An example of myofibrils isolated at nearly rest length. Sarcomere length is 2.7 μm. A clear H-zone containing a distinct M-line is evident, but I-bands are not easily discernible. Two bands are seen approximately 220 to 260 nm on either side of the Z-disks. These bands were frequently seen in rest-length myofibrils, but their origin is unclear. B. An example of a slightly contracted myofibril. Sarcomere length is 2.0 μm. No H-zone is present, and the dark band in the centre of the sarcomere suggests that some double overlap of thin filament has occurred. The bands flanking the Z-disk in Fig.2A are absent in this slightly contracted myofibril.

FIG.3 Electron micrographs of myofibrils from *Lethocerus* fibrillar flight muscle after M-line extraction. A. This myofibril, with a sarcomere length of 2.12 μm, has the same dark bands flanking the Z-disk that were observed in Fig.2A. No M-lines are visible in this myofibril, but all other structural features observed in insect fibrillar muscle myofibrils are evident. B. This M-line-extracted myofibril displays the unusual gaps seen in several instances in the M-line region after M-line extraction. Ultrastructure of the remainder of the myofibril seems normal. Sarcomere length is 2.3 μm.

FIG.4 Electron micrographs of myofibrils from *Lethocerus* fibrillar flight muscle after extraction of thick filaments to leave 1Z1 brushes. A. 1Z1 brushes contain Z-disks with sheaves of thin filaments attached to either side, but no thick filaments are seen. The two Z-disks in this micrograph lie 2.4 to 2.6 μm apart. B. The dark bands lying approximately 220 nm on either side of the Z-disk are evident in this micrograph of 1Z1 brushes. The two Z-disks in this micrograph are 2.2 to 2.4 μm apart.

FIG.5 Electron micrographs of the residue remaining after α-actinin extraction of 1Z1 brushes prepared from *Lethocerus* fibrillar flight muscle. A. The residue remaining after α-actinin extraction resembles 1Z1 brushes that have had many of the thin filaments stripped off. The densely staining Z-disk remnants in this micrograph are 105 nm wide, the same width as Z-disks before αactinin extraction. B. Z-disk remnants in this residue remaining after α-actinin extraction have a trough of slightly decreased density running along their centre as though this area had been partly extracted. Z-disk remnants in this micrograph are 95 to 105 nm wide.

FIG.6 Phase (upper) and fluorescence (lower) micrographs of a myofibril from *Lethocerus* fibrillar flight muscle after incubation with anti-α-actinin conjugated with fluorescein. The myofibril is at rest-length as indicated by prominant H-zones. Z-disks (Z) are indicated in both micrographs. Comparison of the phase and fluorescence micrographs shows that anti-α-actinin binds prominently and exclusively to the Z-disk. The very faint fluorescence that can barely be detected on some of the M-line regions is probably due to a very small amount of M-protein contamination in the α-actinin preparation used to elicit the antibodies.

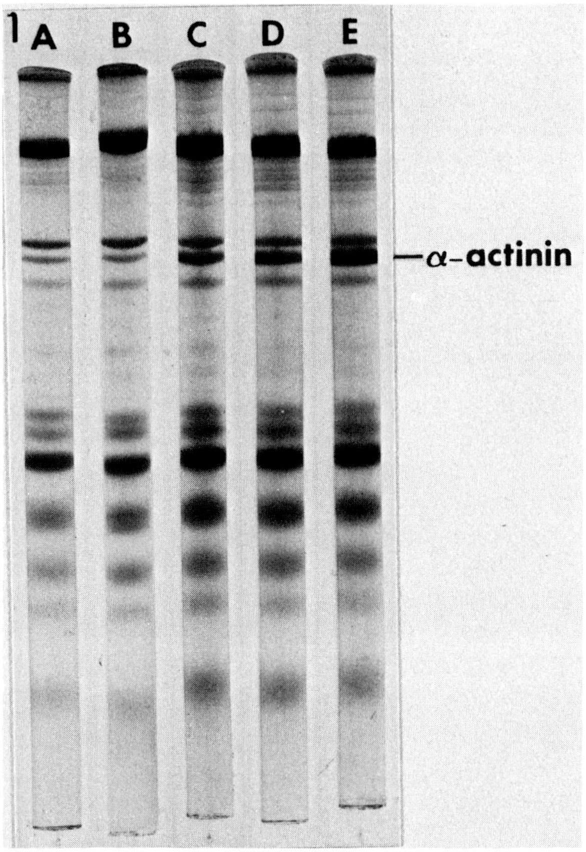

2

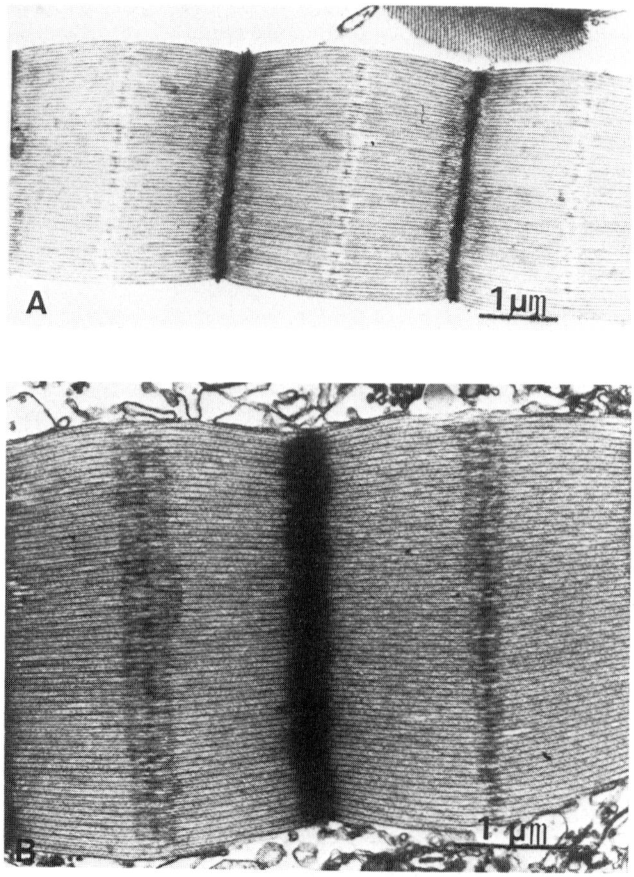

3

4

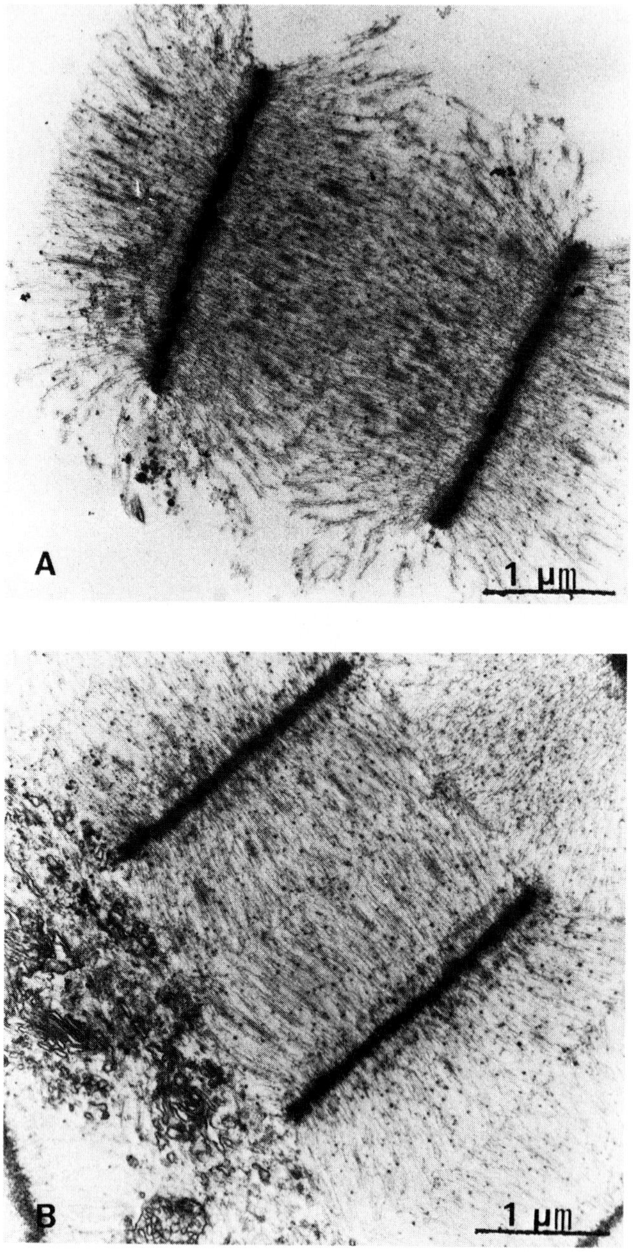

5

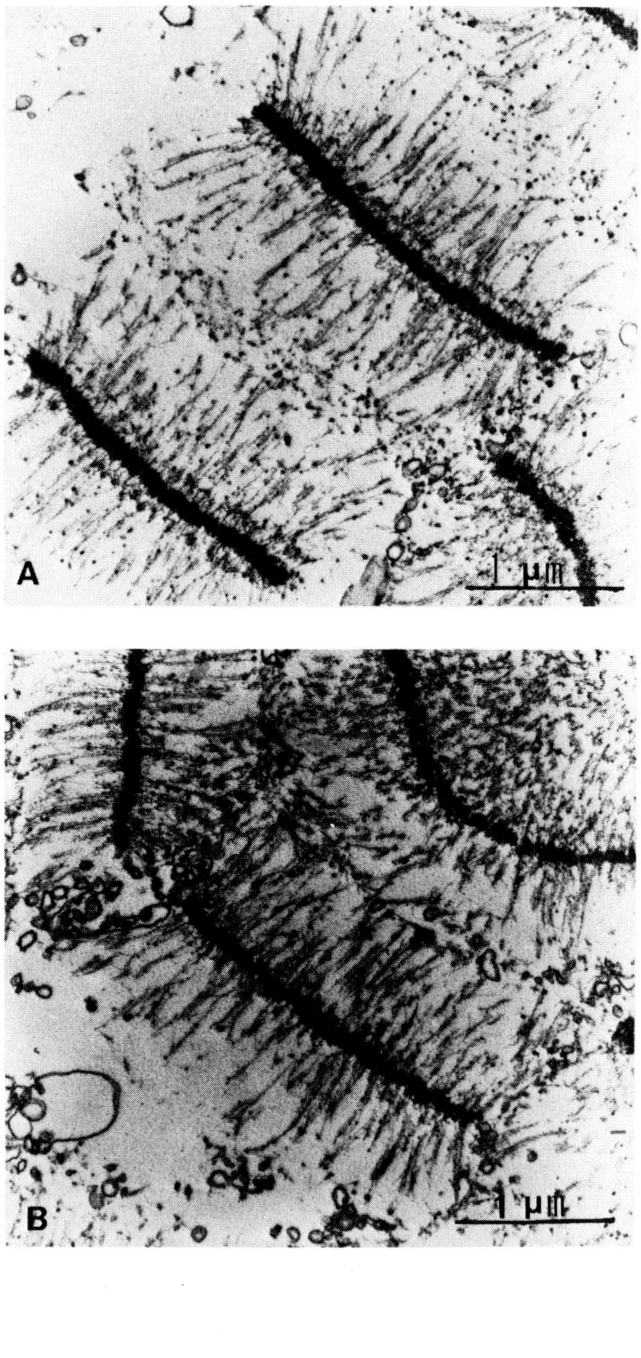

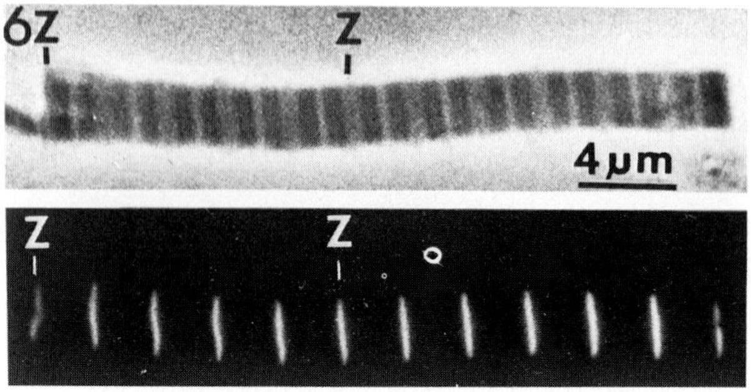
6Z
Z
4 μm
Z
Z

IMMUNOLOGICAL INVESTIGATION OF PROTEINS ASSOCIATED WITH THICK FILAMENTS OF INSECT FLIGHT MUSCLE

B. Bullard, J.L. Bell and B.M. Luke.

A.R.C. Unit, Department of Zoology, South Parks Road, Oxford.

INTRODUCTION

Insect fibrillar muscle has some structural characteristics that distinguish it from other striated muscles. The Z-line is unusually wide, about twice the width of the Z-line in vertebrate striated muscle (Ashhurst 1967) ; the thick filaments are broader than those of vertebrate, though similar to those of other invertebrate striated muscles; and the thick filaments extend the whole length of the sarcomere. These characteristics may be essential for the oscillatory contraction of the muscle.

An important property of fibrillar muscle is the high stiffness in the relaxed state and the most likely cause of the stiffness is mechanical continuity between thick filaments and Z-line (Pringle, this volume). We describe here an investigation of two structural proteins associated with the thick filaments: paramyosin and a protein that may connect thick filaments with the Z-line.

METHODS

Most of the techniques used have been described by Bullard *et al.*, (1977). The methods used for the fractionation of antiserum and determination of antibody specificity are given in that paper.

Dung beetle α-actinin was prepared as described by Hammond and Goll (1975) and purified by two DEAE columns (Goll *et al.* 1972).

Antigen was removed from dung beetle α-actinin preparations by adding the protein to Sepharose-4B which was coupled with the γ-globulin from antiserum to paramyosin. The CNBr Sepharose-4B was obtained from Pharmacia. γ-globulin was obtained from antiserum by precipitating with 40% saturated $(NH_4)_2SO_4$. Unreacted groups on the Sepharose were blocked with 1 M ethanolamine (pH 8). The Sepharose (1g) coupled with γ-globulin (10 mg) was equilibrated with 0.4 M KCl, 0.01 M K-phosphate buffer (pH 7.4); α-actinin (4.7 mg in 1 ml of the same buffer) was mixed with the Sepharose in a beaker and incubated at 4°C for 7 hr. Unbound protein was removed by suction through a sintered glass funnel. The yield of unbound protein from the Sepharose was 3.6 mg.

Immunoelectrophoresis was by the method of Laurell (1965). Glass plates (5 x 5 cm) were coated with 4 ml of 1% oxoid agar in 0.06 M Na-barbitone (pH 8.6), 1 mM NaN_3. Holes were made in the agar close to one edge of the slide and antigens (3-50 μg) were electrophoresed from the holes at 30-40 V from negative to positive for 3 h. Strips about 1 cm wide were cut in the agar such that each strip contained an electrophoresed antigen. Each strip was placed along one edge of another plate (5 x 5 cm.) Antiserum to paramyosin was mixed with 1% agar, 0.06 M Na-barbitone (pH 8.6), 1 mM NaN_3 at $50^{\circ}C$ to a final concentration of 0.5 to 1%. The mixture was poured onto the plates beside the strips of agar containing electrophoresed antigen. The antigen was then electrophoresed at right angles to the first electrophoresis,into the agar containing antibody. The plates were washed in 1% NaCl then distilled water and dried before staining with 1% Amido black in 50% methanol, 7% acetic acid.

Myosin was extracted from fly(*Sarcophaga bullata*) myofibrils as described by Bullard and Reedy (1972). The extracted myofibrils were washed twice in extracting solution and once in 0.1 M KCl, 0.01 M K-phosphate buffer (pH 7.0). They were incubated for 12 h at 4° suspended in the γ-globulin fraction of antiserum from which the anti-paramyosin had been removed by a Sepharose paramyosin column (Bullard *et al.* 1977). The γ-globulin was in the 0.1 M KCl solution. The myofibrils were washed in the 0.1 M KCl solution and a small lump fixed and embedded for electron microscopy.

RESULTS AND DISCUSSION

The site of paramyosin

In a survey of invertebrate muscles Winkelman (1976) found there was no simple correlation between the diameter of the thick filaments and the weight ratio of paramyosin to myosin; Levine *et al.*(1976) found the same for molluscs and arthropods whose muscles did not contain exceptionally large amounts of paramyosin. The flight muscles in different insects also show no correlation between thick filament diameter and the proportion of paramyosin. The thick filaments of all insect fibrillar muscles are about the same diameter (180 Å) but the amount of paramyosin is variable; in transverse sections the filaments often appear to have a hollow core. Calculation using Squire's (1971) model shows that in *Lethocerus* flight muscle, which has a high proportion of paramyosin (6.7% of the total myofibrillar protein), the protein would exactly fill the thick filament core. It follows that in insects such as flies in which less than 1% of the myofibrillar protein is paramyosin the core of the filaments cannot be filled with paramyosin. There appears to be some correlation between the wing beat frequency of insects and the amount of paramyosin in the flight muscle; small insects with high wing beat frequencies have less paramyosin than large ones with low wing beat frequencies (Bullard *et al.* 1973).

Antibody specific to paramyosin binds to the thick filaments of *Lethocerus* myofibrils in the centre of the sarcomere only (Plate 1(b) and Bullard *et al.* 1977). Even in myofibrils from which myosin has been extracted, anti-paramyosin binding is most concentrated in the centre of the sarcomere. Although we have not established whether or not paramyosin extends the entire length of the core of the thick filaments, the antibody binding pattern does suggest there may be more paramyosin inside the thick filaments in the M-region.

This is consistent with results from the nematode *Caenorhabdites elegans* (Dr. W. Waterson, personal communication). The wild-type nematode has muscles that contain paramyosin and the thick filaments are less hollow in the M-region than along the rest of the filament. In a mutant having no paramyosin the thick filaments are hollow in the M-region, suggesting that paramyosin may be confined to the M-region in the wild-type nematode.

The properties of native paramyosin

Some molluscan paramyosins are partially degraded during purification unless proteolytic enzymes are inhibited by EDTA (Stafford and Yphantis, 1972; Yeung and Cowgill, 1976; Melson and Cowgill, 1976). The native α-paramyosin of *M.mercenaria* has a molecular weight of 207,000 and degraded β-paramyosin one of 200,000; removing a short length of the molecule increases the solubility (Merrick and Johnson, 1974; Yeung and Cowgill, 1976). Insect paramyosin is also degraded from an α form to a lower molecular weight β form if proteolysis during preparation is not prevented by EDTA. There is some evidence that in insects the α form may be more soluble than the β form.

α-paramyosin was distinguished from β-paramyosin by comparing the mobility on SDS gels with that of α-actinin. α-paramyosin had a lower mobility than α-actinin and appeared as a discrete band, while β-paramyosin had the same mobility as α-actinin and could not be resolved from it.

The existence of a higher molecular weight α-paramyosin in insect muscle became apparent when purified dung beetle α-actinin, in solutions of low ionic strength, was found to cross react with antiserum to paramyosin. Immunoelectrophoresis of paramyosin against antiserum produced a strong precipitin arc (Plate 2(a)); an α-actinin preparation gave an arc in the same position as the paramyosin arc (Plate 2(b)), showing the antigen had the same electrophoretic mobility as paramyosin. A mixture of the two proteins gave a single arc (Plate 2(c)) which proved the antigens were identical. The most likely explanation of this result was that the α-actinin preparation contained paramyosin. In SDS gels of whole myofibrils paramyosin has a lower mobility than α-actinin (Bullard and Sainsbury, 1977). Although purified α-actinin showed only slight traces of a band of lower mobility (Plate 3(b)), a sample of α-actinin at an intermediate stage of purification had about equal amounts of α-actinin and a protein of higher subunit weight (Plate 3(c)). The protein with the higher subunit weight was selectively removed by adding the sample to Sepharose linked to anti-paramyosin (Plate 3(d)); the remaining protein did not form a precipitin arc with antiserum. This is consistent with the higher subunit weight protein being paramyosin and the antigenic reaction of α-actinin being due to a paramyosin contaminant.

Paramyosin was an unexpected contaminant of α-actinin because α-actinin samples were dissolved in 20 mM Tris-Cl (pH 7.5) and purified insect paramyosin is not soluble at this low ionic strength at pH 7.5. The paramyosin in synthetic mixtures of purified paramyosin and α-actinin was not soluble at low ionic strength and SDS gels of the mixture showed that the paramyosin had a mobility that was the same as that of α-actinin (Plate 3(e)). This shows that purified paramyosin is degraded to β-paramyosin the loss of a small part of the molecule may reduce the solubility at low ionic strength but no systematic tests of paramyosin solubility have been carried out. The α-paramyosin contaminating α-actinin was probably protected from proteolysis during preparation because the extraction solution contained EDTA. A sample of α-paramyosin having a higher subunit weight than β-paramyosin was prepared from *Lethocerus* muscle by including EDTA in all the solutions during preparation, (Plate 3(a)).

Clearly the solubility of α-paramyosin should be investigated further because the formation of thick filaments in the myofibril is probably a consequence of the relative insolubility of the proteins concerned, and paramyosin as a core protein would be expected to be insoluble at physiological ionic strength.

Protein connecting myosin filaments to Z-line

In the course of preparing antibody specific to paramyosin we separated antibody to another protein which was at the ends of the thick filaments and extended from the thick filaments to the Z-line in stretched muscle (Bullard *et al.* 1977). The antigen in this region of the myofibril is likely to produce the mechanical continuity between thick filaments and Z-line and will be referred to as connecting protein.

Antiserum prepared by immunizing rabbits with β-paramyosin containing small amounts of impurities had antibodies both to paramyosin and to connecting protein. A simple way of removing paramyosin antibodies is to absorb the antiserum with α-paramyosin. Even high concentrations of α-paramyosin did not give the precipitin arc characteristic of connecting protein (Bullard *et al.* 1977), which showed that α-paramyosin was less contaminated by this protein than β-paramyosin. Absorbing antiserum with α-paramyosin therefore neutralises antibodies to paramyosin and not those to connecting protein. Plate 1(c) shows strong antibody binding at the ends of the sarcomere in a *Lethocerus* myofibril labelled with absorbed antiserum. It is possible that connecting protein is exposed only at the ends of the sarcomere in intact myofibrils because the thick filaments mask the antigen further in. Myosin can be removed nearly quantitatively from fly myofibrils (Bullard and Reedy, 1972) and the extracted myofibrils were labelled by antibody to connecting protein in the same place as unextracted ones (Plate 4). This shows that more connecting protein was not exposed by removing myosin; nor was the connecting protein at the end of the sarcomere removed with myosin and therefore it is probably confined to a small area either side of the Z-line in the intact myofibril.

The antigenic protein at the ends of the sarcomeres has not been isolated yet, and the antiserum may have antibodies to more than one protein in this region, but there is evidence for an antigen with solubility similar to β-paramyosin and a subunit weight of about 180,000 to 200,000, which may be connecting protein.

The myofibrillar proteins forming a precipitating complex with the antiserum were detected by SDS gel electrophoresis of the immunoprecipitate obtained when the γ-globulin fraction of antiserum was incubated with an extract of dung beetle myofibrils containing all the myofibrillar proteins. Plate 5(a) shows that paramyosin was the major precipitating antigen but a protein of subunit weight 180,000 to 200,000 and small amounts of one of 150,000 were also precipitated. Antiserum from which anti-paramyosin had been removed precipitated predominantly the protein of subunit weight 180,000 to 200,000 (Plate 5(b)); the small amount of protein with the subunit weight of paramyosin was probably due to incomplete removal of anti-paramyosin from the antiserum. The binding pattern of the antiserum to myofibrils showed that at least two passes through a Sepharose-paramyosin column were necessary to completely eliminate antibody to paramyosin.

The high molecular weight protein precipitated by the antiserum had a subunit weight close to that of the heavy chain of myosin; however immunological tests (Bullard *et al.* 1977) showed that the antiserum did not combine with myosin and antiserum absorbed with myosin bound to the same regions of the myofibril as unabsorbed antiserum. Therefore the high molecular weight antigen is not myosin but an unidentified protein. Further work is needed to establish whether or not this antigen is a connecting protein.

Organisation of proteins near the Z-line

The proteins of Z-line and thick filaments appear to be closely associated in the myofibril and connecting protein may also be associated with Z-line proteins. Z-discs isolated from fibrillar muscle show evidence of remaining thick filament attachments; this is seen in SDS gels of Z-discs (Bullard and Sainsbury, 1977) which have minor bands of paramyosin and myosin. There is also a minor protein of subunit weight 180,000 and, as connecting protein extends to the Z-line, this may be from attachments of connecting protein to the Z-discs.

The exposure of antigenic sites on connecting protein to antibody varies with sarcomere length. Plate 6 shows that slightly shortened sarcomeres bind more antibody than those at rest length; in extreme cases of contracture, less antibody is bound (Bullard *et al* . 1977). The functional significance of these changes in the arrangement of the proteins is not clear but similar alterations may occur during oscillatory contraction.

CONCLUSIONS

The two proteins investigated here are associated with the thick filaments and both are likely to have a structural function. Paramyosin is probably not distributed uniformly down the core of the thick filaments and its function may be rather different in insects with a high proportion of the protein and insects with a very low proportion. The interaction between the native α-paramyosin, which is probably more soluble than β-paramyosin, and myosin should be investigated.

There is a protein at the ends of the myosin filaments that probably joins them to the Z-line. The connecting protein is insoluble at physiological ionic strength, as is myosin, and the two would be expected to associate. The method of attachment of the protein to the Z-line is not known; possibly connecting protein forms part of the Z-line lattice and is not labelled by antibody at the Z-line because it is masked by other Z-line proteins.

ACKNOWLEDGEMENTS

We are grateful to Ms. K.S. Hammond for performing the immunoelectrophoresis and for her part in the earlier stages of this work.

REFERENCES

Ashhurst, D. (1967). *J. Mol. Biol.*, 27, 385.

Bullard, B., Hammond, K.S. and Luke, B.M.(1977). in preparation.

Bullard, B., Luke, B.M. and Winkelman, L. (1973). *J. Mol. Biol.*, 75, 359.

Bullard, B., and Reedy, M.K. (1972). *Cold Spring Harb. Symp. Quant. Biol.* 37, 423.

Bullard, B. and Sainsbury, G.M. (1977). *Biochem. J.* 161, 399.

Goll, D.E., Suzuki, A., Temple, J. and Holmes, G.R.(1972). *J. Mol. Biol.* 67, 469.

Hammond, K.S. and Goll, D.E. (1975). *Biochem. J.*, 15, 189.

Laurell, C.B. (1965). *Anal. Biochem.* 10, 358.

Levine, R.J.C., Elfuin, M., Dewey, M.M., and Walcott, B.(1976). *J. Cell. Biol.* 71, 273.

Melson, G.L. and Cowgill, R.W. (1976). *Comp. Biochem. Physiol.*, 55B, 503.

Merrick, J.B. and Johnson, W.H. (1974). *Fedn. Proc. Fedn. Am. Soc. Exp. Biol.* 33, 1521.

Stafford, W.F. and Yphantis, D.A. (1972). *Biochem. Biophys. Res. Comm.* 49, 848.

Squire, J. (1971) *Nature*, 233, 457.

Winkelman, L.B. (1976). *Comp. Biochem. Physiol.* , 55B, 391.

Yeung, A. and Cowgill, R.W. (1976). *Biochemistry*, 15, 4654.

PLATE LEGENDS

Plate 1. *Lethocerus* myofibrils labelled with antiserum. (a) Normal serum, (b) antiserum to pure paramyosin, (c) antiserum to impure paramyosin absorbed with *Lethocerus* α-paramyosin. Antiserus, 0.25 ml, was absorbed with 0.3 mg *Lethocerus* α-paramyosin, 0.12 ml, by the method described by Bullard *et al.* 1977.

Plate 2. The antigenicity of a dung beetle α-actinin preparation. Immunoelectrophoresis of dung beetle paramyosin and an α-actinin preparation, against antiserum to paramyosin. (a) Paramyosin, 6.4 μg; (b) α-actinin, 47 μg; (c) paramyosin, 3.2 μg, mixed with actinin, 47 μg; concentrations of antiserum: (a) and (c) 1%, (b) 0.5%.

Plate 3. The presence of paramyosin in preparations of dung beetle α-actinin. SDS gel electrophoresis, (a) *Lethocerus* α-paramyosin, 4 μg; (b) dung beetle α-actinin, 5 μg; (c) partially purified dung beetle α-actinin (before DEAE chromatography) 12 μg; (d) same as (c) after incubating with Sepharose linked to anti-paramyosin, 10 μg; (e) dung beetle α-actinin, 5 μg, and dung beetle β-paramyosin, 5 μg. A = actin.

Plate 4. Antibody labelling of fly myofibril with myosin removed. Extracted myofibrils were incubated with antibody to connecting protein (see Methods).

Plate 5. Reaction of antiserum with myofibrillar proteins. SDS gel electrophoresis of precipitates obtained by incubating dung beetle myofibril extract with antibody. (a) γ-globulin fraction of antiserum to impure paramyosin, 0.12 ml, with myofibril extract, 0.75 mg; (b) γ-globulin fraction of antiserum to impure paramyosin, from which anti-paramyosin was removed (Bullard *et al.* 1977), 0.25 ml, with myofibril extract, 1.0 mg; (c) normal serum, 0.25 ml; with myofibril extract, 1.0 mg. HC and LC = heavy and light chains of immunoglobulin.

Plate 6. *Lethocerus* myofibril labelled with antibody to connecting protein. Same antiserum as Plate 1 (c).

1

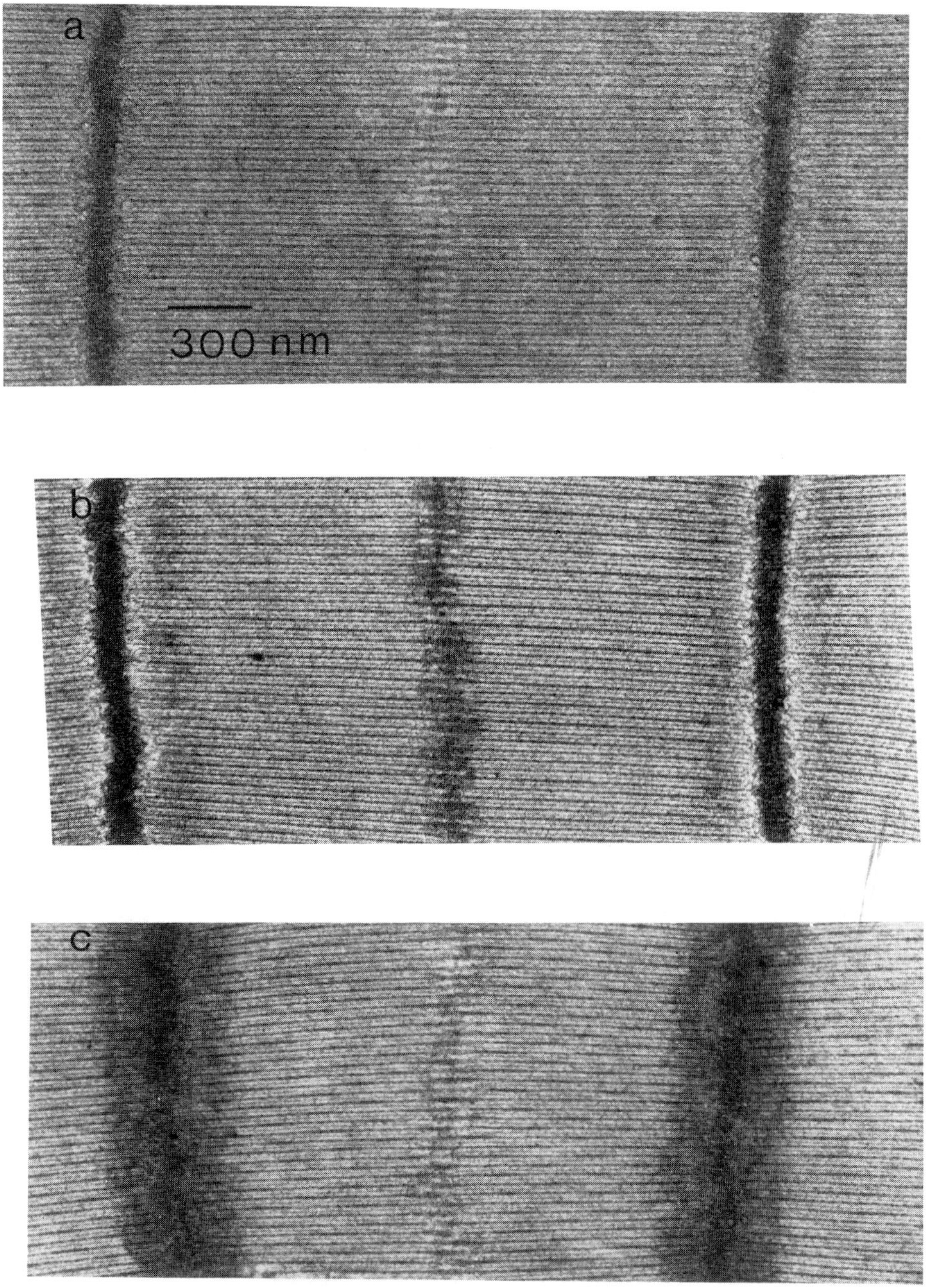

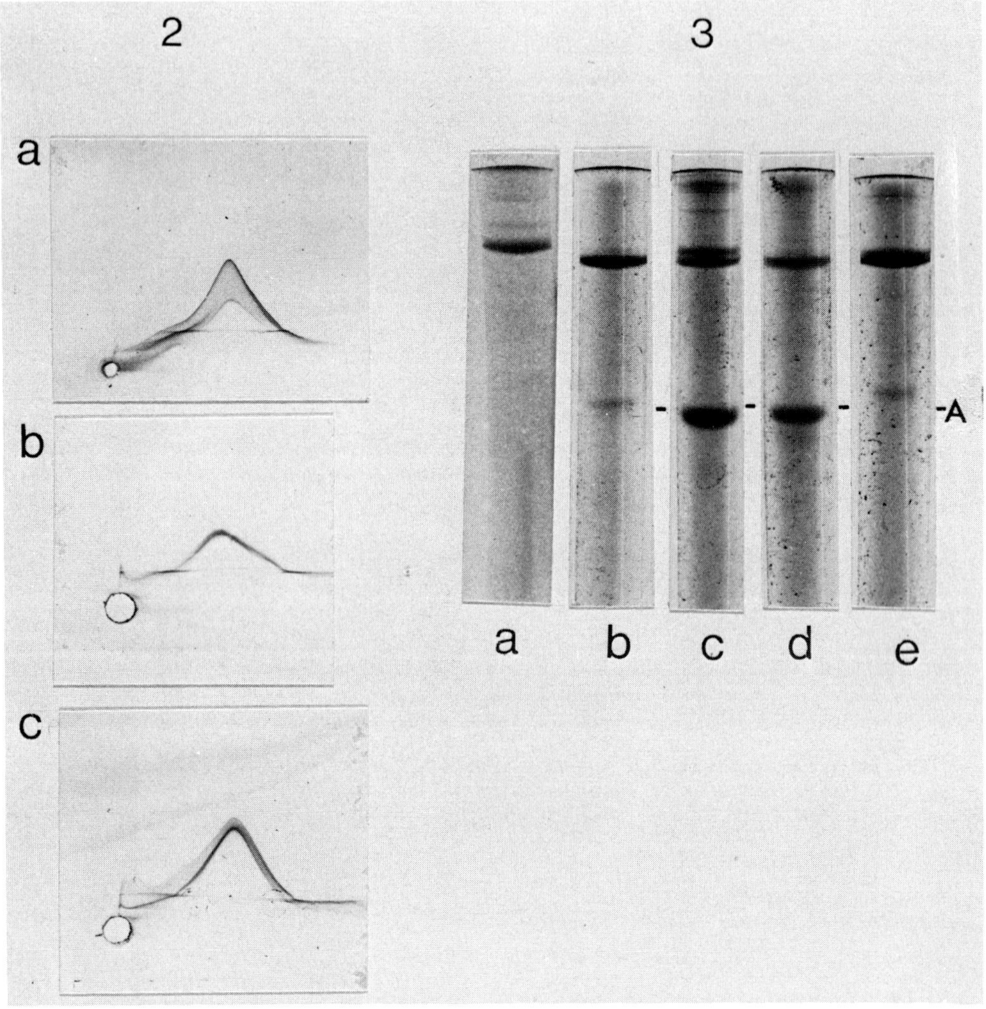
2
3
a
b
c
a
b
c
d
e
A

4

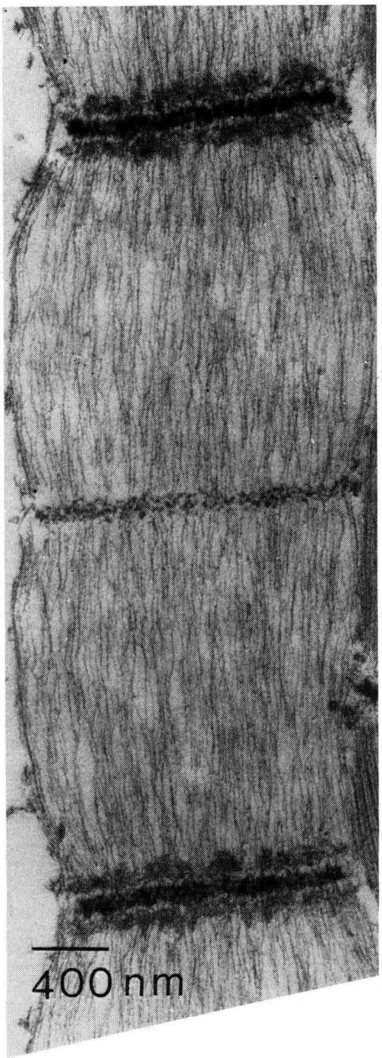

5

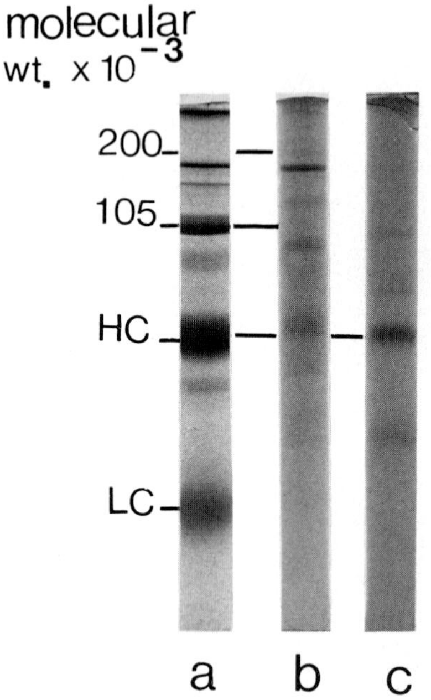

6

THE PARAMYOSIN CONTENT AND LOCALISATION IN THE HONEY BEE

C. Trombitas
Anna Tigyi-Sebes
G. Pallai

Central Laboratory, Medical University, 7643 Pécs Pf.99, Hungary.

We have some preliminary results of the paramyosin content and its location in the bee flight muscle. We prepared paramyosin from the thoracic muscle of the bee according to the method of Bullard and co-workers (1973). We found that this muscle contained a considerable amount of paramyosin. The bee paramyosin was contaminated with one higher and one lower molecular weight band after 10 precipitations, as seen on SDS gels, (Fig.1.).

Rabbits were immunised with this relatively pure paramyosin. Immunodiffusion of antiserum and paramyosin as well as antiserum and total myofibrillar proteins gave one precipitin band. The immune globulin was fractioned with ammonium sulphate. Fluorescein-isothiocyanate was coupled to immunoglobulin by the method of Wood and co-workers (1965). Myofibrils were prepared from glycerinated muscle in rigor according to the method of Ernst and co-workers, (1956) and stained according to the procedure described by Pepe (1966). In these preparations the I-bands only contained C-filaments (Fig.3.). These myofibrils were stained with fluorescein-labelled immunoglobulin from antiparamyosin. The I-bands were stained strongly and the elongated thick filaments in the middle of the A-bands weakly. The overlap zones could not be stained.

In the electron microscope preparations the antiparamyosin was mainly located on each side of the Z-line in a narrow band, as was found by Bullard (1974) and slightly along the M-line. The elongated thick filaments in the middle of the H-band were also stained with antiparamyosin. Our findings resemble those of Bullard *et al.* (1977) when impure paramyosin was employed.

REFERENCES

Bullard B. (1974) In : Pringle J.W.S. *The resting elasticity of insect flight muscle.* In: *Symposium on the muscle.* Biró E.N.A. and Garamvölgyi, N. (Ed). 67.

Bullard B., Luke B. and Winkelman L. *J. Mol. Biol.* 75 (1973) 359.

Bullard B., Bell J. and Luke B.M. (1977) This volume.

Ernst E., Béleczky I. and Nagy J. *Acta Physiol. Hung.* 9 (1956) 7.

Pepe F.A. *J. Cell.Biol.* 28 (1966) 505.

Wood, B.T., Thomson S.H. and Goldstein G. *J. Immunol.* 95 (1965) 225.

FIGURE LEGENDS

Fig.1. Sodium dodecyl sulphate gel electrophoresis of the purified paramyosin and the normal myofibrillar proteins.

Fig.2. Immunodiffusion of purified paramyosin and anti-paramyosin as well as normal myofibrillar proteins and anti-paramyosin.

ap = anti-paramyosin

pp = purified paramyosin

mp = normal myofibrillar proteins.

Fig.3. Isolated myofibrils, stretched in rigor condition. Scale 10μm.

Fig.4. Isolated myofibrils, stretched in rigor condition and stained with fluorescein-labelled anti-paramyosin immunoglobulin. Scale 10μm.

Fig.5. Isolated rest length myofibril stained with anti-paramyosin. Scale 1μm.

Fig.6. Stretched myofibril stained with anti-paramyosin. The anti-paramyosin is located in the elongation zone in the middle of the A-band. Scale 1μm.

1

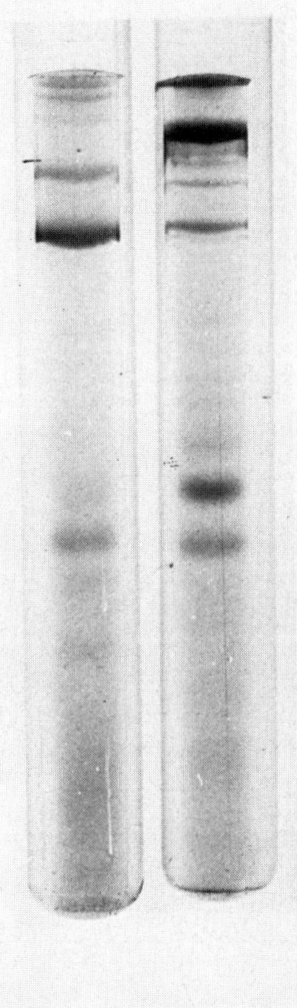

2

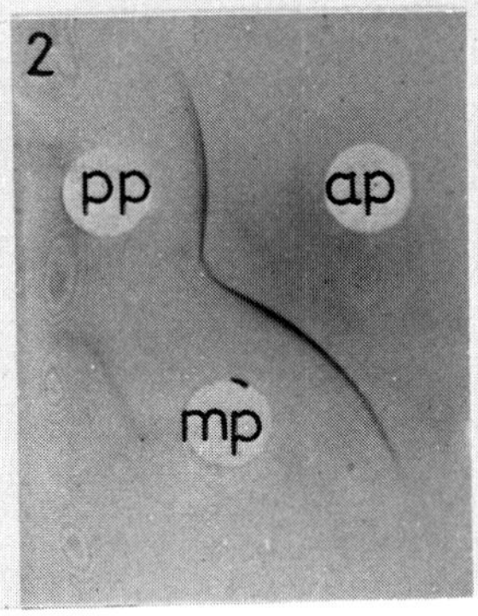

3
4
5
6

THE Z-LINE:
ITS STRUCTURE AND EVIDENCE FOR THE PRESENCE OF CONNECTING FILAMENTS

Doreen E. Ashhurst

Department of Anatomy, Medical School, University of Birmingham, Birmingham B15 2TJ, England.

INTRODUCTION

The Z-lines, which are usually seen in electron micrographs as an electron dense band, demark the boundaries of the sarcomeres along the myofibrils and serve essentially as a firm anchorage for the I-filaments of the adjacent sarcomeres. While it was easy to establish the filament lattices of the A- and I-bands of striated muscles, the Z-lines have proved somewhat refractory to study. The close proximity of the filaments within the Z-line and the electron dense, amorphous Z-line material together combine to obscure much of the inherent structure. In the fibrillar flight muscle of insects these problems are further compounded by the very narrow I-band, which is often partly obliterated by the dense Z-line material. A further complication is provided by the suggestion that the A-filaments are extended into the Z-line by connecting, or C-filaments.

In this paper the ultrastructural evidence on which several models of the Z-line of fibrillar muscle are based will be reviewed, and then an assessment will be made of the available evidence for the presence of C-filaments.

THE Z-LINE

Several models for the Z-line of fibrillar muscles have been proposed which are based on observations of transverse sections of myofibrils in which it is possible to trace the prolongations of the I-filaments into the Z-line. In insects, such as the water bug, *Lethocerus* spp., the myofibrils are up to 4 μm in diameter and in slightly oblique sections, it is possible to trace the progression from the filament lattice of the A-band of one sarcomere, through the I-band to the Z-line and then into the lattice of the I-band of the adjacent sarcomere and finally into the A-band of that sarcomere. Thus the lattice lines of the two adjacent sarcomeres can be traced into the common Z-line.

It is from sections such as this that the models have been derived. The first model was proposed by Auber and Couteaux (1963) from observations of the Z-lines of the Diptera, *Calliphora erythrocephala* and *Eristalis tenax*, the second was proposed by Ashhurst (1967) for the Hemiptera, *Lethocerus maximus* and *L. cordofanus* and the third by Saide and Ullrick (1973) for the hymenopteran, *Apis mellifica*. A fourth model for the honey-bee Z-line has recently been proposed by Trombitás and Tigyi-Sebes (1975), but as this differs fundamentally from the others, it will be discussed separately.

The models suggested by Auber & Couteaux, Ashhurst, and Saide and Ullrick are essentially similar and differ only in minor aspects of detail. The important common feature is that the I-filament lattices of adjacent sarcomeres are prolonged into the Z-lines where they overlap and interdigitate to produce the Z-line lattice. A feature of the filament lattice of fibrillar muscle is that alternate filament rows differ, one consists entirely of I-filaments, while the next has alternating A- and I-filaments. Thus in the I-band, the alternating rows are easy to recognise since one has twice as many filaments as the other; the possible contribution of the A-filaments, or their prolongations, to the I-band will be considered later. The rows of I-filaments from adjacent sarcomeres can be followed into the Z-line where their precise interdigitation can be seen; it is from sections such as Plate 1 that Ashhurst produced the model illustrated diagrammatically in Fig.1. The Z-line in transverse section appears to be pierced by a number of holes which are surrounded by six thin filaments forming a hexagon (Fig.2A). The Z-line material is preferentially associated with the filaments giving the hexagons a dense border and it is also found in the spaces between the hexagons; it does not appear to penetrate the hexagons and thus their centre has less electron density. It should be stressed that while the filament arrangement is clearly hexagonal, the actual shape of the holes in the Z-line is somewhat blurred so that perfect hexagons are rarely visible.

The ability to trace the lattices from two adjacent sarcomeres into the Z-line also removes any ambiguity about the alignment of the filament lattices of adjacent sarcomeres relative to each other; the lattices of the adjacent sarcomeres in the water bug flight muscle are rotated relative to each other, so that the A-filaments of one sarcomere occupy the triagonal positions in the A-filament lattice of the next sarcomere (see Fig.5).

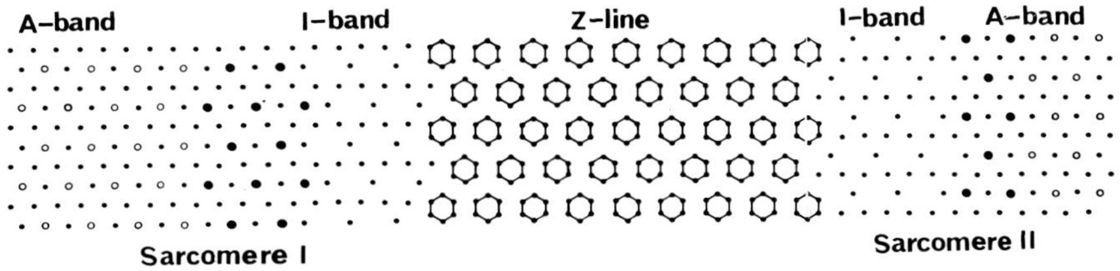

Fig.1. Diagram to illustrate the precise interdigitation of the I-filament lattices of adjacent sarcomeres which produces the Z-line lattice.

Auber *et al's.*, model was derived from fly muscle where the myofibrils are of smaller diameter (1-2μm) and it is not possible to see the lattices of the adjacent sarcomeres in their micrographs. Thus, while they were able to trace the I-filament lattices from one sarcomere into the Z-line, they were not able to account for the appearance of six filaments surrounding the holes in the centre of the Z-line, or to determine the relationship of the lattices of the adjacent sarcomeres unequivocally. They proposed two models, which are illustrated in Fig.2. C & D, to account for the presence of the six filaments. Firstly, they suggest that the I-filaments of adjacent sarcomeres might be exactly opposite each other and that they bifurcate and fuse in the centre of the Z-line (Fig.2.C), or secondly, that the I-filaments are staggered and that these bifurcate and join with the bifurcations of two different I-filaments from the adjacent sarcomere (Fig.2.D). Both these models require a deflection in the position of the filaments which surround the holes in the Z-line. Examination of Auber *et al's.*, micrographs and of my own micrographs of *Calliphora* Z-lines (Plate 2A) suggests that there is <u>no</u> deflection of the I-filament lattice as it enters the Z-line proper, and therefore it appears most likely that the dipteran Z-line is produced in exactly the same manner as that of the water bugs. In a later paper, Auber (1969) confirms that the filament lattices of adjacent sarcomeres are aligned relative to each other as in *Lethocerus*, but he says that the method of association of the thin filaments remains to be determined.

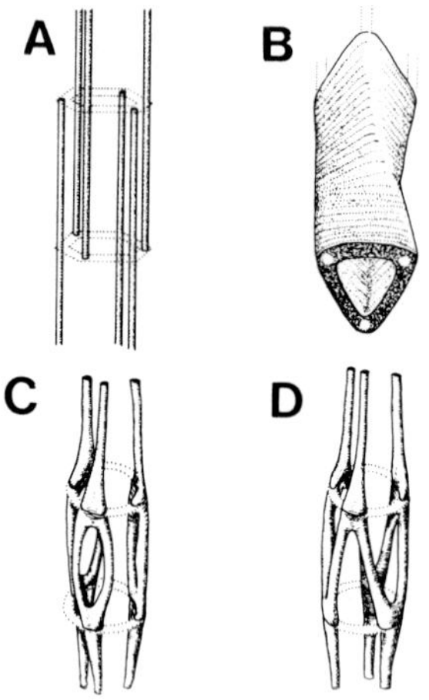

Fig. 2 A. Diagrammatic representation of the arrangement of the I-filaments around the "holes" in the Z-line of *Lethocerus*.
B. Drawing of the twisted Z-line tubes described by Saide *et al.* in *Apis*. (Reproduced by permission of Drs. J.D. Saide and W.C. Ullrick.)
C and D. Diagrams to illustrate the two possible methods of bifurcation and fusion of the I-filaments of adjacent sarcomeres suggested for *Calliphora* by Auber *et al.* (Reproduced by permission of Profs. J. Auber and R. Couteaux.)

The model for the honey-bee Z-line proposed by Saide *et al* is again similar. Here the authors can see the I-filament lattices of the adjacent sarcomeres approaching the Z-line in the manner proposed by Ashhurst (1967), but they found that the dense rims of the tubes passing through the Z-line obscured the position of the filaments. They do, however, on other evidence conclude that the Z-line lattice of the honey-bee is generated in exactly the same manner as that of the water-bugs (Saide *et al.* 1973; Fig. 2) . In the micrograph of a honey-bee Z-line (Plate 2B) it is possible to trace the I-filament lattices into the Z-lines and thus to confirm that the lattice is exactly similar to that of *Lethocerus*.

The holes which penetrate this Z-line most commonly appear to be triangular, rather than circular or hexagonal (Saide *et al.*, 1973; Ashhurst, unpublished observations), and triangular formations of dense amorphous material are visible peripheral to the Z-line proper. The triangles are in the positions where three I-filaments are closely associated, and thus due to the geometry of the lattice association, the triangles on one side of the Z-line will have a different orientation to those on the other side (Fig. 2B). Since the axis is straight and the lumen continuous throughout the Z-line, Saide *et al.* suggest that twisted triangular tubes, with prolongations of the I-filaments in their walls (Fig. 2B) occur in the honey-bee Z-line (Saide *et al.*, 1973, Fig. 2).

Fibrillar muscle also occurs in the Coleoptera, but so far there is no published description of the Z-line of a beetle muscle. The micrograph (Plate 2C) shows a slightly oblique section of a Z-line of the fibrillar muscle of *Scarabeus femoralis*, the dung beetle. Comparison of this micrograph with those in Plates 1 and 2A and B, indicates that the beetle Z-line is similar to those previously described; clear evidence for the interdigitation of the adjacent I-filament lattices can be seen.

How can we account for the relatively small differences in the appearance of these Z-lines described by the various authors? Two points are immediately apparent. Firstly there are differences in the fixatives used in the preparation of the tissue for electron microscopy. Auber *et al.* fixed the fly muscles in buffered osmium tetroxide only, Ashhurst fixed bee, fly, scarab beetle and water bug muscles in gluteraldehyde in cacodylate buffer and postfixed in aqueous osmium tetroxide, while Saide *et al.* fixed bee muscle in glutaraldehyde in phosphate buffer containing calcium chloride and postfixed in osmium tetroxide. The obscuring of the Z-line lattice mentioned by Saide *et al.* could be due to their use of phosphate buffer in the fixative; in my own experience more material is preserved after the use of phosphate-buffered glutaraldehyde than after cacodylate-buffered glutaraldehyde and it may be that less Z-line material is preserved by cacodylate-buffered fixatives and hence less material is present leaving the details of the Z-line visible. A second series of differences occur in the diameter of the myofibrils which mean that fly myofibrils must be cut more obliquely than the myofibrils of other insects to show the progression from the A-band into the Z-line and this leads inevitably to some distortion of the lattices. One difference which does appear to be real and unequivocal is that the I-filaments of the Diptera are enlarged and oval in cross-section just before they enter the Z-line.

It therefore seems reasonable to conclude that such differences as do occur in the structure of the Z-lines of fibrillar muscle from the four orders of insect in which they occur are small. The most important point is that the Z-line lattice is generated in precisely the same way in all these muscles. Differences in the shape of the "holes" in the Z-line appear to be due simply to the amount and disposition of the dense Z-line material.

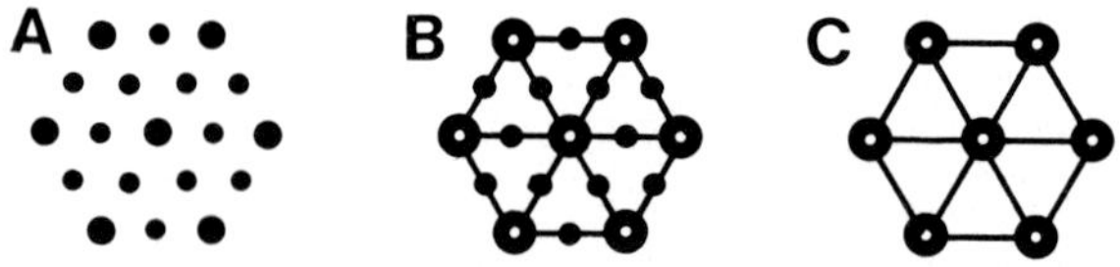

Fig. 3 Diagrams illustrating the Z-line model for *Apis* postulated by Trombitás *et al*
(Reproduced by permission of Drs. C. Trombitás and A. Tigyi-Sebes).

More recently, Trombitás and Tigyi-Sebes (1975) have proposed another model for the Z-line of the honey-bee (*Apis mellifica*) from their observations of slightly stretched, osmium fixed material. They consider that the A-filaments are continuous into the Z-line and that in the Z-line these extensions of the A-filaments from adjacent sarcomeres are continuous with Z-filaments, that is, the A-filaments of one sarcomere are joined to those of the adjacent sarcomere through the Z-filaments. The authors stress that the centre to centre spacing of the Z-filaments is exactly the same as that of the A-filaments (Fig. 3). Another point stressed by these authors is that the I-filaments cannot be seen in the Z-line. The micrographs of transverse sections used to support this model are of relatively thick, very contrasty sections; such sections could not be expected to show details of the I-filament lattice of the Z-line.

The essential feature of the model illustrated in Figure 3 is that the Z-line lattice is regular and that the hexagonal patterns produced by the material linking the Z-filaments are exact. A detailed examination of the micrographs used as the main evidence for this model (Trombitás *et al.*, 1975, Fig. 4) reveals that the hexagons are not regular. Alternate triangular sections are not only larger, but also less electron dense than the intervening triangles. The explanation of this phenomenon is simple and is illustrated in Figure 4; the lighter triangles are the holes in the Z-line surrounded by six I-filaments described by the previous authors, the dense lines are the accumulations of the Z-line material in association with the I-filament prolongations, and the more electron dense triangles are the spaces between the holes where there is less Z-line material. There is thus no unequivocal evidence for this model.

The model above is at complete variance with the models discussed earlier. It does not take into account the clear evidence presented by the previous authors for the precise interdigitation of the I-filament lattices of the adjacent sarcomeres, and the precise alignment of the lattices of the adjacent sarcomeres so that the A-filaments occupy a triagonal position relative to those of the next sarcomere. Indeed, in Trombitás *et al's* model, the lattices of the adjacent sarcomeres must be in the identical orientation, otherwise there could be no continuity of the A-filaments from one sarcomere to the next.

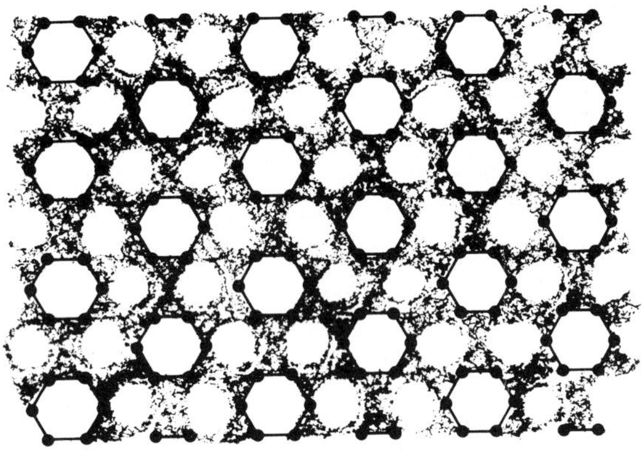

Fig. 4 Diagram to illustrate the explanation of Trombitás *et al's* model for the honey-bee Z-line. The hexagonal "holes" of the Z-line lattice are indicated. The other holes are produced by the non-uniform distribution of the Z-line material between the hexagons; in the honey-bee this results in less density in precisely the position where the A-filaments are if the complete lattices of two sarcomeres are superimposed (see Figure 5). Each of these areas is surrounded by a hexagonal array of six light areas, three like itself, and three "holes" of the Z-line lattice.

CONNECTING FILAMENTS (C-filaments)

In their 1963 paper, Auber *et al.* deduced from the transverse sections of the I-band of *Eristalis tenax* that the pointed end of the primary or A-filaments "est reliée à la strie Z par une formation moins dense que le filament primaire lui-même, mais qui conserve la même orientation que lui". The prolongations of the A-filaments are illustrated in Auber *et al's* Plate VIIb; close examination of this plate reveals amorphous material in the positions in the I-band lattice occupied in the A-band by the A-filaments. The material has no defined shape or structure; sometimes it appears to be in the shape of a star, the points of which extend towards the six surrounding I-filaments. The point which must be stressed here is that Auber *et al.* do not describe a filament joining the A-filaments to the Z-line, though their Figure 4 might suggest that they do.

The idea that the A-filaments might be joined to the Z-line in fibrillar muscle has some attraction. These muscles have peculiar elastic properties; for example, *Lethocerus* muscle in the resting state develops tension when stretched up to 9% of its resting length, but above 9% the tension does not increase further and the stretch is irreversible (Pringle, 1967). The proposed connections between the A-filaments and Z-line were invoked as the structural correlate of the required elastic element, and were henceforth termed "connecting" or "C"-filaments (Pringle, 1967). Thus Auber *et al's* star-shaped formations became "thin connecting filaments" (Garamvölgyi, 1965a, p.409; Zebe, Meinrenken and Rüegg, 1968, p.603). The existence of the C-filaments has become somewhat controversial and the evidence pertaining to them will now be critically reviewed.

Since the I-band is normally so narrow in resing fibrillar muscle, it is immediately apparent that ny slight contraction of the muscle during fixation could lead to the penetration of the ends of the A-filaments into the peripheral regions of the Z-line. The amorphous, electron dense, Z-line material does not have a regular distribution; in fact an examination of any muscle reveals that the boundary between the Z-line and the I-band is very difficult to determine, especially when critical measurements are required. It is therefore possible, and in my view probable, that the appearance of amorphous material forming an irregular shape in the I-filament lattices of fibrillar muscles is due to the distribution of the Z-line material, which will tend to protrude into the larger spaces produced in the I-band lattice by the absence of the A-filaments. When slight contraction occurs on fixation, the A-filaments will abut onto this material. In this context, it is significant that it is much more difficult to hold and thus prevent the contraction of the short flight muscle fibres of bees and flies than the long muscle fibres of water-bugs during fixation. It is, therefore interesting that Ashhurst (1967) and Cullen (1971) found no evidence from transverse sections of *Lethocerus* and other Hemipteran muscles for the presence of C-filaments; the I-band lattice consists only of I-filaments (Plate 3). In *Apis, Calliphora* and *Scarabeus* muscle, however, sections prepared in exactly the same manner as those of *Lethocerus* show amorphous material similar to that described by Auber *et al* in the I-filament lattice (Plate 2 A,B and C) but it is considered that this appearance is explained by the argument above.

Some authors have, however, suggested that I have interpreted the appearance of the I-band in my published micrographs of *Lethocerus* muscle incorrectly (Ashhurst, 1967). They insist either that they can count the same number of filaments in both the I- and A-band lattices (Ernst, Kovács, Metzger-Török and Trombitás, 1969), or that amorphous material is present in the appropriate C-filament position (Saide *et al.*, 1973). Perhaps I do not have such a good imagination, but the published micrographs are only two of many, of both fresh and glycerinated muscle, which show a clear I-band lattice, devoid of A-filaments or any continuation of them.

Some other workers have claimed to see C-filaments in longitudinal and transverse sections of *Apis, Bombyx,* and *Lethocerus* muscle. Saide *et al;* (1973,

Plate III) counted the number of filaments in marked areas of the A- and I-band lattices of *Apis* and concluded that the same number of filaments occur in both lattices; it is, however, difficult to see any precise regularity in this I-band lattice, but where some regular areas occur, spaces corresponding to the position of the A-filaments are visible. The transverse sections used by Garamvölgyi (1965) to illustrate the occurrence of C-filaments in *Bombyx* muscle lack contrast and are extremely difficult to interpret, while Zebe *et al.*'s (1968) *Lethocerus* muscle seems to be contracted since the I-band is obliterated and the solid tapering ends of the A-filaments are present in the lattice up to the Z-line. These authors took their main evidence for the presence of C-filaments from longitudinal sections. Zebe *et al.* suggest that the A- and I-filaments pass into the Z-line, where the A-filaments can be followed to the starting point of an I-filament, and *vice versa*. Garamvölgyi, on the other hand, suggests that A-filaments can be seen crossing the Z-line from one sarcomere to the next. Both these interpretations of longitudinal sections are in conflict with the models for the Z-line discussed earlier; it is implicit in the geometry of this model that no filament of one sarcomere is exactly opposite another in the adjacent sarcomere.

The explanation of Garamvölgyi's and Zebe *et al's* findings is found in the superimposition of the filaments within the thickness of the section. This is apparent from Figure 5, in which the filaments which will be included in a 48nm thick section are indicated. In sections 1 and 2, the A-filaments of one sarcomere could appear continuous with the A- or I-filaments respectively of the next sarcomere, while in section 3, there will be no filament continuity across the Z-line.

There is, therefore, no unequivocal evidence for the presence of connecting filaments from studies of unstretched fresh or glycerinated fibrillar muscle.

Some completely distinct evidence for the presence of connections between the A-filaments and Z-line has been obtained by the examination of longitudinal sections of fibrillar muscle stretched in rigor to 115 to 200% of rest length. Under these conditions, the I-filaments are pulled away from the Z-line leaving a clear zone. Thin filaments which emanate from the ends of the A-filaments can be seen crossing this light zone to insert into the Z-line (Plate 4A) (Cullen, 1971; Garamvölgyi, 1969; Reedy, 1971; Trombitás and Tigyi-Sebes, 1974; White, 1967; White and Thorson, 1973). In addition, White *et al.* have transverse sections (Plate 4B) through this region which show that the C-filament lattice has the same organisation and dimensions as the A-filament lattice of fibrillar muscle. Thus in stretched muscle, there is continuity between the A-filaments and Z-line, but it must again be emphasised that these filaments appear entirely different to the formations described by Auber *et al.* (1963). Why these filaments are visible only after stretching muscle fibres in rigor remains, as yet, a mystery.

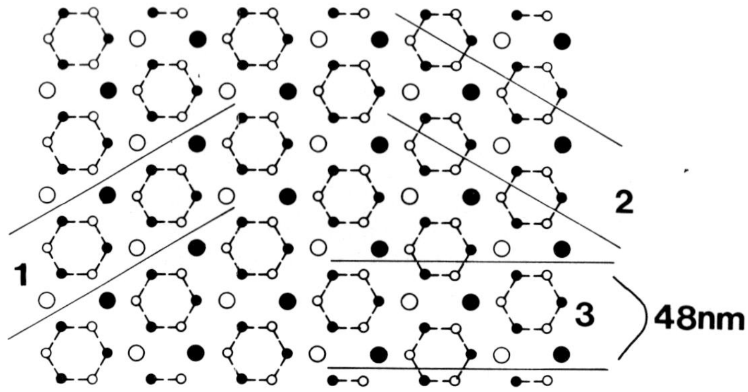

Fig. 5 In this diagram, the complete filament lattices of two sarcomeres are superimposed so that the I-filaments form the hexagonal lattice seen in the Z-line. The circles O represent the filaments from one sarcomere, ● from the other; the A-filaments are shown larger. The lines inserted into the lattice indicate the filament composition of 48 nm thick longitudinal sections of this Z-line. See text for further explanation.

There is now evidence accumulating for the presence of a narrow band of material extending outwards from the Z-line to enclose the ends of the A-filaments. The presence of a protein peculiar to this region has been found immunologically, (Bullard,personal communication). It is suggested that, on stretching, the molecules of this material become more highly organised and orientated to form filaments and that this extension, resulting in a more precise molecular orientation, alters the staining properties of the protein so that it is now stained by the routine electron microscopical methods (Pringle, 1974).

This hypothesis offers the most satisfactory explanation of the anomalies discussed above which is presently available. In conclusion, Pringle (1974) suggests that the term "connecting" or "C"-filament should now be abandoned and instead the term "connecting material" should be employed for the material responsible for the phenomenon of resting elasticity.

CONCLUSIONS AND DISCUSSION

The main conclusions which can be drawn from the preceding discussion are

1. The Z-line of fibrillar flight muscle is produced by a precise interdigitation of the I-filament lattices of the adjacent sarcomeres and it follows from

the geometry of this interdigitation that the filament lattices of the adjacent sarcomeres are aligned so that the A-filaments of one sarcomere occupy a tri-agonal position with respect to those of the adjacent sarcomeres. The nature of the amorphous Z-line material is unknown.

2. The evidence for the presence of connecting or C-filaments linking the A-filaments to the Z-line is still equivocal. True filaments in this position are seen only when a myofibril in rigor is stretched and the I-filaments are pulled away from the Z-line.

The Z-line is still a source of controversy and many questions remain to be answered. For instance, the nature of the prolongations of the I-filaments within the Z-line is known, though, since Bullard and Sainsbury (1977) found that actin is one of the major Z-line proteins in *Lethocerus cordofanus*, these filaments may be composed of actin throughout their length. Tropomyosin and α-actinin are also major protein constituents; the former may play a part in forming the lattice, while the latter may be a cross-linking protein (Bullard *et al.*, 1977). The presence of the minor components, myosin and paramyosin may be due to remnants of the A-filaments adhering to the isolated Z-discs, but small amounts of a further unidentified protein of molecular weight 180,000 were also found in *Lethocerus* Z-lines. This may be similar to the polypeptide of molecular weight 175,000 isolated by Saide and Ullrick (1974) from honey-bee Z-discs. These authors also found three other polypeptides of molecular weights 87,000, 113,000 and 158,000; these were not found in water-bugs by Bullard *et al.* (1977). Earlier studies had suggested that since the Z-line was removed by lipase digestion, it must contain a lipoprotein (Garamvölgyi, 1965b), but the later workers have attributed this digestion to impurities in the lipase (Bullard *et al.*, 1977).

Despite investigations such as those just described, we still do not know the composition of the amorphous Z-line material, and whether this, or some other protein is responsible for cementing the filaments within the Z-line together. The distance between adjacent filaments around the holes in the Z-line lattice is about 14nm (this approximation is derived from Figure 1 and the measurement of the centre to centre spacing of the myosin filaments in *Lethocerus* of 48 nm (Reedy, 1967)), and whatever is responsible for holding these filaments in position, is also responsible for the integrity of the myo-fibril. Without a firm anchorage for the I-filaments in the Z-line, the filament lattices of the sarcomere could not exist.

The nature of the association between the A-filaments and Z-line also remains to be elucidated. The recent work in which the presence of the pro-tein peculiar to the I-band has been established (Bullard, personal communication) would seem at present to offer an attractive solution if the protein proves to possess the necessary properties which could account for the filamentous connections being visible only when a myofibril in rigor is stretched. Further work should show whether this is so.

ADDENDUM

Sainsbury and Hulmes (this volume) describe the optical diffraction pattern produced from an electron micrograph of a transverse section of the *Lethocerus* Z-line lattice. The pattern indicates that the spacings of the I-filament prolongations into the Z-line are not uniformly spaced around the holes. The filaments are in pairs, one from each adjacent sarcomere; the distance between their centres is approximately 14.4 nm, while the distance between neighbouring pairs is approximately 16 nm. This paired arrangement would account for the triangular appearance of the holes, which is most pronounced in the bee and beetle Z-lines.

The next problem is how and where this deflection of the I-filament lattices occurs. The irregular spacings cannot be accounted for by a slightly different rotation of the filament lattice as this would distort the lattice. Therefore, a slight deflection of the I-filaments must occur in the Z-line.

REFERENCES

Ashhurst, D.E. (1967) *J. Mol. Biol.*, 27, 385 and 441.

Ashhurst, D.E. (1971) *J. Mol. Biol.*, 55, 283.

Auber, J. (1969) *J. Microscopie*, 8, 197.

Auber, J. and Couteaux, R. (1963) *J. Microscopie*, 2, 309.

Bullard, B. and Sainsbury, G.M. (1977) *Biochem. J.*, 161, 399.

Cullen, M.J. (1971) D.Phil. thesis, University of Oxford.

Ernst, E., Kovács, K., Metzger-Török, G., Trombitás, C. (1969) *Acta Biochim. Biophys. Acad. Sci. Hung.*, 4, 177.

Garamvölgyi, N. (1965a) *J. Ultrastructure Res.*, 13, 409.

Garamvölgyi, N. (1965b) *J. Ultrastructure Res.*, 13, 425.

Garamvölgyi, N. (1969) *J. Ultrastructure Res.*, 27, 462.

Pringle, J.W.S. (1967) *Progr. Biophys. Mol. Biol.*, 17, 3.

Pringle, J.W.S. (1974) *Symp. Biol. Hung.*, 17, 67.

Reedy, M.K. (1967) *Amer. Zool.*, 7, 465.

Reedy, M.K. (1971) In Contractility of muscle cells and related processes. Ed. R.J. Podolsky; Prentice-Hall, Englewood Cliffs, N.J. p. 227.

Saide, J.D. and Ullrick, W.C. (1973) *J. Mol. Biol.*, 79, 329.

Saide, J.D. and Ullrick, W.C. (1974) *J. Mol. Biol.*, 87, 671.

Trombitás, K. and Tigyi-Sebes, A. (1974) *Acta Biochim. Biophys. Acad. Sci. Hung.*, 10, 83.

White, D.C.S. (1967) D.Phil thesis, University of Oxford.

White, D.C.S. and Thorson, J. (1973) *Prog. Biophys. Mol. Biol.*, 27, 173.

Zebe, E., Meinrenken, W. and Rüegg, J.C. (1968) *Z.Zellforsch*, 87, 603.

PLATE LEGENDS

Plate 1. Electron micrograph of a transverse section of a myofibril of *Lethocerus maximus* at the level of the Z-line. The I-filament lattices of the sarcomeres I and II can be traced into the Z-line where they interact to form the Z-line lattice.

Plate 2. Electron micrographs showing the filament lattices of the Z-lines and adjacent sarcomeres of:

A. *Calliphora erythrocephala*

B. *Apis mellifica*

C. *Scarabeus femoralis*

Plate 3. Electron micrograph showing the I-band of *Lethocerus maximus*. Spaces in the lattice corresponding to the position of the A-filaments in the A-band can be clearly seen (arrows).

Plate 4. Electron micrographs of myofibrils of *Lethocerus cordofanus* stretched in rigor. The clear zones are visible.

A. Longitudinal section which shows thin filaments emanating from the A-filaments and crossing the clear zone to the Z-line.

B. Transverse section showing the A-band lattice and part of the clear zone. In the latter filaments which are arranged in a lattice identical to that of the A-filaments are seen.

Micrographs provided by Dr. D.C.S. White.

1
I
II
100 nm

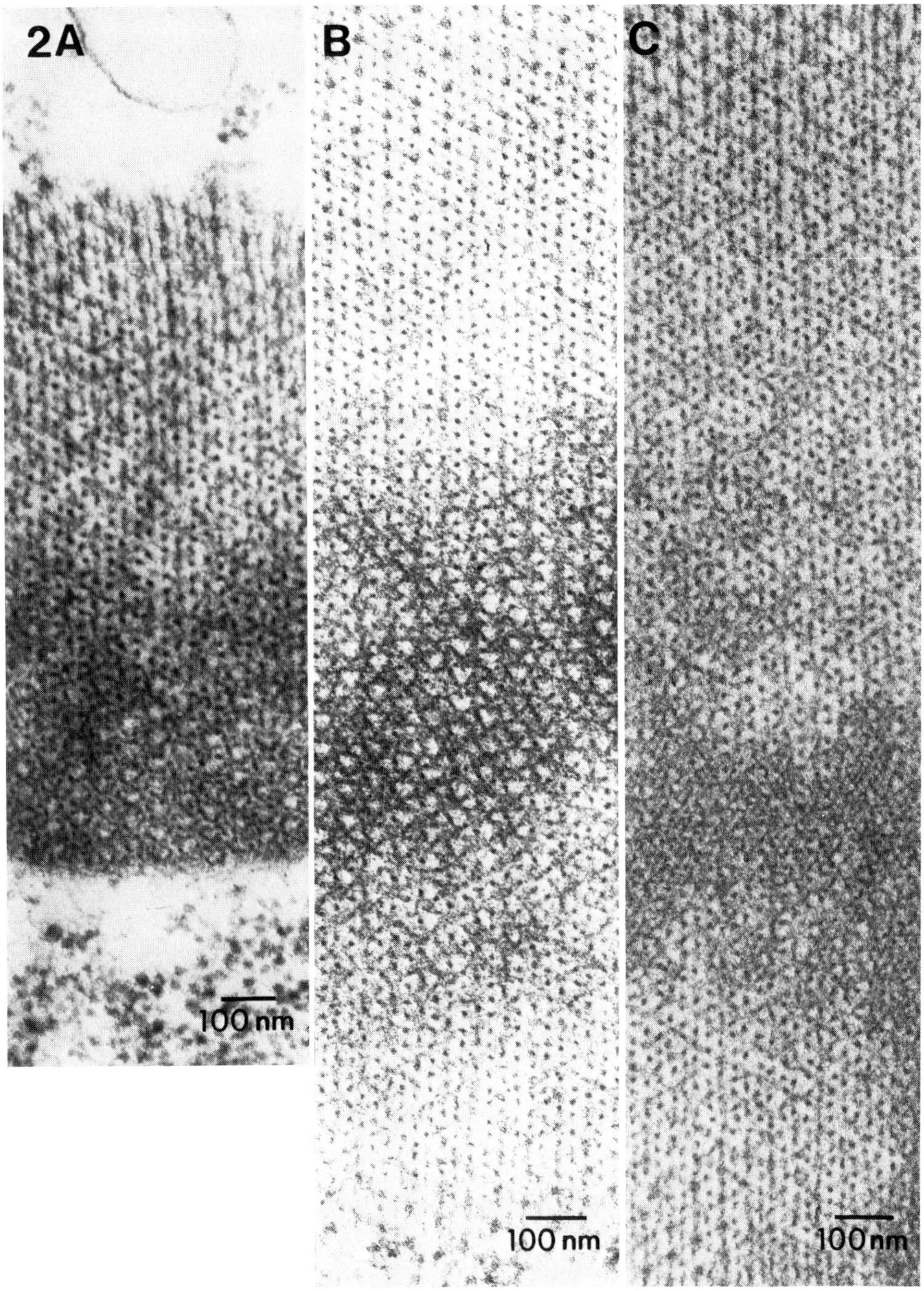
2A
B
C
100 nm
100 nm
100 nm

3
100 nm

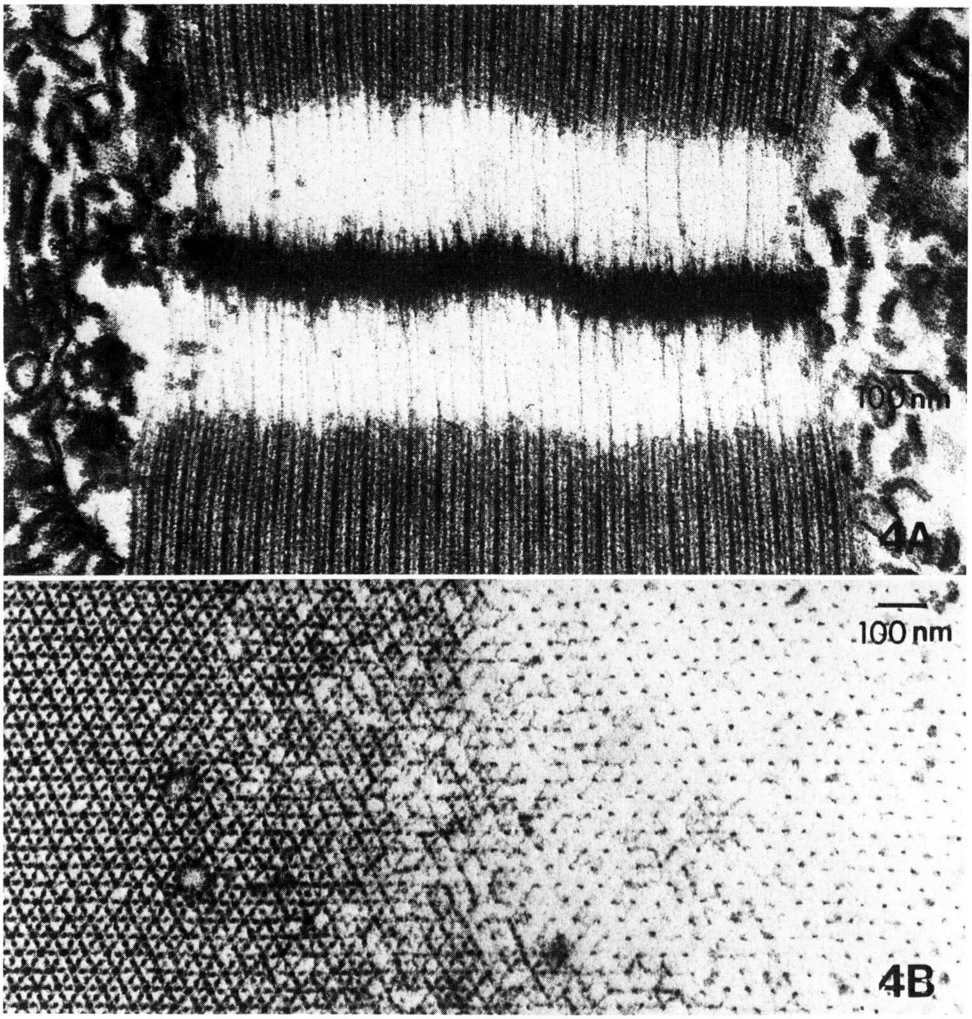
100 nm
4A
100 nm
4B

NOTES ON THE STRUCTURE OF THE Z-DISC OF INSECT FLIGHT MUSCLE

Gillian M. Sainsbury* and David Hulmes+

*A.R.C. Unit, Department of Zoology, Oxford.
+Molecular Biophysics Laboratury, Department of Zoology, Oxford.

Optical diffraction patterns from electron micrographs of transverse sections of *Lethocerus* flight muscle were obtained using a diffractometer from Polaron Instruments Ltd., London. An optical reconstruction derived from a diffraction pattern is shown in Plate I b); Plate I a) shows the original micrograph from which the diffraction pattern was made. The optical reconstruction substantiates the model of the Z-disc described by Ashhurst (1967, 1971), in which groups of three I-filaments from one sarcomere interdigitate with three I-filaments from an adjacent sarcomere to form a hexagonal lattice. In this model no filament of one sarcomere is opposite a filament from the adjacent sarcomere and A-filaments are absent. The optical reconstruction (Plate I b) shows clearly that each hexagon formed by the dense I-filaments from adjacent sarcomeres is composed of three pairs of I-filaments. The two I-filaments forming a pair are derived from adjacent sarcomeres. The pairs are linked at the perimeter of the hexagon by a structure of less dense material, with which they form a triangle. This less dense material is in turn linked by one or two strands to two different pairs of linked I-filaments in the adjacent hexagon. Links of less dense material may also be seen on the inner border of each hexagon, forming a circle around a white central space. Each pair of I-filaments in a hexagon is in the order of 14.4 nm apart, from centre of filament to centre of filament, whereas unlinked I-filaments in the hexagon are in the order of 16 nm from centre to centre. This implies that the linking process leads to a slight deflection of I-filaments as they pass through the disc.

Plate II shows an electron micrograph of a transverse section through a Z-disc which has been isolated from the contractile elements of the whole myofibril, as described by Bullard and Sainsbury (1977). This micrograph demonstrates that it is possible to isolate Z-discs using solutions of high ionic strength without substantially disrupting the Z-line structure. This micrograph can be compared with that in Plate I a) in which the section is cut from an intact myofibril.

We are grateful to Mrs B.M. Luke for the electron microscopy.

REFERENCES

Ashhurst, D.E. (1967) *J. Mol. Biol.* 27, 385.

Ashhurst, D.E. (1971) *J. Mol. Biol.* 55, 283.

Bullard, B. and Sainsbury, G.M. (1977) *Biochem. J.* 161, 399.

PLATE LEGENDS

Plate I a) Electron micrograph of a transverse section through the Z-disc of *Lethocerus* flight muscle. Scale bar 200 nm.

b) Optical reconstruction from transverse section micrograph of *Lethocerus* flight muscle. Scale bar 200 nm.

Plate II Electron micrograph of a transverse section through an isolated Z-disc of *Lethocerus* flight muscle. Scale bar 200 nm.

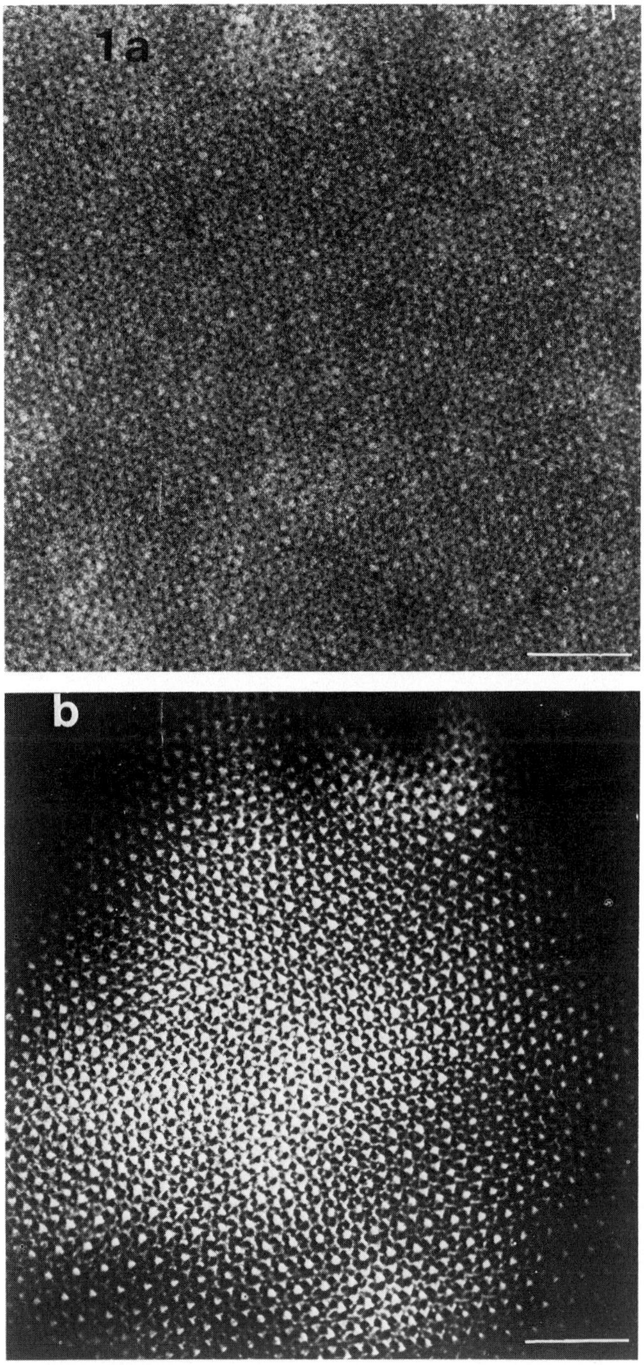
1a
b

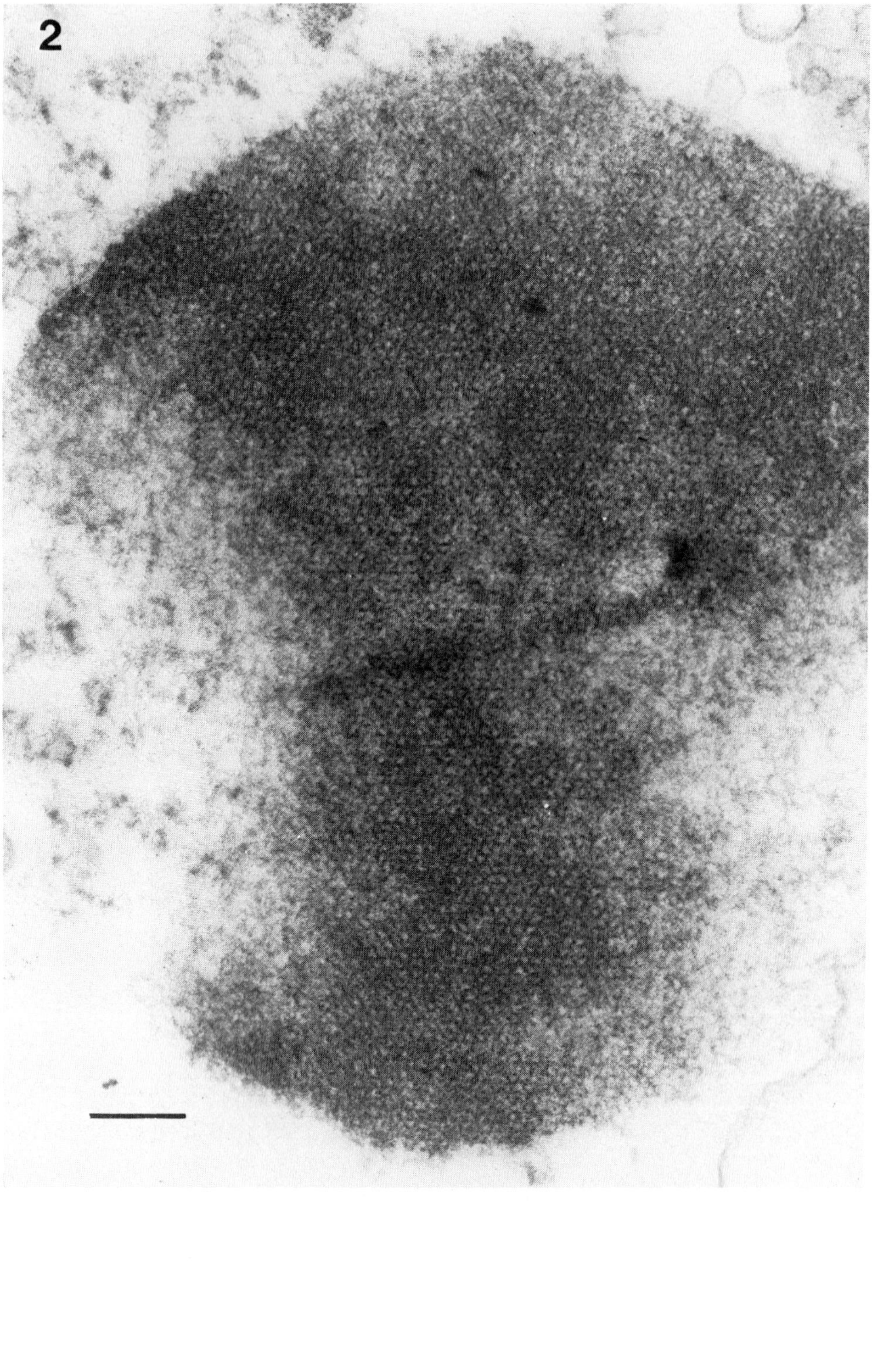

FINE STRUCTURE AND MECHANICAL PROPERTIES OF INSECT MUSCLE

Charles Trombitás and Anna Tigyi-Sebes

Central Laboratory, Medical University, 7643 Pécs, Pf. 99., Hungary

INTRODUCTION

It is a well-known fact that insect flight muscles are generally very inextensible even in a relaxed state. This means that these muscles cannot withstand an extension of more than about 10% without tearing or stretching irreversibly; at this extension overlap has only been reduced by a similar percentage, as was reported by Hanson, 1956; Pringle, 1967; and White, 1967. On the other hand, Garamvölgyi, 1965 showed that honey-bee flight muscle has similar extensibility to vertebrate muscle if the stretch is performed slowly. This result was supported by Saide and Ullrich, 1973 and Pringle, 1974. However, a relatively fast stretch causes a change in the extensibility of bee muscle. A part of the muscle remains at its original extensibility, while the other part of the same muscle becomes inextensible like other insect flight muscles. Irreversible extension of fibrils causes the appearance of several different sarcomere structures which may be used to examine the problems of resting elasticity and inextensibility of flight muscle. They may also be used to verify the existence of C-filaments and the localisation of protein within difficult regions of the sarcomere.

METHODS

In our experiments fibre bundles and isolated myofibrils of honey-bee flight muscle (*Apis Mellifica*) were used.

(a) Fibre bundles. The bees were prepared according to the method of Garamvölgyi and Belágyi (1968) at room temperature. The two dorso-ventral flight muscles were left *in situ* in the thorax and all other connections between the top and bottom of the thorax were cut. The wings as well as the mesothoracic legs were bound to a glass stick by means of a thin thread of cotton. The muscle was stretched relatively fast, the stretch being applied in about two or three seconds by hand. The extension of the sarcomeres varied from a few per cent to about 200 per cent. The complete preparation of the bee was carried out in about two minutes. During the preparation no bathing solution was used. Then the muscle was either

immediately fixed. A double fixation with glutaraldehyde and osmium

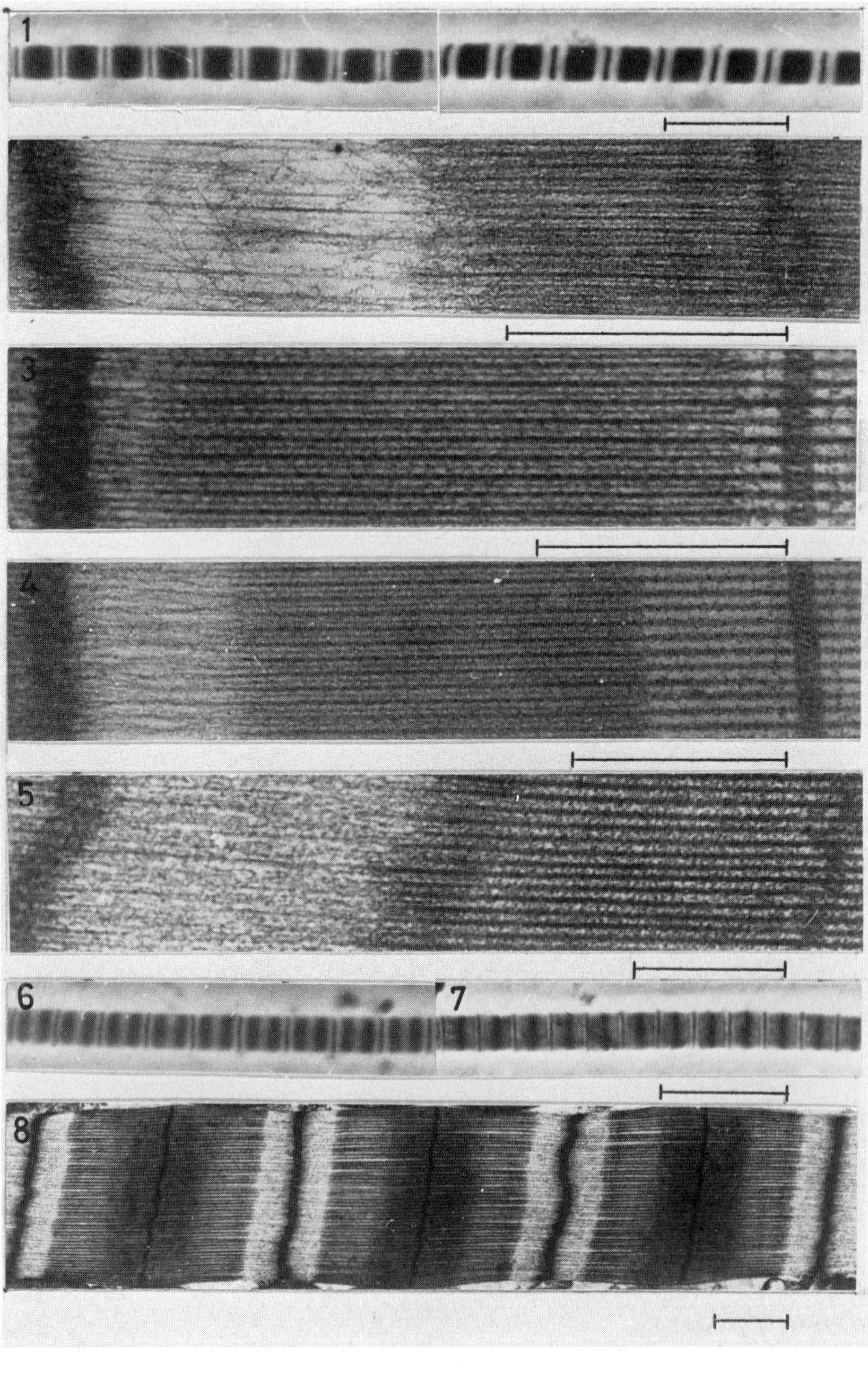
1
2
3
4
5
6
7
8

tetroxide was used in the usual manner.

<u>or</u>

extracted with glycerol. After three weeks or longer storage at -20°C the muscle was

i. fixed in the above manner

ii. treated with rigor solution and released before fixation

iii. treated with rigor solution and stretched again before fixation

iv. treated with rigor solution, stretched again and treated with activating solution of ITP-containing solution before fixation.

Dehydration and embedding were performed in the usual way. The sections were cut so that the fibre axes were parallel to the knife edge.

TABLE I Solutions (mM)

	ATP	ITP	$MgCl_2$	EGTA	Ca-EGTA	KCl	PO_4	References
Relaxing	5	-	5	5	-	50	20	White, 1970
Activating	5	-	5	5	5	50	20	White, 1970
Rigor	-	-	5	5	-	50	20	White, 1970
ITP-activating	-	5	5	-	5	50	20	Zebe *et al.*, 1968

All solutions were at pH 7.1.

(b) <u>Isolated myofibrils</u> The myofibrils were isolated according to the method of Ernst *et al*,(1956) by disintegration of the fibres between needles on a slide in a drop of either Pringle's solution or relaxing solution. In these preparations myofibrils of different striated patterns were present. The individual myofibrils were examined in the phase contrast microscope and were then fixed and processed in situ on the slide, using the reagents described above. The myofibrils were embedded in capsules containing Araldite which were inverted

---

Fig.1 Stretched fresh myofibrils in Pringle's solution on a slide. H-zones are not seen. Scale 10 μm

Fig.2 Electron micrograph of fresh myofibril stretched in Pringle's solution on a slide. The A-band contains both the thick and thin filaments as the muscle is stretched in rigor. Scale 1 μm

Fig. 3-5 Muscle stretched in life; varying degrees of stretch. The regular striated pattern as well as both the thick and the thin filaments are observable. Scale 0.5 μm

Fig.6 Stretched fresh fibril in relaxing solution on a slide, showing normal striated pattern. Scale 10 μm

Fig.7 Myofibril from a muscle stretched in life, after glycerination and homogenisation. Dense M-bands accompanied by very short I-bands. Scale 10 μm

Fig.8 Muscle stretched in life; "reversed striation" A-bands. Scale 1 μm

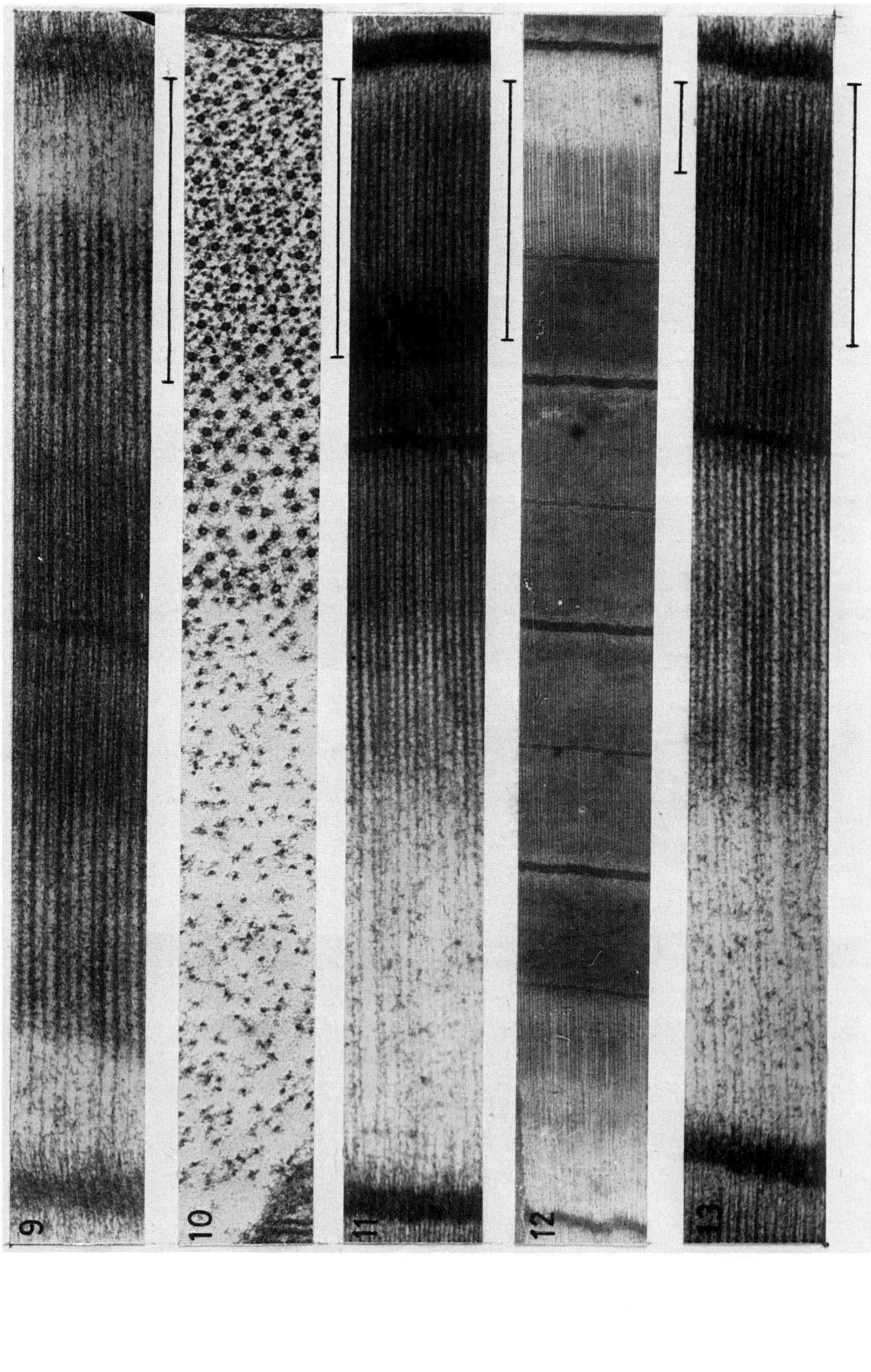
9
10
11
12
13

over the slide surface. After polymerisation the blocks were peeled off along with the embedded myofibrils from the slide surface, thin sections were made and the individual myofibrils were examined in an electron microscope.

RESULTS AND DISCUSSION

Extensibility of the fresh muscle

Hanson (1956) reported that myofibrils of blowfly isolated in 0.1 M phosphate buffer were only able to elongate a few per cent reversibly, and further stretch produced very light irreversible bands on one or both sides of the Z-line. Similar clear I-bands were observed by Garamvölgyi and co-workers (1964) in isolated myofibrils of bee stretched in Pringle's solution on a slide, but they did not consider the apparent absence of the H-zones as an irreversible change (Figure 1). Our micrographs taken from these isolated and stretched fresh myofibrils show that H-zones really are absent. The whole A-band contains thick and thin filaments alternately, as if the myofibrils had stretched in rigor condition (Figure 2). It seems that there is no essential difference between the extensibility of the blowfly myofibrils and that of the bee myofibrils when the solution does not contain ATP, in spite of the fact that the intact flight muscle of bee is very extensible (Garamvölgyi, 1965; Saide and Ullrick, 1973; Trombitás and Tigyi-Sebes, 1975). It is very probable that these solutions dilute the ATP of the myofibrils enough for them to behave as if they were in the rigor state. But when the fresh bee myofibrils were stretched in relaxing solution, they showed the normal striated pattern (Figure 6). This phenomenon suggests that the extensibility of the myofibril depends fundamentally on the state of the cross bridges.

It is usually assumed that in the normal external environment of the body fluids and the absence of nervous excitation, an intact muscle is in a fully relaxed state. In spite of this fact, when the fresh flight muscle of the bee was stretched relatively fast in the thorax, the individual sarcomeres in the same fibre did not elongate uniformly. A part of the muscle elongated in a

---

Fig. 9 Muscle stretched in life: the dense zone in the middle of the A-band contains thin filaments C filaments connect the thick filaments to the Z-line. Scale 1 µm.

Fig. 10 Muscle stretched in life: transverse section. The end of the A-band does not contain thin filaments. Scale 0.5 µm

Fig. 11 Glycerinated muscle, stretched in rigor solution and treated with activating solution. The broken thin filaments are moved across the M-line. Scale 1 µm

Fig. 12 and 13 Glycerinated muscle, stretched in rigor solution and treated with ITP-activating solution. In the stretched sarcomeres the broken thin filaments are almost totally moved from the stretched part of the sarcomere to the contracted side. Scale 1 µm

14

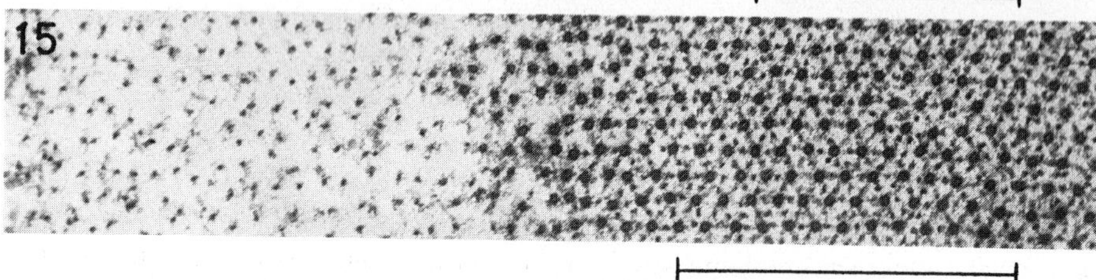
15

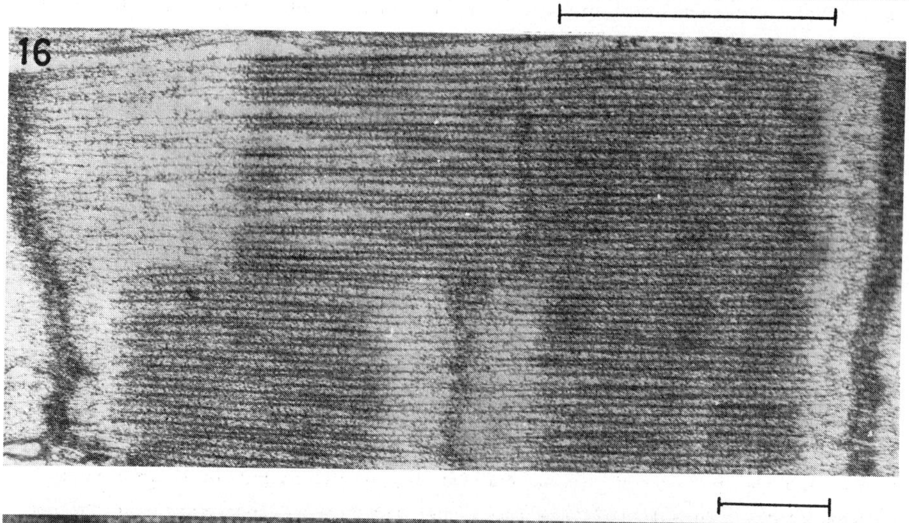
16

17

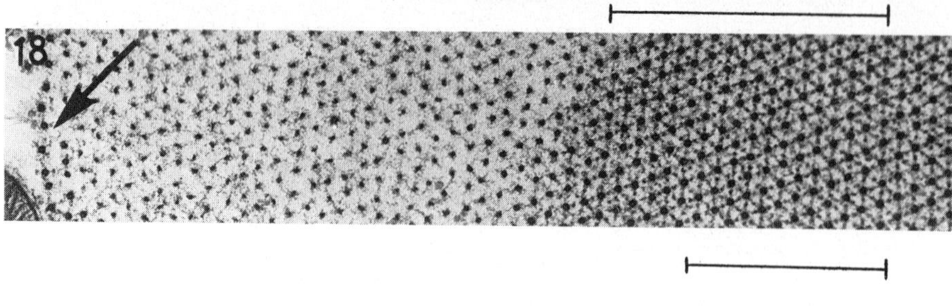
18

regular manner, i.e. the elongation of a sarcomere was accompanied by an increase of the I-bands together with a corresponding widening of the H-zone and a decrease of the length of the overlap zones. This process was observed over the whole range of elongation, up to a 200% extension , (Figures 3, 4 and 5), and the length change was always reversible. After the release of the muscle the thin filaments returned to their original position, while the I-bands and H-zones disappeared.

On the other hand in a large part of the muscle the sarcomeres remained in a shortened state or elongated irreversibly. In the latter case we observed different fibril structures. The most frequent and striking phenomenon was the appearance of the "reversed striation" in the A-bands, where the middle of the A-band had a dense zone and both ends of the A-band became lighter than the overlap zones, (Figure 8). This phenomenon could also be observed in the phase contrast microscope when muscles were first stretched and then glycerinated before homogenisation (Figure 7). The central dense zone of the A-band contains all of the actin filaments, as can be seen in the electron microscope (Figures 9 and 10). In this case clear evidence for the presence of connecting filaments could be seen. Since both ends of the A-band contained no actin filaments, the continuity of the thick and connecting filaments was clearly visible. This has only been demonstrated previously by White (1967) and Reedy (1972) under special conditions, namely in muscles stretched in the rigor state. A possible explanation of this phenomenon is as follows. Immediately upon stretch the sarcomeres become activated and further stretch evokes "stretch activation", the maintained tension of which inhibits the sliding movement of the filaments. As a result the actin filaments are broken and moved towards the M-line. This idea could be supported by experimental data. Similar fibril structure was developed when the glycerinated muscle stretched in rigor solution was treated with activating

---

Fig. 14 Muscle stretched in life: the A-band contains the interlocked A- and I-filaments as in the rigor muscle. C-filaments connect the thick filaments to the Z-line. Scale 0.5 μm

Fig. 15 Muscle stretched in life: the frequency of the thin filaments in the I-band correspond to the frequency of the thick filaments in the A-band. Scale 0.5 μm

Fig. 16 The elongation of the thick filaments in the middle of the H-band increases the length of the A-band. Muscle stretched in life.

Fig. 17 Muscle stretched in life: the arrangement of the filaments in the case of the elongated H-band. Scale 0.5 μm

Fig. 18 Muscle stretched in life: the tapered thick filaments in the elongated zone maintained their original positions. The peripheral thick filament layers are thicker (arrow). Scale 0.5 μm

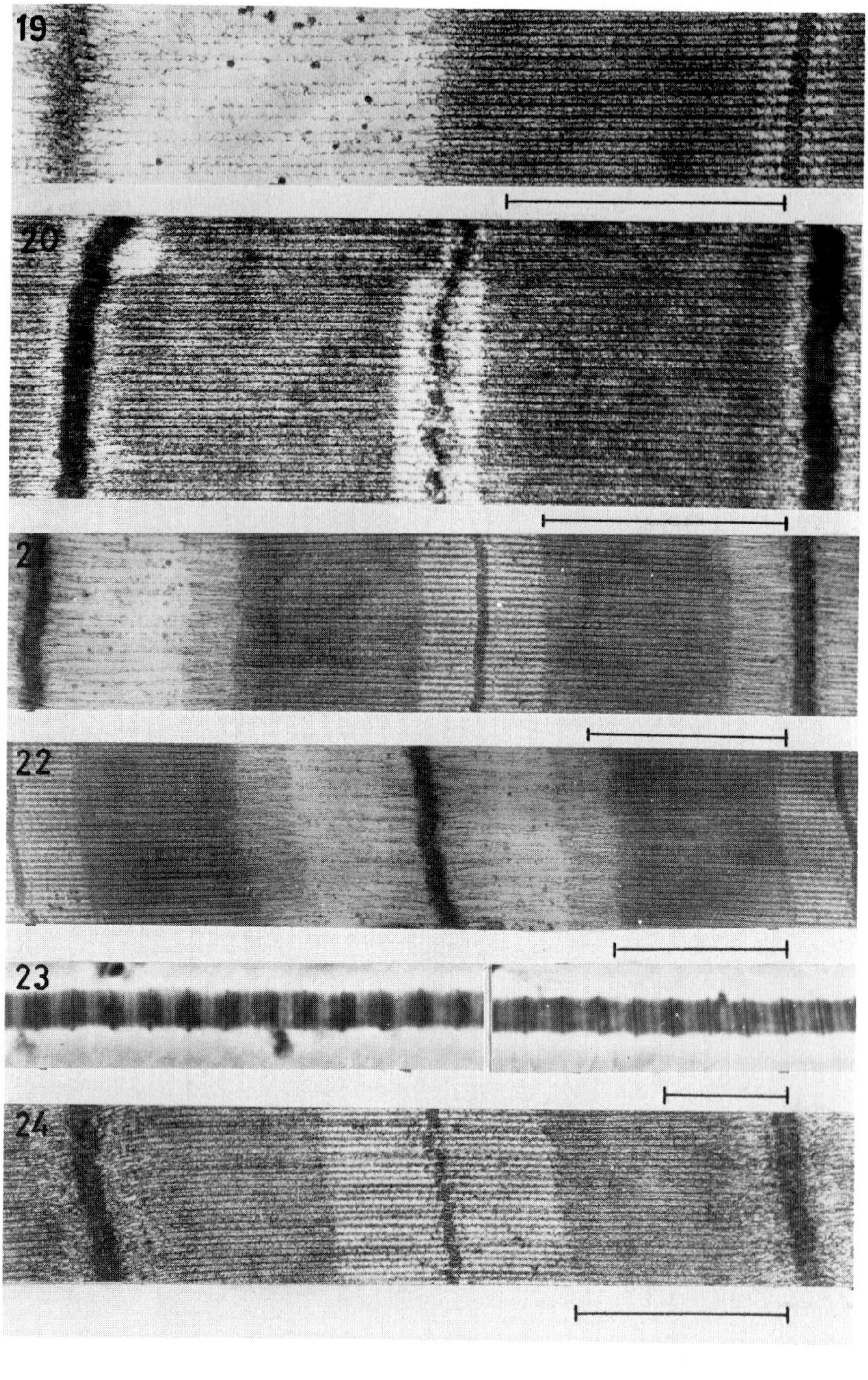
19
20
21
22
23
24

solution or ITP-activating solution (Fig.11). In the ITP-activating solution the actin filaments occasionally moved from one half of the sarcomere into the other (Figs. 12 and 13).

We have infrequently observed fibrils in which the actin filaments remained in their original position in spite of the stretch. In this case either the I-filaments broke or the A-filaments elongated as in muscles stretched in rigor (see the following paragraph. Figs. 14 and 15 show examples of the breaking of the actin filaments and Figs. 16 and 17 show examples of the elongation of the thick filaments in the middle of the A-band. In this case the thick filaments became plastic on both sides of the M-line. Elongation of the thick filaments was always irreversible, the increase of the sarcomere length remaining after release of the muscle. In cross section the tapered thick filaments can be seen in the elongated zone (Fig.18). These elongated zones have a peculiar property; the thick filament layer placed on the surface of the fibril never elongates. Since in the rigor muscle the fixed cross-bridges are responsible for similar changes, it is very probable that irregular cross-bridge interactions play an important role in the inextensibility of the bee flight muscle in the living state.

Extension of the Insect Flight Muscle in Rigor

Muscle stretched in rigor is often used for mechanical, X-ray diffraction and electron microscopic experiments. Although Reedy and coworkers (1965) did not find any significant change in the cross-bridge arrangement when the muscle was stretched in rigor by up to 5 per cent over its resting length, the stretch caused marked morphological changes in the muscle structure when the muscle extension was more than 10 per cent. It is theoretically predicted that both tension and strain are minimal at the ends

---

Fig. 19 Glycerinated muscle stretched in rigor solution. Scale 1 μm The thin filaments are broken off from the Z-line.

Fig. 20 Glycerinated muscle stretched in rigor solution. Scale 1 μm The thick filaments are elongated on both sides of the M-line.

Figs. 21 and 22 Muscle stretched in life and then again after glycerination in rigor solution. Scale 1 μm

Figs. 23 and 24 Muscle stretched in life and then released after glycerination in rigor solution.

Fig.23. Scale 10 μm
Fig.24. Scale 1 μm

of A- and I-filaments, whereas they are maximal at the Z-lines in the I-filaments and the M-lines in the A-filaments. These are therefore the probable site of microscopic extension. Indeed it was observed first by White (1967) that when fibres were stretched in rigor, the actin filaments break away along the Z-line in some sarcomeres leaving a zone of very low density on one or both sides of the sarcomeres between the Z-line and the inter-locked A- and I-filaments. These "clear zones" contained only connnecting filaments. This was supported by Reedy (1972). We have also observed these "clear zones" (Fig.19). On the other hand Chaplain and co-workers (1969) reported, that "...... above 2.5% fibre extension the H-zone becomes plastic in rigor muscle, the H-zone just extends further and further if the fibres are stretched beyond this point".

On this basis they suggested that non-linear elasticity located mainly in the H-zone might explain "stress relaxation". But they did not give morphological evidence for the H-zone elongation. We have structural evidence of this phenomenon both in fresh and glycerinated muscle (Figs. 16, 20). Moreover, we observed the breaking of the actin filaments and the elongation of the H-zone together in a single sarcomere. An interesting question arises: why do both phenomena occur under the same controlled bio-chemical conditions. To answer this question we extracted both extended and resting length muscle with glycerol. When we stretched the extended muscle further in rigor solution the thin filaments always broke away along the Z-lines (Fig.21). On the other hand only elongated H-zones were produced when the resting length muscle was stretched in the rigor solution. (Fig.20).

The mechanical peculiarities of the filaments

It has generally been assumed that the filaments preserve their original length either in active contraction or during passive stretching. In spite of this assumption, elongation of thick filaments was shown by Garamvölgyi (1965). Indeed, when the sarcomere is stretched above twice its rest length, the thick filaments become plastic on both sides of the A-band. Similar phenomena can be produced in the middle of the A-band in both living and rigor muscles (Figs. 16,17,18 and 20). This means that the thick filaments have a plastic property which may be connected with the paramyosin content of the filaments (Squire 1972, Bullard *et al.*,1973).

Garamvölgyi (1965) provided evidence that the C-filaments had consider-able mechanical strength when he found that the extension of the muscle beyond the double resting length produced elongation of the A-filaments. This finding was supported by White and Thorson (1973) who pointed out that the clear zones on both sides of the Z-line in rigor muscle could not have been produced without mechanical strength of C-filaments, and by Pringle (1974) who emphasized that the maintenance of tension in rigor fibres stretched beyond the plastic limit, implied that the C-filaments had considerable mechanical strength.

Our present evidence confirms this concept. In particular, the production of additional clear zones on a second stretch in rigor after their initial production by a stretch in life (Fig.22) is evidence that the C-filaments retain sufficient strength to break adjacent actin filaments (cf. White and Thorson, 1973). Pringle (1974) supposed that at short extension the substance of C-filaments has to show an elastic behaviour, and a plastic behaviour at greater extension beyond the plastic limit. Since the bee muscle is very extensible, this question can be studied at greater extension of the muscle. When the glycerinated muscle stretched in living condition was released in rigor solution then large H-zones were observed without I-bands in the phase contrast microscope, but the length of the sarcomeres was slightly greater than in the resting state (Fig.23). In the electron microscope, beside large H-zones, short dense I-bands could be observed with folded actin filaments (Fig.24). Since it has been shown that the actin filaments projected out of the A-band remained in their original position and length after breaking in rigor solution (Fig.21), the C-filaments must be responsible for this shortening of the sarcomeres. Similar shortening of glycerinated muscle was observed by Garamvölgyi (1969) in the phase contrast microscope. These data prove that the C-filaments can shorten themselves even at great extension, and exert so much force during shortening that they can fold the actin filaments projecting out of the A-band. Therefore, we think that the connecting material should also show elastic behaviour at greater extension in bee muscle. On the basis of this finding it is very probable that connecting material is responsible for the resting elasticity as was suggested by Pringle (1967) and White (1967).

REFERENCES

Bullard B., Luke B. and Winkelman L. (1973) *J.Mol.Biol.*, 75, 359.

Chaplain R.A., Frommelt B. and Brandt M. (1969) *Kybernetik*, 5, 177.

Ernst E., Béleczky I. and Nagy J. (1956) *Acta Physiol., Hung.*, 9, 7.

Garamvölgyi N (1965) *J. Ultrastruct. Res.*, 13, 409.

Garamvölgyi N. (1969) *J. Ultrastruct. Res.*, 27, 462.

Garamvölgyi N. and Belágyi J. (1968) *Acta Biochim. Biophys. Acad. Sci. Hung.*, 3, 195.

Garamvölgyi N., Kerner J. and Cser-Schultz M. (1964) *Acta Physiol. Hung.*, 24, 381.

Hanson J. (1956) *J. Biophys. Biochem. Cytol.*, 2, 691.

Pringle J.W.S. (1976) *The resting elasticity of inner flight muscle. In: Symposium on the muscle.* Birő E.N.A. and Garamvölgyi N. (Ed.) P.67.

Pringle J.W.S. (1967) *Progr. Biophys. Mol. Biol.*, 17.

Reedy M.K. (1972) *Electron microscope observations concerning the behaviour of the cross-bridges in striated muscle. In: Contractility of Muscle Cells and Related Processes.* Podolsky R.I. (Ed.) p.229.

Reedy M.K., Holmes K.C. and Tregear R.T. (1965) *Nature*, 207, 1276.

Saide J.D. and Ullrich W.C. (1973) *J.Mol.Biol.*, 79, 329.

Squire J.M. (1973) *J.Mol.Biol.*, 77, 291.

Trombitás K. and Tigyi-Sebes A. (1975) *Acta Biochim. Biophys. Acad. Sci. Hung.*, 10, 83.

White D.C.S. (1967) *Ph.D. Thesis, Oxford University.*

White D.C.S. (1970) *J. Physiol.*, 208, 583.

White D.C.S. and Thorson I. (1973) *Progr. Biophys. Mol. Biol.*, 27, 173.

Zebe E., Meinrenken W. and Rüegg J.C. (1968) *Z.Zellforsch . Mikrosk. Anat.*, 87, 603.

THE STRUCTURE OF INSECT THICK FILAMENTS

John M. Squire

Biopolymer Group, Department of Metallurgy and Materials Science, Imperial College of Science and Technology, LONDON S.W.7.

INTRODUCTION

Of the two principal filament types which occur in muscles, it is the thick, myosin-containing, filaments which are the less well understood. The general form of the thin actin filaments is now reasonably well characterised (see review by Squire, 1975).

Insect thick filaments resemble those in vertebrate skeletal muscle in being roughly cylindrical in shape (Fig.1.) and having tapered ends. They also carry projections along all of their length except in the central part of the filament now termed the M-region (Sjostrom and Squire, 1977; previously known as the pseudo H-zone). But insect thick filaments are longer (2.2 μm in *Lethocerus* flight muscle) and greater in backbone diameter (180 to 200 Å in *Lethocerus* flight muscle) than their counterparts in vertebrates (1.57μm and 120 to 140 Å respectively). In addition, in transverse section, they frequently appear to be hollow, the hollow being approximately 100 Å in diameter (Fig.2.(a)).

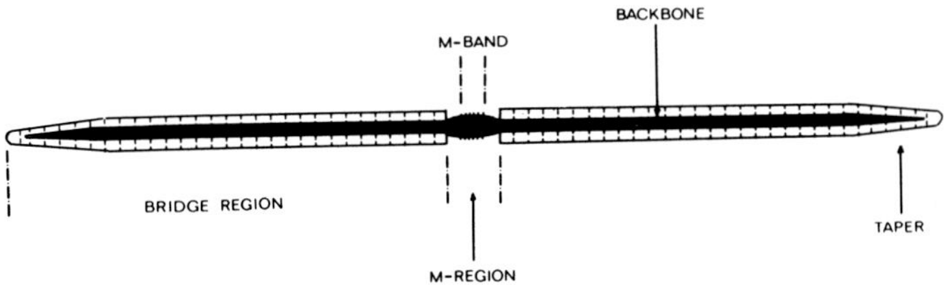

Fig.1. Schematic view of the thick filaments in insect flight muscle. The black central line represents the cylindrical filament backbone and the shaded regions the positions of the myosin projections. The thick filament backbone widens at the M-region and there are lateral M-bridges (forming the M-band) in this region. The filament tapers at each end.

The myosin molecules in insects are typical in consisting of an α-helical rod portion about 1500 Å long at one end of which are two globular heads (Fig.3.(b)). These heads carry the ATPase and actin binding properties of the whole molecule. Such molecules pack together so that the rods form the shaft of the myosin filaments and the heads form projections on the filament surface. It is these projections, normally termed the myosin crossbridges, which are thought to interact with actin during muscular contraction.

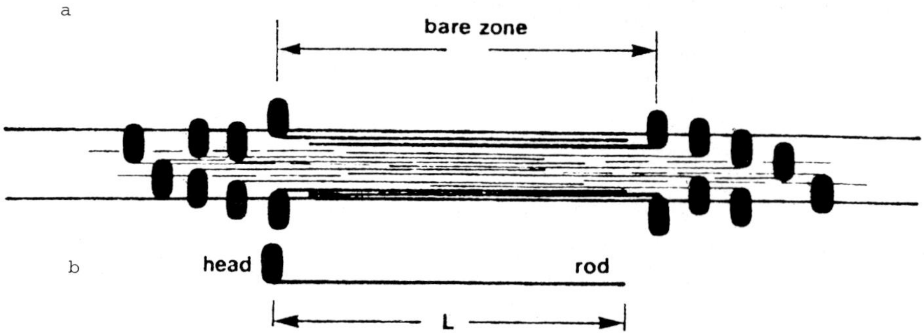

Fig.3. Scheme for the packing of myosin molecules (b) in the bare zone of bipolar myosin filaments according to Huxley (1963). The bare zone is the region of the thick filament which forms the M-region in the intact A-band. 'Head' in (b) refers to the location of the two terminal globular heads of the myosin molecule.

Fig.2. (a) Transverse section of *Lethocerus cordofanus* treated with relaxing solution. (Glutaraldehyde fixed; uranyl acetate and lead citrate stained; bar represents approximately 1500 Å). This shows the hollow circular profiles of the thick filaments in the bridge region. The small dots are actin filaments.

(b) Transverse section through the M-band of *L. cordofanus* treated with Tris.Cl. relaxing solution; otherwise treatment as in (a). Bar represents 1300 Å. Note the variety of orientations of the oval shaped thick filaments.

(c) Slightly oblique transverse section through I-band/Z-band region of *L. cordofanus*. (Relaxing solution, acrolein dichromate fixative, otherwise treatment as in (a), bar represents 1600 Å). See text for description of appearances.

(d) Nearly transverse section through the same kind of region as (c) but from glutaraldehyde fixed muscle, previously relaxed in Tris. Cl. relaxing solution. (Bar represents 1600 Å).

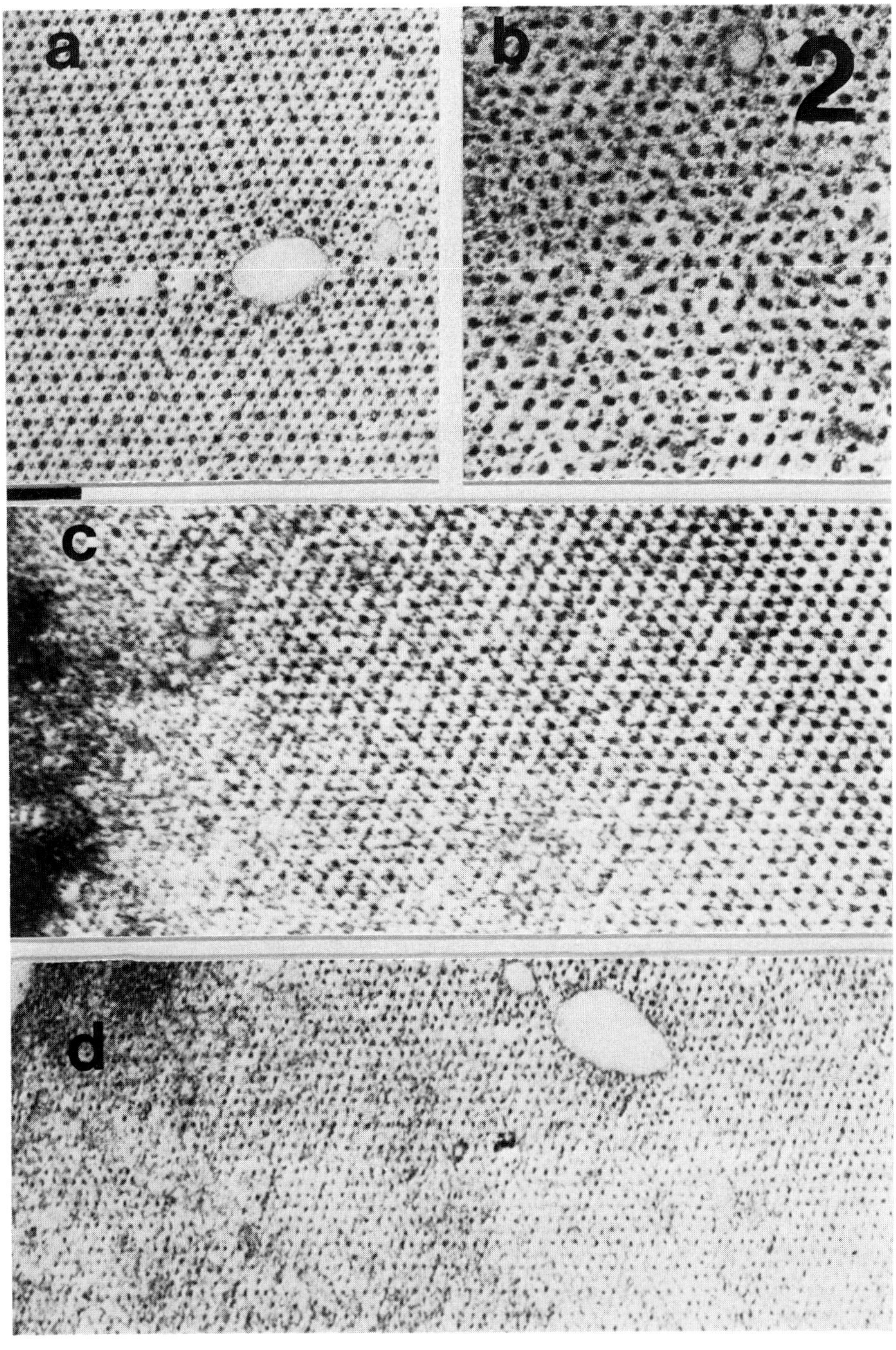
a
b
2
c
d

The packing scheme proposed by Huxley (1963) to account for the bipolarity of the thick filaments in vertebrate striated muscle seems to apply equally well to the insect muscle thick filaments. This scheme, illustrated in Fig.3(a), packs the rods of the myosin molecules in an anti-parallel fashion to form the central bipolar M-region, while on each end of this assembly myosin molecules add on by parallel aggregation to complete the whole bipolar filament structure.

Myosin is not the only component of the thick filaments. Bullard *et al.*, (1973) showed that another common component is the protein paramyosin which was previously known to occur in abundance in molluscan muscles where it serves as the backbone structure of the molluscan thick filaments (Elliott and Lowy, 1970). It is also clear that the M-band, which serves to cross-link adjacent thick filaments at the centre of the M-region, must comprise a number of other proteins but these have not yet been identified.

With these basic ideas on thick filament structure in mind, it is clear that analysis of further structural details can be divided into a number of separate but related questions:

(i) What is the arrangement of the myosin heads on the surface of the thick filament backbone?

(ii) How are the myosin rods packed in three-dimensions to give rise to the available details on filament shape and size mentioned above and to account for the crossbridge array determined in (i)?

(iii) What are the structures of the M-region and the regions where the thick filaments taper?

(iv) What is the location and role of paramyosin in the thick filaments?

These questions will be considered here and a number of answers given. As is usual in discussions of thick filament structure, most of these answers will be tentative. But fortunately, it is now becoming clear that at least some reasonably firm conclusions about thick filament structure can be reached. This is particularly true for the thick filaments in the flight muscles of insects such as the giant water bugs (*Lethocerus*) which have been more extensively studied than any other insect muscles. Because of this, the discussion given here is primarily concerned with the myosin filaments in insect flight muscles. The structures of other insect thick filaments will be considered briefly in the final section.

MATERIALS AND METHODS

Electron micrographs of two types of relaxed preparation of the flight muscles of *Lethocerus cordofanus* are included in this review. Both types of preparation were from muscles which were glycerinated and stored in the deep freeze (Jewell and Rüegg, 1966).

Glycerol was washed out of the muscles using standard salt. The muscles were then treated with relaxing solution (5 mM ATP, 5 mM $MgCl_2$, 5 mM EGTA, 50 mM KCl, 20 mM Phosphate buffer). They were then fixed using either 0.1 M glutaraldehyde or using acrolein/dichromate according to the method of Robison and Lipton (1969). Each type of preparation was post fixed with osmium tetroxide, dehydrated in acetone, embedded in araldite, sectioned and stained in uranyl acetate and lead citrate.

A second type of relaxed preparation used Tris.Cl buffer (5 mM ATP, 50 mM KCl, 5 mM $MgCl_2$, 4 mM EGTA, 10 mM Tris.Cl). Subsequent treatment was as above. This preparation may have removed part of the M-protein.

Ultrathin sections were cut on a Huxley ultramicrotome and were viewed on an AEI EM 6 or PHILIPS EM300 electron microscope operating at 80 to 100 kV and using a primary magnification of about 10000 to 12000.

THE MYOSIN CROSSBRIDGE ARRAY IN INSECT FLIGHT MUSCLE

Early evidence from electron microscopy.

The first clear indications of the nature of the crossbridge array in insect flight muscles were obtained in an elegant series of experiments by Reedy (1967, 1968). He found that in ultra-thin transverse sections of glycerinated muscle, regular arrangements of crossbridges from the thick to the thin filaments could be observed. The crossbridges on each thick filament seemed to be grouped in two diametrically opposed pairs, each pair separating and linking to two different actin filaments. This configuration, which Reedy called the 'flared X', is illustrated schematically in Fig.4.

That these appearances could be seen so clearly was mainly due to the fact that over quite an extensive area of a myofibril all of the flared Xs pointed in the same direction. In such a case reinforcement of the image by eye is automatically accomplished and the structure involved tends to be readily apparent. Some variations of the basic flared X structure were occasionally observed as illustrated in Fig.4(b).

Slightly oblique sections showed that at different levels along the thick filaments the flared X images systematically pointed in different directions thus indicating that the sections were sampling different levels of a helical array of crossbridges. Longitudinal sections of rigor insect flight muscle also showed a regular appearance. This was characterised by the existence of axial repeats of 385, 235 and 145 Å as revealed by optical diffraction analysis of the micrographs. These repeats are all orders of a basic repeat of about 8 x 145 Å or 1160 Å. Reedy therefore suggested that the crossbridge array could be described as a two stranded helix of true repeat

a

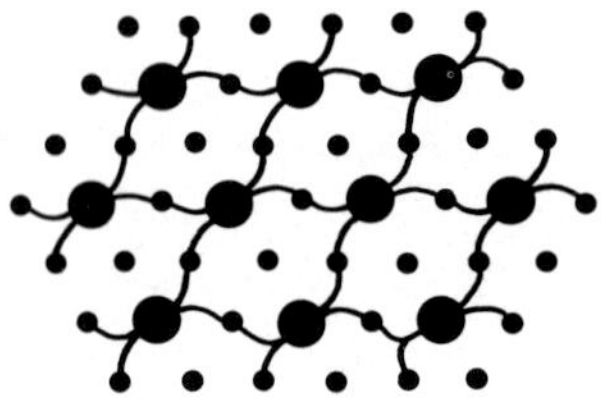

b

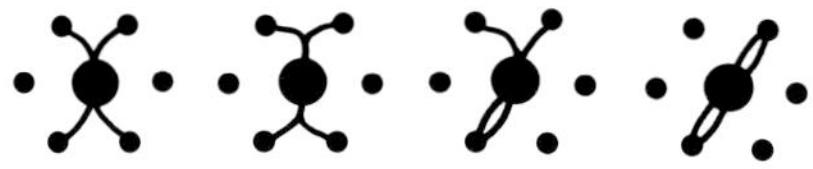

Fig.4. (a) Diagrammatic representation of the flared X crossbridge configurations seen by Reedy (1967, 1968) in thin transverse sections of insect flight muscle. (Large circles, thick filaments; small circles, actin filaments).

(b) Representations of some slight variations of the flared X crossbridge configuration seen by Reedy.

1160 Å and subunit axial translation 145 Å (Fig.5.(a)). Each strand in the helix would have a pitch of 770 Å (2 x 385 Å) and each lattice point on each helical strand would comprise two arms of the flared X configuration.

The appearance of Reedy's micrographs also led him to suggest that the six actin filaments surrounding a single thick filament were themselves arranged on a helix. He found that the actin ends of the crossbridges were arranged on a helix generated by 60° rotations between adjacent thin filaments. Such a rotation is equivalent to a 128 Å axial translation (i.e. 385 Å/3) between filaments. Since the actin repeat in insect muscle is 385 Å the six actin filaments described a two-strand 6/1 helix of myosin attachment points (target areas) with a subunit axial translation of 128 Å and a true repeat of 385 Å (Fig.5.(c)). Although Miller and Tregear (1972) have since shown that such an arrangement of thin filaments can only be developed in a statistical way over a large area of the myofibril, the existence of this basic kind of actin arrangement does not seem to be in dispute.

Reedy's final conclusion concerned the screw sense of the helix of flared Xs. He was able to show by a series of independent methods that the flared X helix was left-handed. His observations also confirmed the thin filament screw sense as being right-handed, as originally reported by Depue and Rice (1965).

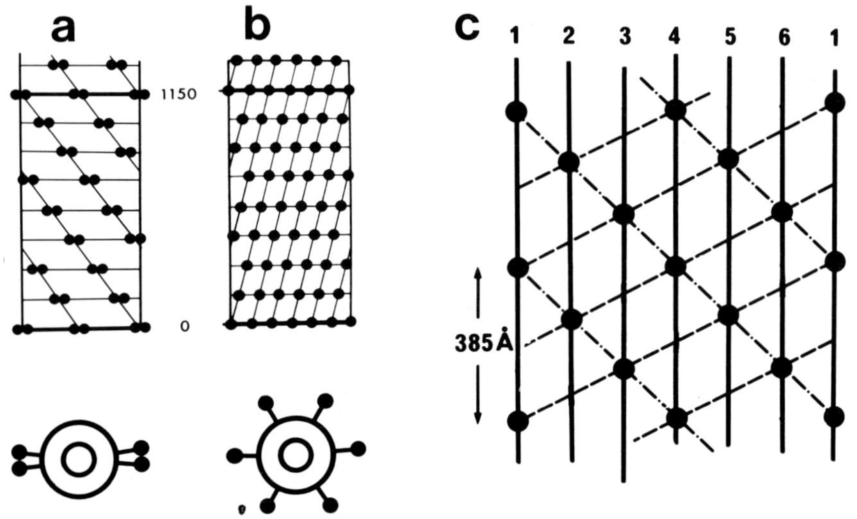

Fig.5. (a) and (b) are radial projections (above) and cross-sectional views (below) of the alternative 2-strand and 6-strand models for the thick filaments in insect flight muscle (Reedy, 1967, 1968; Squire, 1972).
(c) Radial projection of the helix of actin target areas according to Reedy (1968).
Vertical lines; actin filaments numbered sequentially around a single thick filament.

The main uncertainty of these results at the time was taken to be the composition of each arm of the flared X. Likely possibilities seemed to be either that each arm represented one head of the two headed myosin molecule or that the heads of two molecules originated at the same lattice point and that each of the arms of the flared X therefore represented the heads of a different myosin molecule. But the latter seemed more likely to be correct since early estimates of the myosin content in this muscle (Chaplain and Tregear, 1966) indicated that there was more than enough myosin in each thick filament to give four molecules per 145 Å repeat.

It will be shown later that these conclusions about thick filament structure are only part of the truth. But there is no doubt that Reedy's work, summarised briefly here, represented a very major breakthrough in the available knowledge and understanding of muscle structure.

X-ray diffraction evidence

Work carried out in parallel with Reedy's electron microscopy in Professor J.W.S. Pringle's muscle group in the Oxford University Zoology Department was a detailed X-ray diffraction study of insect flight muscle. This work was carried out by Drs. Andrew Miller and Richard Tregear. On the assumption that the myosin crossbridges are not attached to actin in relaxed muscle, it should, in principle, be possible to use the X-ray diffraction evidence from relaxed muscle to determine the geometry of the crossbridge array.

The diffraction patterns from relaxed glycerinated insect flight muscle obtained by Miller and Tregear (1972) included layer lines at axial spacings of 1/389 $Å^{-1}$ and 1/230 $Å^{-1}$ and a meridional reflection at 1/145 $Å^{-1}$. All of these were thought to be due to the ordered helical array of myosin projections on the surfaces of the thick filaments. The form of this diffraction pattern is summarised in Fig.6.(c). Like Reedy, Miller and Tregear (1972) interpreted their results in terms of the two-stranded helix of crossbridges shown in Fig.5.(a). They concluded from modelling studies that such a helix of crossbridges would account reasonably well for the observed distribution of intensity in their diffraction patterns, provided that these crossbridges were situated very near to the myosin filament surface. On this basis Reedy's electron micrography and Miller and Tregear's X-ray diffraction work were both consistent with the two-stranded helical model for the crossbridge array.

But in the early 1970's it became apparent that the structure of one of the observed myosin aggregates in vertebrate smooth muscle (The myosin ribbons: Small and Squire, 1972) was quite different from the two-stranded model for the insect thick filaments and would demand a different kind of myosin packing. It was equally different from the then current two-stranded 6/1 helical model for the thick filaments in vertebrate skeletal muscle (Huxley and Brown, 1967). The possibility was therefore investigated that the structures of these filaments might be much more closely related than had previously been suggested (Squire 1971). It was then found that alternative but closely related filament structures would account equally well for the available X-ray diffraction and electron microscope data. In particular, it was suggested that the crossbridge array in insect flight muscle might consist of the heads of six myosin molecules evenly spaced around the filament circumference at one transverse level (or crown) and that successive levels of bridges 145 Å apart might together form a six-stranded crossbridge helix of the kind shown in Fig.5.(b).

The value of six for the number (N) of crossbridges per crown was based on the evidence that the thick filaments must clearly have a minimum rotational symmetry that is two-fold (as shown by the flared X appearance in Reedy's micrographs) and that there are enough myosin molecules in each thick filament (Chaplain and Tregear, 1966) to allow the number N of crossbridges per crown to be about 5 or 6. A helix with 6-fold rotational symmetry and with N = 6 was the obvious choice.

That such a helix could explain the observed X-ray diffraction results can be seen in the following way. The diffraction pattern of a single continuous helical strand of density of pitch P has the form of a cross of intensity in which the arms of the cross consist of layers of intensity separated axially by n/P from the equator of the diffraction pattern. Here n is any integer, positive, negative or zero (Fig.6(a)). The effect of adding further continuous helical strands coaxial with the first so as to generate an N-fold helix with N-fold rotational symmetry about the helix axis, is to reduce the axial repeat of the structure to 1/Nth of its previous value and this results in the loss of all layers (n/P) except those for which n = kN (k integral) This is indicated in Fig.6(a). The spacing of the layer nearest to the equator in such a diffraction pattern is N/P.

If such an array of continuous helices is then made discontinuous so that the subunits on all N strands lie on a series of planes perpendicular to the helix axis and spaced a distance p apart in an axial direction, then the diffraction pattern changes to the form indicated in Fig.6.(b). This time, in addition to the helix centred on the origin, in which layers occur with spacings kN/P from the equator, there are other identical crosses of intensity which are centred on the axis (meridian) of the diffraction pattern at axial separations from the centre of the pattern equal to m/p where m is any integer. Fig.6(b) shows that the layers in the helix cross centred at a spacing 1/p along the meridian are at positions 1/p - kN/P from the equator.

A diffraction pattern of this kind can be compared with the observed diffraction patterns from relaxed insect flight muscle (Fig.6.(c)). The pattern consists of helix crosses centred at the origin of the pattern and at axial positions + or - 1/145 $Å^{-1}$ along the meridian. This indicates immediately that the value of p is 145 Å and that the structure is helical. It is also clear that any N-stranded helix which satisfies the rules 1/145 - N/P = 1/235 and N/P = 1/385 will account for the observed axial positions of the layer lines (Squire, 1972). It can be readily verified that if N = 2 then P = 770 Å (from both equations) and that if N = 6 then P = 2310 Å. These two

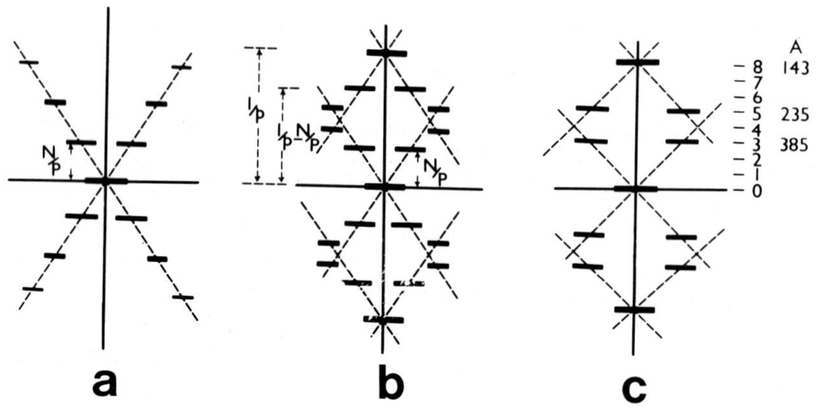

Fig.6. Analysis of the form of the X-ray diffraction patterns from relaxed insect flight muscle (c) in terms of the expected patterns of an N-stranded continuous helix of pitch P (a) and a discontinuous N-stranded helix (b) in which the axial period of the discontinuity is p (Miller and Tregear 1972; Squire 1972). See text for details.

helices are the two shown as (a) and (b) respectively in Fig.5. They can be described as a 2-stranded 16/3 helix of pitch 770 Å and a 6-stranded 16/1 helix of pitch 2310 Å. The notation X/Y helix is used throughout and means there are X repeating units in Y turns of the helix.

From this analysis it is clear that the 6-strand helix accounts as well for the axial periodicities in the observed X-ray diffraction patterns as does the 2-stranded model. It will also account approximately for the radial distribution of intensity along the 385 Å and 235 Å layer lines if the myosin crossbridges are centred at a radius of about 200 Å from the myosin filament axis. A detailed analysis of the observed intensity distribution will reveal how well the 6-stranded model will account for the observations.

Clearly, for the 6-stranded model to be acceptable as an alternative to the model with N = 2, it must also be able to explain the existence of the flared Xs seen by Reedy in micrographs of rigor muscle. A possible explanation for these was put forward by Squire (1972). Stated simply it was suggested that a flared X consisted of four of the six crossbridges in each crown and that the other two crossbridges were not visualised. The reason for this was taken to be that only in the case of the four observed crossbridges were the appropriate actin filaments in the correct orientation for interaction with a crossbridge to occur. In the case of the other two crossbridges the actin filaments would be wrongly oriented for myosin attachment. The unattached bridges would then become disordered during the preparative procedure for electron microscopy (possibly by being plastered back onto the thick filament backbone) and as a result would not be visualised. Evidence to support this suggestion was, in fact, provided by Reedy himself (Reedy, 1971). He found that occasionally in some of his micrographs a position in the hexagonal filament lattice normally occupied by an actin filament was vacant. In such a case, if this filament would normally have been linked to one arm of the flared X, then invariably that arm of the flared X was missing too. The implication is clear: crossbridges not attached to actin filaments are not visualised in electron micrographs.

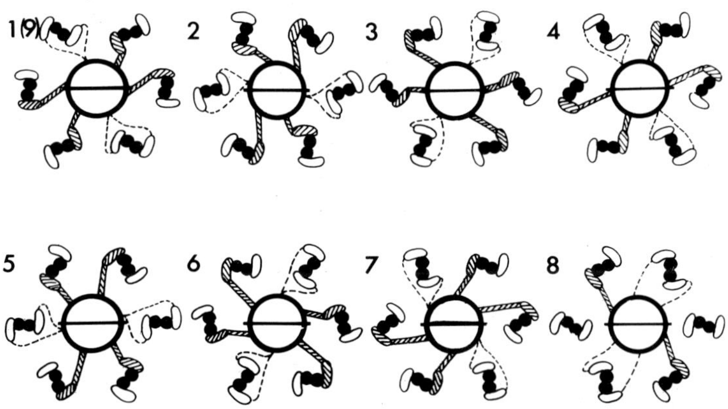

Fig. 7. Generation of flared X patterns from a 6-stranded myosin filament at different axial levels in the A-band separated by 145 Å. The large open circles represent the thick filament backbones and the small solid circles represent the two-strand actin filaments. Good crossbridge interactions with actin are shaded. Poor interactions which require considerable distortion of the crossbridges are shown by dashed lines. The thin filaments are related by 60° rotations, and 128 Å translations as in Fig.5.(c). The different flared X configurations are just as in Fig.6. of Reedy (1968). (After Squire 1972).

Carrying his analysis one step further, Squire (1972) showed that if the actin filaments are arranged on the helix illustrated in Fig.5.(c) then the changes in orientation of the flared Xs in different axial levels along the A-band (Reedy, 1967, 1968) would be generated by a six-stranded myosin filament (Fig.7). In brief, it was found that the positions where attachment of crossbridges to actin was not favourable followed the same helix as the actin filaments. The reasons for this are obvious and this kind of idea is entirely consistent with the idea that there are target areas along the thin filaments where myosin crossbridge attachment is strongly favoured. On this basis the appearance of Reedy's micrographs could be taken to reflect the geometry of the crossbridge interactions with the thin filaments but not the nature (i.e. symmetry and helix sense) of the crossbridge array in relaxed muscle.

Myosin content in the thick filaments.

The early results of Chaplain and Tregear (1966) on the myosin content in insect flight muscle which indicated that N might be 6, clearly favoured a 6-stranded model for the thick filaments. But in the face of the striking flared X images surprisingly little weight was generally attached to these results. However, more recent work by Reedy, Bahr and Fischman (1972) using quantitative microscopy and by Bullard and Reedy (1972) and Tregear and Squire (1973) using quantitative S.D.S. gel electrophoresis has all tended to confirm that N is 6.

For this reason the 6-stranded filament structure seems attractive. But a formal alternative possibility is a 2-stranded helix, such as that shown in Fig.5.(a) but with three myosin molecules at each lattice point on the helix instead of two. Unfortunately the more spread out azimuthally the three myosin molecules are in such a 2-stranded model, the more like a 6-stranded structure the filament will appear. In the analysis which follows true and pseudo 6 foldness would not be distinguishable. No published evidence has yet shown unambiguously which of these two types of structure is correct. But in the following pages some arguments based on the 3-dimensional geometry of the A-band show that it is becoming hard to escape from the conclusion that the thick filaments have at least pseudo 6-fold rotational symmetry.

The structure of the M-region

In transverse sections of insect flight muscle through the M-region, the thick filament profile is clearly elliptical (Fig.2.(b)). As in vertebrate skeletal muscle the thick filaments are cross-linked by an M-band structure. In insect flight muscle (*Lethocerus*) these cross-links run from the thick filaments to the trigonal points of the hexagonal filament array (Fig.8.). There they appear to link to thin longitudinal M-filaments (Ashhurst, 1967; Pringle, 1972).

The fact that the filament profile at the M-region is elliptical (i.e. with a 2-fold rotation axis) rather than say circular, seems at first sight to suggest that the crossbridge symmetry might really be 2-stranded.

a

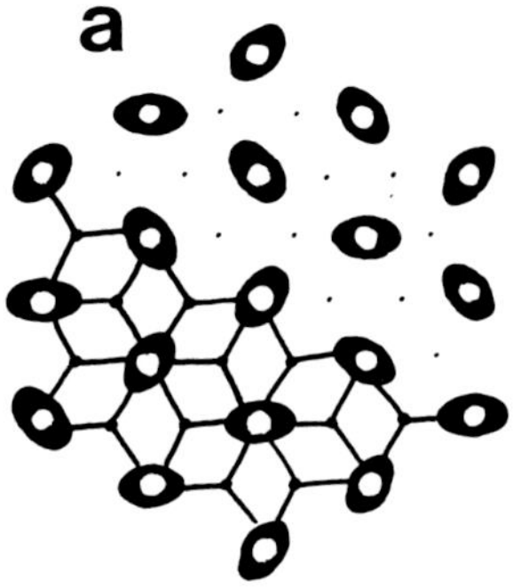

b

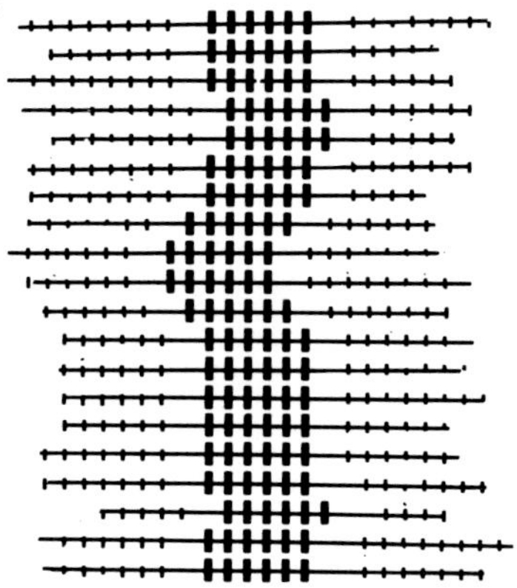

Fig.8. (a) Representation of the M-band appearance as in Fig.2.(b) showing the different orientations of the oval thick filament backbones, and the M-bridge connections reported by Ashhurst (1967a) and Pringle (1972).

(b) Representation of the systematic axial displacement of the M-bridges which sometimes can be seen in longitudinal sections of insect flight muscle (Reedy, 1968, 1971). The thick transverse bands represent M-bridge positions on the thick filament (long thin lines). Displacements are always by small multiples (usually $\pm$ 1) of the 145 Å M-bridge repeat.

But it is clear from transverse sections through this region (Fig.2.(b)) that adjacent thick filaments have markedly different orientations: the ellipses point in one of three different directions about 60° apart as illustrated in Fig.8. But the orientations of adjacent filaments are not systematically related. If the thick filaments can be considered to behave as rigid rods then all the way through the A-band there will occur myosin filaments with three different absolute orientations. In the case of a 2-stranded filament and assuming that all the filaments are in axial register, then the crossbridge origins on adjacent filaments would have to be oriented randomly in one of three positions about 60° apart just as they are in the M-region. The result is shown in Fig.9. But, of course, one of Reedy's results (Reedy 1967, 1968) is that at one level along the A-band all of the flared X crossbridge configurations point in a single direction. One might reasonably ask how crossbridges originating in so many different orientations could interact with actin to produce identically oriented flared X structures. Clearly the actin target areas will still be in the same place, but the crossbridge origins may well be up to 90° away azimuthally. It seems doubtful that there could be such things as target areas on the thin filaments if the myosin crossbridges had sufficient flexibility to allow binding to actin filaments 90° away.

It can readily be seen that the same kind of problem does not occur with the 6-stranded model since a rotation between filaments of any multiple of 60° leaves the positions of the crossbridge origins unchanged. There will always be some of these crossbridges in suitable positions for interaction with the actin target areas to occur and, provided the thin filaments are regularly arranged, a unique flared X orientation at a particular level in the A-band will automatically be produced.

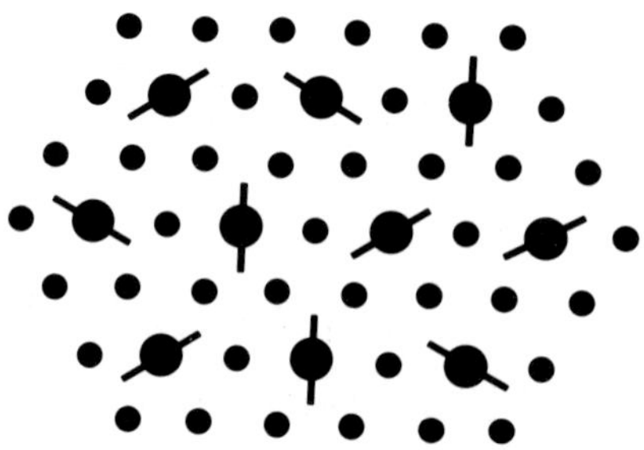

Fig.9. Diagram corresponding to the kind of random thick filament orientation seen in the M-band (Figs. 2 (b) and 8(a)) but now applied to the overlap region of the A-band. The directions of crossbridge origins for a 2-stranded filament are shown. To satisfy Reedy's evidence these would be required to interact with the thin filaments to produce flared X configurations pointing in the same direction (Reedy, 1968).

A similar kind of deduction can also be made from the longitudinal structure of the M-region. Longitudinal sections of the insect M-band clearly show the existence of M-bridges between adjacent myosin filaments (Reedy, 1967; 1968; 1971). On one filament these M-bridges are separated axially by about 140 to 150 Å - a spacing which can be presumed to be equal to the axial subunit repeat of 145 Å in the helix of myosin crossbridges. One interesting feature of these M-bridges is that it is often seen that adjacent thick filaments are displaced from each other axially by one M-bridge repeat of 145 Å. Such displacements seem to occur irregularly through a myofibril and seem to represent simple dislocations (i.e. crystal slip) of the 'crystalline' M-bridge array (Fig.8.(b)). Despite these imperfections in A-band assembly, the rigor crossbridge array revealed by Reedy's micrographs shows no evidence of the thick filament displacements which the M-band shows must occur. If it is accepted that it is only the positions of the actin target areas which determine the nature of the rigor crossbridge array, then provided the I-band is regular so will be the rigor crossbridge interactions.

But, on the basis of a 2-stranded myosin filament, the crossbridge origins on two successive 145 Å spaced crowns are related by a rotation of 67.5°. This means that the thick filaments in a small area of a myofibril, some of which may have relative axial displacements of 145 Å or possibly 2 x 145 Å, will have crossbridge origins related by relative orientations of 67.5°, or even 2 x 67.5°. Yet such filaments still appear to produce flared X crossbridge configurations pointing in the single direction defined by the actin target areas. As mentioned above, either the concept of actin target areas must be questioned, or the 2-stranded thick filament model is inadequate.

One more point needs to be made. It is crucial to question whether or not the relative rotations and relative axial shifts of the thick filaments (illustrated in Figs.8.(a) and (b)) are related in such a way that the perfection of the array of crossbridge origins in the A-band is maintained. If they were, the kind of deductions given above would be invalidated. Fortunately it does seem that these two types of A-band imperfection are not systematically related. In the first place it appears from the frequency of undulations in the M-band structure seen in longitudinal sections, both that axial shifts between adjacent filaments are the exception rather than the rule and that when they do occur they will tend to be systematically related. On the other hand, the orientations of the oval filament profiles in transverse sections through the M-band (Fig.2(b)) are such that it is relatively rare to find neighbouring filaments pointing in the same direction.

A detailed investigation of the M-region in insect flight muscle is now in progress and it is, therefore, hoped that this question will be decided in the near future.

ROD PACKING IN THE MYOSIN FILAMENT BACKBONE

Once the geometry of the crossbridge array on the thick filament surface has been determined unambiguously it is a relatively easy matter to determine the few most likely packing schemes for the myosin rods in the filament backbone. Deciding which of these models is correct is a much more formidable task.

On the assumption that the myosin heads are equivalently placed on the thick filament surface and that likewise the myosin rod packing is such that all molecules are equivalent, then two classes of rod packing scheme are possible (Squire, 1973; 1975). In one, the myosin rods are close packed to form a continuous cylindrical layer of molecules (Fig.10(a); Squire, 1973). The second class of model first packs the myosin rods into sub-filaments and then packs these sub-filaments to form a cylindrical shell (Fig.10.(b)). As a general rule if the myosin heads form a regular N-stranded helix, then there could be b x N sub-filaments of equivalent molecules in the backbone, where b is integral. Each sub-filament would follow a helical path; the helix being one that can be drawn through the lattice points of the appropriate crossbridge array - in this case that in Fig.5(b) . A very large number of different sub-filament models can be envisaged, each corresponding to one of the many families of helices that can be drawn through the crossbridge lattice. But fortunately, there are good reasons for limiting the possibilities to one or two likely alternatives.

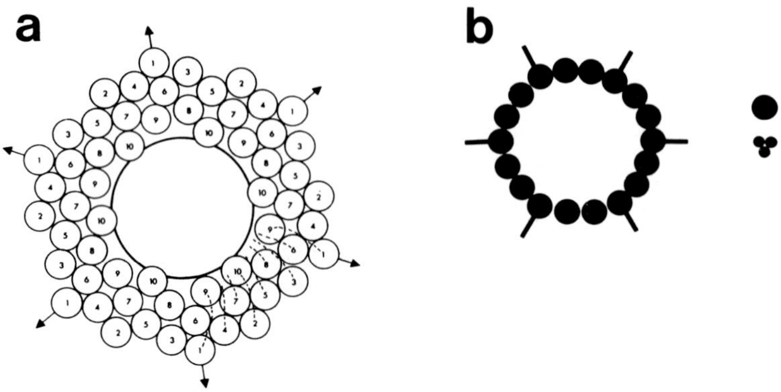

Fig.10. (a) Diagram showing a cross-sectional view of the continuous layer model for myosin rod packing in the 6-stranded thick filament of insect flight muscle (after Squire, 1973). Each circle represents a cross-section through one myosin rod. The numbers in the circles represent the different axial levels (in multiples of 145 Å) at which the myosin heads on the ends of those rods will be located. For details see text.

(b) Cross-section of an 18 sub-filament model for myosin rod packing in the 6-stranded filament. This is the only sensible packing scheme which includes an interaction between molecules staggered axially by 3 x 145 Å. But even this gives an outer diameter which is much too great ($\sim$ 330 to 340 Å). For details of model see text.

The structures of either the continuous layer (Fig.10.(a)) or the sub-filament model (Fig.10.(b).) are uniquely determined by the axial staggers which occur between interacting myosin rods. Some likely values for these staggers have been determined from studies of synthetic aggregates of myosin and its proteolytic sub-fragments (Lowey, *et al.*, 1967; 1969, Cohen, *et al.*, 1970; Harrison, *et al.*, 1971; Kendrick-Jones, *et al.*, 1971) and of the properties of myosin in solution (Burke and Harrington, 1971; Harrington, *et al.*, 1972). From these studies it appears very likely that a parallel axial stagger of 3 x 145 Å is involved, together with a stagger of a low multiple of 145 Å (other than 3n x 145 Å where n is integral). Possible staggers in addition to 430 Å are therefore 145 Å, 290 Å, 580 Å and 725 Å It was shown by Squire (1973) that in the case of the continuous layer model the most sensible kind of structure was one with staggers of 3 x 145 Å and 5 x 145 Å or possibly 3 x 145 Å and 2 x 145 Å; the former being slightly preferred in the case of the ribbons in vertebrate smooth muscle. A similar model can be derived for the thick filaments in insect flight muscle using 3 x 145 Å and 5 x 145 Å staggers as shown in Fig.10.(a). This gives a filament diameter of about the right value (180 to 200 Å, Reedy, 1967) if the myosin rods are assumed to be uniform cylindrical rods of diameter 20 Å (Elliott *et al.*, 1968). It also gives a core diameter of about 100 Å; a value which agrees well with the appearance of these filaments in transverse sections (Fig.2.(a)). In Fig.10(a) the number in any circle represents the axial level at which the crossbridge on the end of that particular rod will be located. For example the circle marked 8 means that the crossbridge on the end of that rod will appear on a crown (8 - 1) x 145 Å above the arrowed circles marked 1. The path which one myosin rod traces as it travels from the filament surface to the core is indicated by the dotted lines. This corresponds to an axial movement of about 1500 Å between the crossbridge and the other end of the myosin rod, and a radial movement of about 40 Å. Each myosin rod is tilted slightly both towards and around the filament axis. In Fig.10.(a) the dotted lines join circle numbers 1 and 9 etc. to show that the filament repeat is 8 x 145 Å (1160 Å).

Another feature of Fig.10.(a) is that the circles marked 1 and 6 are in close contact, indicating that an axial stagger of 5 x 145 Å may occur as well as the 3 x 145 Å interaction. But it should also be noted that circles 1 and 3 are very close and that very slight adjustment of the packing scheme could bring these circles into contact. It is, therefore, difficult to decide whether the important interactions in such a scheme would be multiples 3 and 5 or 3 and 2 of the basic 145 Å repeat.

Turning now to the sub-filament models and assuming that a 3 x 145 Å stagger between molecules must occur, then the only sensible structure is an 18 sub-filament arrangement with a 3 x 145 Å stagger between molecules in each sub-filament, about three molecules in any sub-filament cross-section and a stagger between sub-filaments of 145 Å. Clearly if the 3 x 145 Å stagger was not involved several other schemes would be possible.

The 18 sub-filament model can be tested if it is assumed that each myosin rod is 20 Å in diameter and behaves as a solid cylinder. The three molecule cross-section of each sub-filament would then have a mean diameter of about 43 Å. It can easily be shown that a cylindrical shell comprising 18 sub-filaments of this diameter would itself have an outer diameter of 330 to 340 Å and an inner diameter of 150 to 160 Å. It is clear that these figures are well above the range of experimentally observed values (e.g. outer diameter 180 to 200 Å; Reedy, 1968; Cullen, 1971) and therefore that either the sub-filament model is wrong or that the 435 Å stagger must go. Further evidence is needed for a more definite conclusion, but taken at its face value existing evidence does seem to favour the continuous layer structure illustrated in Fig.10.(a).

CONTENT AND ROLE OF PARAMYOSIN IN THE A-BAND

Bullard, Luke and Winkelman (1973) isolated and purified paramyosin from a variety of insect muscles. It was also shown by S.D.S. gel electrophoresis that in the flight muscles of *Lethocerus* there is about enough paramyosin (11% of the thick filament) to fill the thick filament core. Since paramyosin is known to have such a role in the thick filaments in other muscle types (e.g. molluscan muscle) this possibility has its attractions.

However, various lines of evidence suggest that the role of paramyosin is really much more subtle. Firstly there is only about 2% paramyosin in the myosin filaments of the blowfly (Bullard, *et al.*, 1973). Secondly it has been observed that filaments near to the middle of a myofibril sometimes appear more solid than those at the periphery (Auber, 1969; Reedy, *et al.*, 1972). Lastly Levine, *et al.*, (1976) have correlated paramyosin content in a variety of invertebrate muscles with the length of the appropriate thick filaments, but not with their diameter. It therefore seems that the role of paramyosin is not merely to fill up the core of the insect myosin filaments. It might possibly be associated with the nucleation of filament growth at the M-region and also with the thick filament length determining mechanism. In such a case one might expect to find paramyosin in the middle part (M-region) of the thick filaments. Antibody labelling studies by Bullard (this volume) have indeed shown that paramyosin is present in the M-region and also in the proximal parts of the bridge regions.

One intriguing feature of the M-region is that despite the apparent 6-fold nature of the crossbridge array, the thick filament profile in the M-region only has 2-fold rotational symmetry. At the same time it seems that each thick filament can make M-bridge cross-links to six other thick filaments via the trigonal points of the myosin filament array. How can the thick filaments have a 6-fold crossbridge array and yet sometimes show much less symmetry? One possible answer may be that the paramyosin in the filament core in the M-region is itself assembled into a structure with two-fold rotational symmetry. The myosin crossbridge array would then build up from two nucleating points until the full six-fold crossbridge array had been developed. Further studies on the M-region may help to clarify this point. If it is correct it means that the crossbridge array may have true 6-fold rotational symmetry but that the filament as a whole would only possess pseudo 6-fold symmetry.

THE THICK FILAMENT TAPER AND TIP REGION

Is has been suggested in several papers (Auber and Couteaux, 1963; White, 1967; Reedy, 1971) that connecting (C-) filaments link the tips of the thick filaments to the Z-band. The most striking evidence for this is from longitudinal and transverse sections of highly stretched rigor muscle. In such sections continuations of the thick filaments into the Z-band can be clearly seen (White, 1967; Reedy, 1971). But Ashhurst (1967) studied transverse sections through the Z-band region of unstretched sarcomeres (such as that in Fig.2(d)) and found no evidence for continuity of the thick filaments.

Fig.2.(c) shows a similar section to Fig.2.(d) except that it is slightly more oblique and is from a muscle that was fixed using the acrolein/ dichromate procedure which often causes the myosin components of the section to be exceptionally densely stained (e.g. Small and Squire, 1972). In Fig.2.(c) it is clear that the myosin filaments are very heavily stained and that the thin filaments hardly show up at all.

In Figs.2 (c) and 2(d) a number of features are worth noting:

(i) The thick filaments often appear hollow near their tips.

(ii) As the thick filament diameter reduces in the tapered region (Fig.1.) the diameter of the hollow core also reduces.

(iii) In parts of the tapered region the thick filaments have an essentially triangular profile (Fig.2.(c) centre). This is yet another indication that the filament possesses both 3-fold and 2-fold rotational symmetry which, taken together, must mean that the filament is 6-stranded.

(iv) Some continuation of the thick filaments into the Z-band does seem to occur in Fig.2.(c). But in Fig.2.(d) which is from a glutaraldehyde fixed preparation the myosin positions in the open hexagonal actin array become vacant before the Z-band is reached (Ashhurst, 1971).

Taken together, these results seem to suggest that C-filament connections if they occur cannot penetrate far into the thick filament core -

otherwise at the filament tip these filaments would not appear hollow. The only alternative possibility seems to be that for some reason they stain extremely weakly so that the thick filament core appears empty and they are not normally seen between the filament tip and the Z-band. One might then reasonably ask why they show up in sections of stretched muscle. Bullard (this volume) has isolated a new protein component from *Lethocerus* which does seem to be located in the vicinity of the Z-band. Further investigations of this protein may help to clear up the C-filament story.

CONCLUSION

The most crucial conclusion to be reached about thick filament symmetry is that the crossbridge arrangement does appear to be 6-standed. It must therefore be described by a 6-stranded 16/1 helix of pitch 2310 Å with one myosin molecule at each lattice point (Squire, 1972). It has also been suggested that this crossbridge arrangement is built up from a paramyosin core which has 2-fold rotational symmetry. In a similar way it seems possible that the crossbridge array may taper down to a structure with 3-fold rotational symmetry, possibly having only three myosin heads per crown at the tip. Further evidence for this latter point will be presented elsewhere. On the basis of current evidence it is also reasonable to conclude that the myosin rods pack as a continuous layer structure (Fig.10.(a)) just as seems to occur in the thick filaments of vertebrate skeletal muscle (Squire, 1975).

Concerning the structures of myosin filaments in insect muscles other than flight muscles, there is at present so little evidence on them that it is not possible to reach firm conclusions. Two results are worth noting. One is that Tregear and Squire (1972) using quantitative S.D.S. gel electrophoresis estimated that there is sufficient myosin in *Lethocerus* leg muscle to allow at least 5 myosin molecules per crown on the thick filaments. The second is that Rodger (1973) studied a number of insect muscles using X-ray diffraction and found that the diffraction patterns from many muscles, although often poorly developed, at least showed similar axial periodicities to the *Lethocerus* flight muscles. Apart from these results, there is such a diversity in thick filament length and M-band appearance (Pringle, 1972) that it is not possible to come to any further conclusions. However, the general model of myosin filament structure (Squire 1971; 1972; 1973) does seem to have survived the tests so far applied to it. On that basis one can suggest that all insect thick filaments of the same diameter may have the same number of crossbridges per crown. If they also have the same axial repeat then the helical symmetry of the crossbridge lattice and the nature of the rod packing arrangements would also be the same.

ACKNOWLEDGEMENTS

I am indebted to Miss Sheila Lauchlan for technical assistance and to Dr. A. Freundlich, Mr. P.K. Luther and Mr. P.M.G. Munro for discussions. I would also like to take this opportunity to thank Professor J.W.S. Pringle for providing facilities and support for me at the Zoology Department, Oxford, during a difficult stage in my career in early 1972. The work reported here was supported by a project grant from the Medical Research Council.

REFERENCES

Ashhurst D.E. (1967a) *J. Cell.Sci.,* 2, 435.

Ashhurst D.E. (1967b) *J.Mol.Biol.,* 27, 385.

Auber J. (1969) *J. Micr.,* 2, 325.

Auber J. and Couteaux R. (1963) *J. Micr.,* 2, 309.

Bullard B., Luke B. and Winkelman L. (1973) *J. Mol. Biol.,* 75, 359.

Bullard B. and Reedy M.K. (1972) *Cold Spring Harbor Symp.Quant.Biol.,* 37, 423.

Burke M. and Harrington W.F. (1971) *Nature,* 233, 140.

Chaplain R.A. and Tregear R.T. (1966) *J.Mol.Biol.,* 21, 275.

Cohen C., Lowey S., Harrison R.G., Kendrick-Jones J. and Szent-Gyorgyi A.G. (1970) *J. Mol.Biol.,* 47, 605.

Cullen M.J. (1971) *D.Phil.Thesis, Oxford University.*

Depue R.H. and Rice R.V. (1965) *J.Mol.Biol.,* 12, 302.

Elliott A. and Lowy J. (1970) *J.Mol.Biol.,* 53, 181.

Elliott A., Lowy J., Parry D.A.D. and Vibert P.J. (1968) *Nature,* 218, 656.

Harrington W.F., Burke M., Barton J.S. (1972) *Cold Spring Harbor Symp.Quant. Biol.,* 37, 77.

Harrison R.G., Lowey S. and Cohen C. (1971) *J.Mol.Biol.,* 59, 531.

Huxley H.E. (1963) *J.Mol.Biol.,* 7, 281.

Huxley H.E. and Brown W. (1967) *J.Mol.Biol.,* 30, 383.

Jewell B.R. and Rüegg C. (1966) *Proc.Roy.Soc.B,* 164,428.

Kendrick-Jones J., Szent-Gyorgyi A.G. and Cohen C. (1971) *J.Mol.Biol.,* 59, 527.

Levine R.J.C., Elfvin M., Dewey M.M. and Walcott B. (1976) *J. Cell.Biol.,* 71, 273.

Lowey S., Goldstein L., Cohen C. and Luck M. (1967) *J.Mol.Biol.,* 23, 287.

Lowey S., Slayter H.S., Weeds A. and Baker H. (1969) *J.Mol.Biol.,* 42, 1.

Miller A. and Tregear R.T. (1972) *J.Mol.Biol.*

Reedy M.K. (1967) *Am. Zool.,* 7, 465.

Reedy M.K. (1968) *J.Mol.Biol.,* 31, 155.

Reedy M.K. (1971) *In Contractility of Muscle Cells and Related Processes,* ed. R.J. Podolsky Prentice Hall. 229.

Reedy M.K., Bahr S.F. and Fischman D.A. (1972) *Cold Spring Harbor Symp.Quant.*

Biol., 37, 397.

Robison W.G. and Lipton B.H. (1969) *J.Cell.Biol.*, 43, 117a.

Rodger C. (1973) *D.Phil. Thesis, Oxford University.*

Sjostrom M. and Squire J.M. (1977) *J. Mol.Biol.*, 109, 49.

Small J.V. and Squire J.M. (1972) *J.Mol.Biol.*, 67, 117.

Squire J.M. (1971) *Nature*, 233, 457.

Squire J.M. (1972) *J.Mol.Biol.*, 72, 125.

Squire J.M. (1973) *J.Mol.Biol.*, 77, 291.

Squire J.M. (1974) *J.Mol.Biol.*, 90, 153.

Squire J.M. (1975) *Ann. Rev. Biophys. Bioeng.*, 4, 137.

Tregear R.T. and Squire J.M. (1973) *J. Mol. Biol.*, 77, 279.

White D.C.S. (1967) *D. Phil. Thesis, Oxford University.*

CROSSBRIDGE CONFORMATION

ELECTRON MICROSCOPE STUDIES OF *LETHOCERUS* FLIGHT MUSCLE IN RIGOR

Michael K. Reedy

William E. Garrett, Jr.

Department of Anatomy, Duke University Medical Center, Durham, North Carolina 27710.

From the lore of early molecular genetics, word has come down of "Hershey Heaven", so-called after A.D. Hershey's description of scientific happiness: "To have one experiment that works, and keep doing it all the time". (Benzer, 1966). Examining thin sections of rigor *Lethocerus* flight muscle in the electron microscope is something like that. The experiment works often, if not always. The sections are very pretty, and they give one rather pretty ideas.

It seems to help the experiment to work if one makes X-ray diffraction pictures of the fiber bundle during preparation. The patterns are always lovely. The bad experiments have always turned up in the lot that didn't get X-rayed. Good specimens seem to be almost uniformly good, both at first sight and after much searching; poorer rigor specimens, with soft or fuzzy or heavily lumpy periods, never seem to have nice areas. As for relaxed muscle, we've only met 3 or 4 good specimens in about 4 dozen efforts. So we are happy to describe the electron microscopy of rigor muscle here to-day. A good deal of the electron microscopy and optical diffraction analysis of micrographs will be summed up using diagrams and computer-filtered images. Near the end, we discuss filtered images from relaxed muscle, valuable for what they tell us (and don't tell us) concerning the surface lattice of insect thick filaments.

In order to get the best possible orientation of fibers, we use a "dry, flat" embedding procedure modified from H.E. Huxley's routine of inverting a capsule full of Araldite over a small fiber bundle laid on a stiff 2 mm polypropylene sheet. We take advantage of the fact that resin-infiltrated fibers no longer face any risk of drying. We separate the bundle into individual fibers and push them well away from all excess resin, to make monolayer rafts of 6-15 single fibers which cling closely parallel to one another and to the polypropylene surface, held by surface tension of a very thin layer of resin. This manipulation, which uses specially sharpened needles, is very fussy, rather like knitting mittens for a mosquito, but it pays off. After oven curing just enough

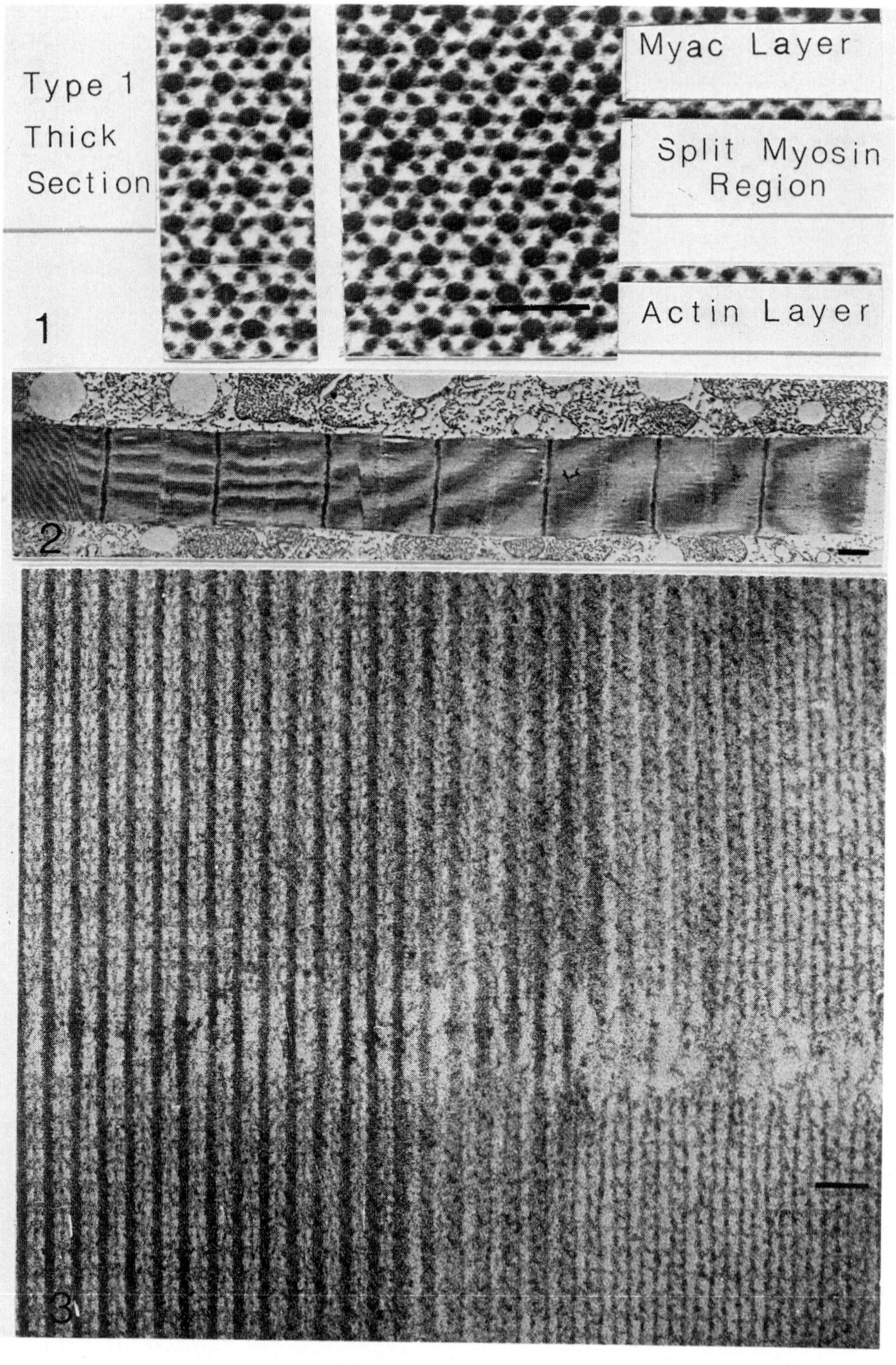
Type 1
Thick
Section
Myac Layer
Split Myosin
Region
Actin Layer
1
2
3

to stabilize these rafts, we invert resin-filled lidless BEEM capsules over them and cure again. The resulting block has a flat surface (gelatin capsules make it undesirably concave) ideal for knife alignment, yielding perfectly oriented longitudinal sections within 5-10 μm of the original surface.

The most useful thin sections are type 1, i.e. parallel to 1,0 layer planes (Fig.1), easily recognized at low magnification by a more-like succession of myac and actin layers (Fig.2). We search the changing moire pattern of successive sarcomeres until a widespread myac or actin layer occurs and then photograph that. A slight rotation of the filament lattice in successive sarcomeres and the slight bowed curvature of filaments nearer the fibril surface guarantee a constantly-changing moire-like effect.

The widest extent of myac or actin layers is sought because only one-third to one-half of such an area includes a single filament layer that is well centered between the two section surfaces. The remainder shows faint or strong diagonal striping, betraying a more or less one-sided view of the filament layer while sampling the crossbridges in the region between a myac and an actin layer (Fig.3).

These diagonally-striped split myosin regions seemed at first the most interesting parts of type 1 thin sections, once we were past the distraction of making initial observations about crossbridge angle (Reedy, Holmes and Tregear, 1965). The striping clearly indicated the presence, lateral register, and (2 x 385 Å) pitch of a two-stranded crossbridge helix around the thick filaments. They were called split myosin regions, rather than split myac layers, because our initial pre-occupation was more with their meaning for the cross-bridge surface lattice on the myosin filament shaft than with their meaning for the actin "target" lattice.

Fig.1 This micrograph shows the rigor myofilament lattice in a thick transverse section labelled to demonstrate how type 1 thin (200-250 Å) sections showing single filament layers are perpendicular to type 1 thick (1250 Å) sections. Scale bar 100 nm.

Fig.2 Moire-like effect in type 1 thin sections guides the search for widest areas of single filament layers. Scale bar 1 μm.

Fig.3 This type 1 thin section exhibits myac layer on left, actin layer on right, and transitional split myosin region in between. Crossbridges appear as double chevrons on left, as beads and rungs on right, and as diagonal stripes running 10:30 o'clock to 4:30 o'clock in middle. Scale bar 100 nm.

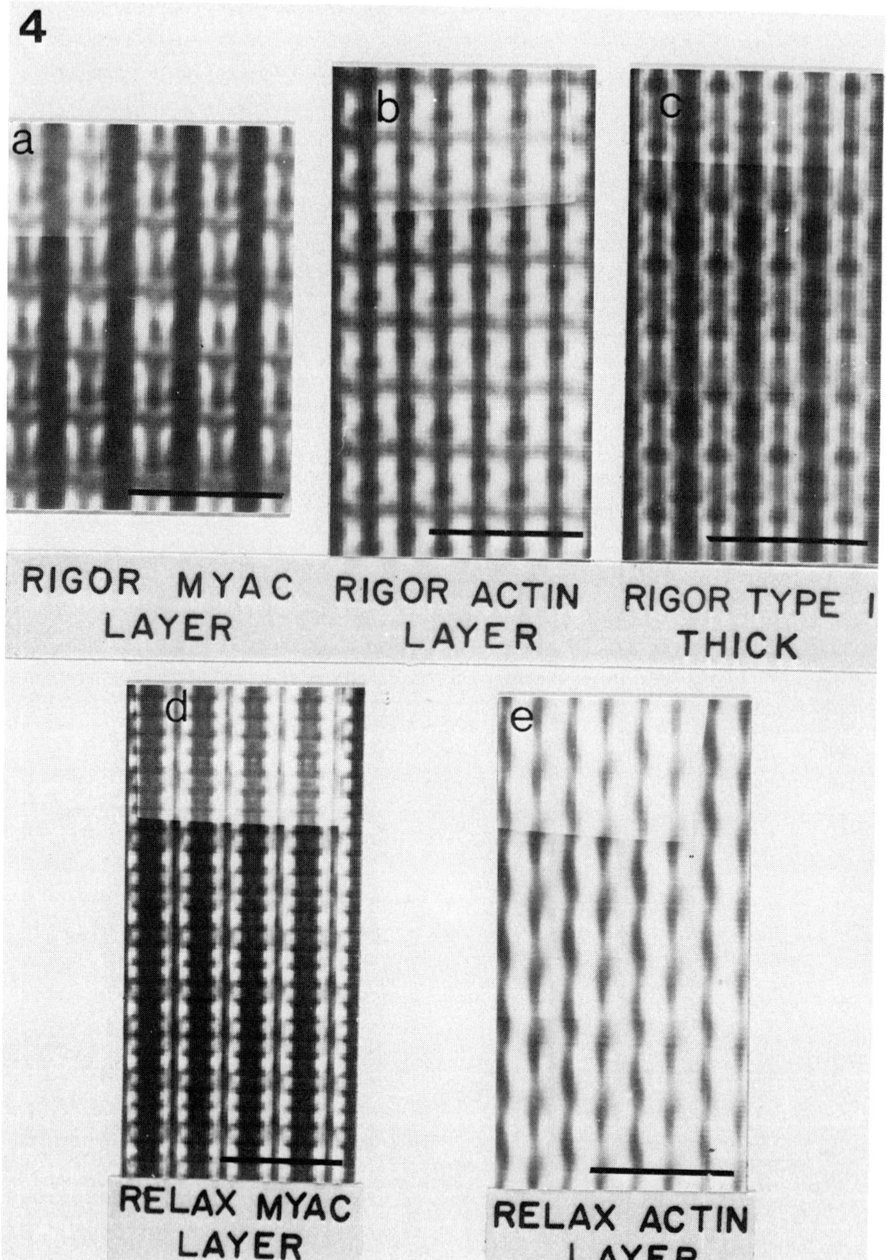
4
a
b
c
RIGOR MYAC LAYER
RIGOR ACTIN LAYER
RIGOR TYPE I THICK
d
e
RELAX MYAC LAYER
RELAX ACTIN LAYER

Using serial sections, it was possible to establish that the diagonals running from 10:30 o'clock to 4:30 o'clock (filaments vertical) as in Fig.3 were "scraped off" the near surface of a myac layer, indicating a left-handed helix. A simpler diagnosis of the helix sense required single oblique sections, cut about 40° to the fiber axis to permit direct sighting along the helical gyres of one side of each thick filament (Reedy, 1967, 1968). Later it turned out that proper longitudinal sections, both thick and thin, can present the effect clearly when tilted 30° - 50° about a transverse axis. When vertical filaments are tilted top end toward the viewer, the sighting effect produces strong periodic projections on the right side and spread-out diffuse projections on the left side of each filament, consistent with the left-handed screw sense.

The most direct and accessible electron microscope view of cross-bridges occurs in myac layers of type 1 thin sections, where the bridges extend between thick and thin filaments directly along the plane of the section. The general angling of rigor bridges around 45° to form actin-centered chevrons pointing toward the M-region, first noted by Reedy *et al* (1965) is evident in computer filtered images from such electron micrographs (Fig.4a). The double chevron grouping, composed of 2 nested chevrons comprising 4 bridges, marking each 385 Å axial repeat, is clear in the filtered image. Finally a modulation in the grouping, such that every third 385 Å repeat of chevrons tends to appear more single than double, expresses the 2 x 1155 Å period required for an exact repeat of the match between a myosin filament and one of the adjacent filaments. An exact repeat of the axial and azimuthal match requires 16 x 145 Å on the thick filament and 42 x 55 Å on the thin filament. An 1155 Å period did not show up clearly enough to deserve comment in the filtered images of other rigor views. The exact three dimensional crossbridge arrangements which express this long period are difficult to specify, so let us for the moment excuse our ignorance by calling it a defect structure brought about by matching the 145 Å period to the 385 Å actin helix repeat.

The chevron grouping may not seem convincingly doubled in flat views, but sighting transversely across a myac layer brings out unmistakably the doublet character of the rows of bridges. In fact, we can do even better, for an exact edge-on view is presented by type 1 thick sections (Fig. 4c), where the apparent thick and apparent thin filaments are actually edge-on views

---

Fig.4 Computer filtered images from digitized electron micrographs of *Lethocerus* flight muscle (Garrett, 1976). A single value for amplitude and for phase was chosen from each lattice point on the computed Fourier transform, and used in computing Fourier synthesis. Resynthesized images therefore show maximum averaging. Scale bars 100 nm.

5

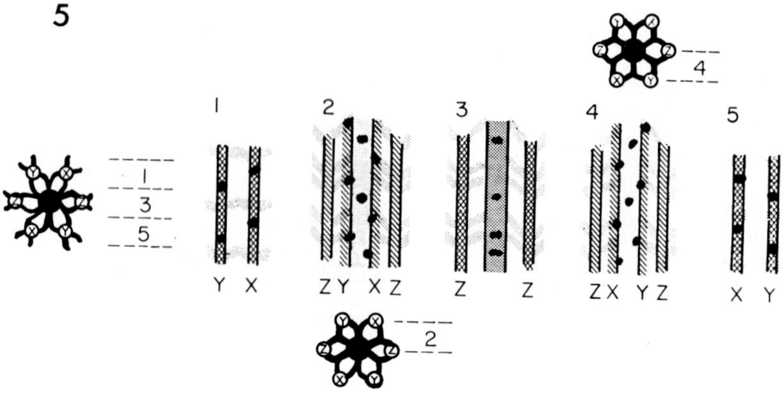

6

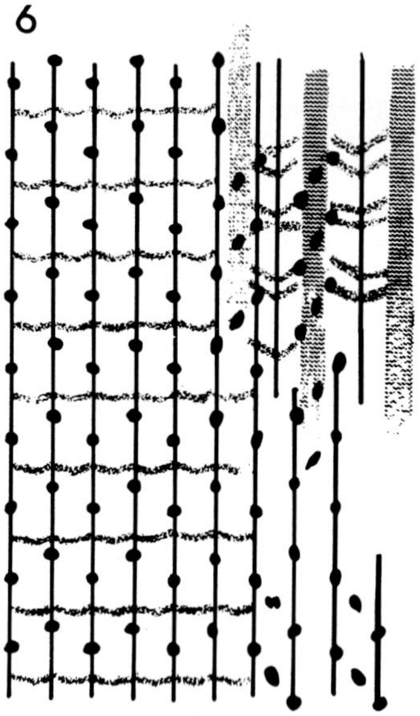

7

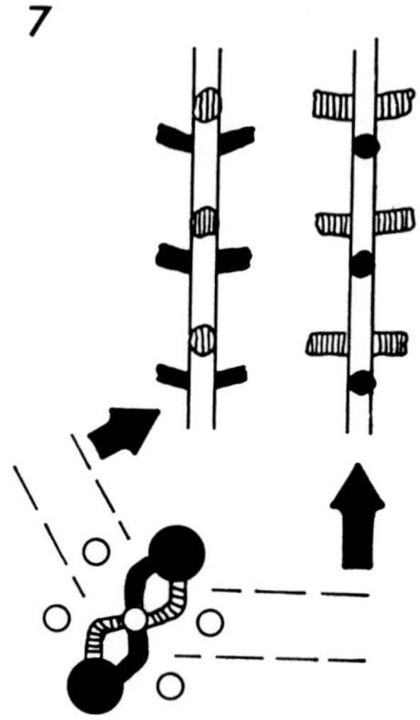

of myac and actin layers. 385 Å doublet beading along apparent thick filaments verifies the edge-on chevrons as doubled. Singlet beading along the apparent actin filaments occurs one-half period out of register with the myac doublet beading. Likewise, returning to the flat view of type 1 thin sections, there is a 385 Å singlet runging in actin layers which is staggered one-half period in register with the double chevron groups found in adjacent parallel myac layers.

The lateral and axial register among the features of type 1 thin sections is diagrammed in Figs.5 through 9. The pattern of beads and rungs along thin filaments of an actin layer appears in a filtered image (Fig.4b) exactly as diagrammed in Figs. 6 and 9. Alternate thin filaments are just alike, but the beads are staggered from one actin to the next. When actin layers are tilted 30° to 60° about a longitudinal axis, some beads become rungs and some rungs become beads (Fig.7), confirming that beads and rungs are just attached ends of crossbridges projecting at different azimuths from the thin filaments

---

Fig.5 Appearance of type 1 thin rigor section depends upon level of section in filament lattice. Axial crossbridge period of 385 Å is marked by double chevrons in myac layers (3), by {bead, rung} grouping in actin layers (1 and 5) and by beaded diagonal striping in split myosin regions (2 and 4). Overall register among these elements is as displayed here and in Fig.6; for example, actin layer rungs are midway between level of double chevrons and myac layer, and are aligned with middle of diagonal stripe which overlies each myosin in split myosin region. Y-shaped bridges in cross section view are derived from flared X patterns (Fig.11). Bridges appear as lateral projections and as dense beads, depending on the angle of view. Note that views 1 and 2, above myac layer, are mirror images to corresponding views 5 and 4, below myac layer.

Fig.6 Diagram shows appearance of type 1 thin section. Lateral register shown among various patterns is as described for Fig.5. Adjacent, thick or thin filaments in myac layer are essentially identical, but adjacent thin filaments in actin layer are unlike, while alternate thin filaments are identical. One myac layer is represented in upper part of drawing, while the influence of a second myac layer appears in lower part, where diagonal stripes show location of unseen thick filaments. Note that successive layers of thick filaments occupy alternating positions with respect to corridors of the intervening actin layer. (See Fig.9).

Fig.7 Along a single filament in an actin layer, 60° tilt in viewpoint or section plane causes beads to become rungs and rungs to become beads. Curved, angled bridges represent end-on view of one double chevron set, as reconstructed with flared X-bridge groups. Note that one more 60° tilt beyond the two portrayed is possible. It would present this same filament as seen in a myac layer, and would show all 4 bridges as rungs (i.e. as a double chevron), rather than as beads. Such changes have been verified by tilting actin layer sections +60° and -60°.

8

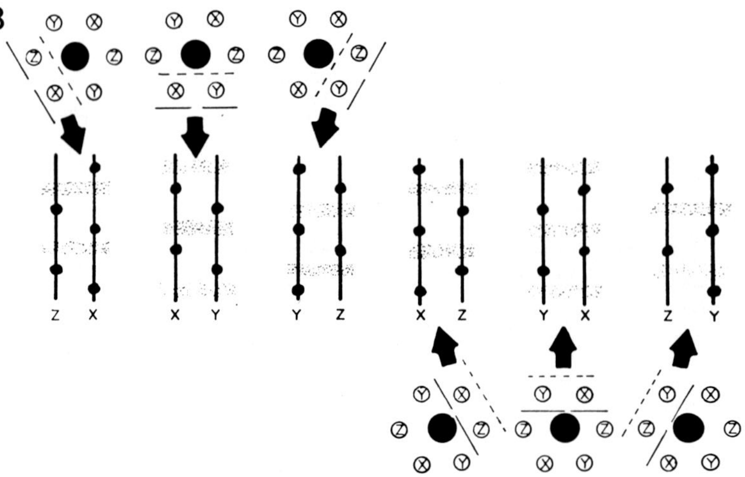
Z X
X Y
Y Z
X Z
Y X
Z Y

9

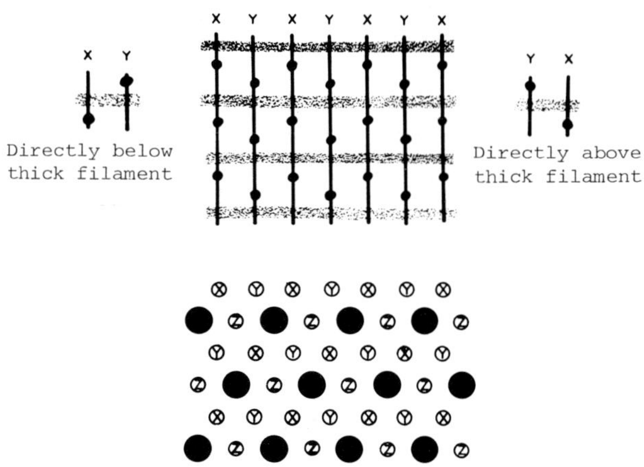
X Y
Directly below thick filament
X Y X Y X Y X
Y X
Directly above thick filament

(see flared X discussion below). With this in mind, it is clear that the 385 Å repeat is marked along each thin filament by a doublet grouping (1 bead plus 1 rung), then a gap, then doublet, then gap, quite like the double chevron groupings along actin filaments in myac layers. In either presentation, the attached bridges mark two out of three equal subsegments of each 385 Å repeat along each filament, leaving the third subsegment empty. If we combine two adjacent filaments in an actin layer, the axial stagger of bead positions from one to the other fills in the missing third, so that the three levels marked by projecting the rung, bead, bead repeat on to the common meridian account directly for the 385/3 Å or 128 Å meridional reflection in optical transforms of actin layer images.

The 1/3 stagger between adjacent filaments of an actin layer needs only to be repeated 6 times to reconstruct the whole envelope of 6 actins around one thick filament. This process is diagrammed in Fig.8. Remembering that each actin filament is shared by 2 myosins and is thus shared between 2 six-fold actin envelopes, the extension of this lattice regularity throughout the array of actins within one fibril can be reconstructed (Fig.9). All of the thick filaments are in helical register across the whole fibril, while the thin filaments fall into 3 sets of helical register, labelled x,y and z in the diagrams, each set rotated 1/3 with respect to the other 2.

In the thinnest transverse sections of rigor muscle the flared X was discovered. These sections had to be 200 Å or less; with thickness less than 1.5 x 145 Å, the section would include a single crown of bridges more often than two successive crowns along each thick filament. {We might say 1.5 x 128 Å with even more justification, since the average actin attachments are spaced

---

Fig.8 The setting of an actin layer pattern in a type 1 thin section gives enough information to indicate how to build up crossbridge pattern of complete six-fold actin envelope around thick filament. Six hypothetical actin layer sections are presented here in counterclockwise sequence. Each is shown as viewed flat from the upper side (marked by short dashed line). Change in appearance of filament X in going from section Z-X to section X-Y coincides with 60° change in viewing angle, as described in Fig.7. Once you see that this is all it takes to get from Z-X to X-Y, the rest of the diagram is simply redundant. Note that Z-X, X-Y and Y-Z are viewed from inside the envelope, and that mirror images of each pattern are provided (respectively) by X-Z, Y-X and Z-Y, viewed from outside the envelope.

Fig.9 Actin layer pattern can be built up from the two different {bead, rung, bead} groupings seen on either side. Once six-fold actin envelope is built up from actin layer pattern, as in Fig.8, then extended actin lattice can be developed as shown in the cross sectional image, from the relationship between actin envelopes around any two adjacent thick filaments.

10

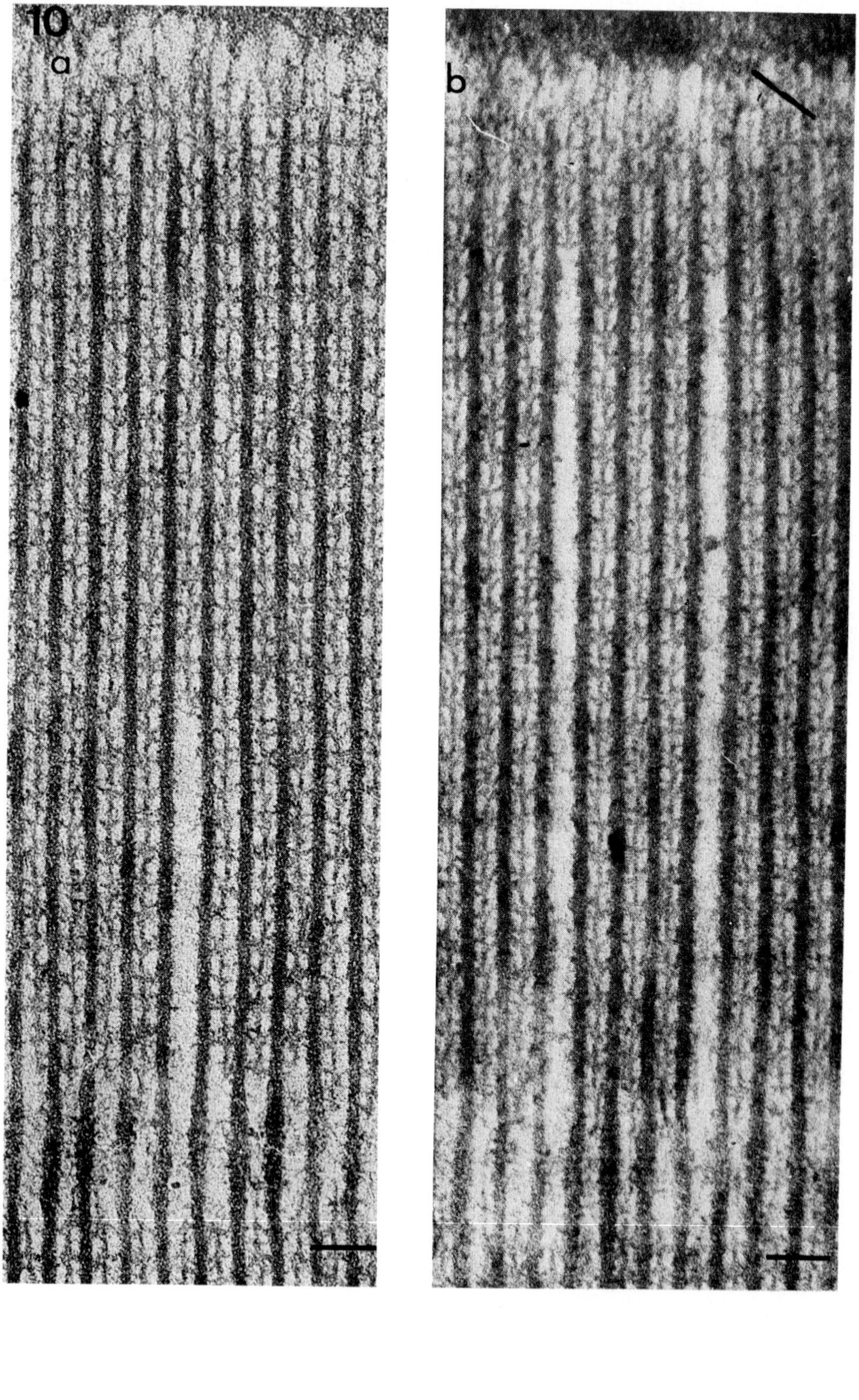

every 128 Å. The beating between 145 Å (myosin ends of bridges) and 128 Å (actin ends of the same bridges) produces the 1155 Å period. The periods match every 1155 Å, and they also go through an awkward patch of maximum mismatch every 1155 Å. We can't at the moment recognise any specific variations on the flared X form as more characteristic of the awkward patch than of the easy patch, even though Reedy has presumed to suggest such variations in a published diagram (Fig.6 of Reedy, 1968).} In any case the single crown of insect rigor turned out to include 4 bridges, the flared X, whose gracefully arching arms explained the bead,rung appearance of double chevrons when they were tilted 60° to the section plane, as in actin layers (and in Fig.7). In binding 4 and missing 2 of the 6 actins around each myosin, the flared X exhibits 2/3 filling of the azimuth, matching the 2/3 marking of 385 Å axial periods which we have pointed out in the myac and actin layer images. We have not yet found a section which included so perfect a transverse orientation of flared Xs. Instead a slight obliquity sampled successive levels of crowns in crossing the fibril, recognized by a succession of fields of differently-rotated flared Xs, changing by 60° anti-clockwise with each step along a system of left-handed helices (See fig. 11k and plates VII and X in Reedy, 1968).

Can we preserve and recognize the bridges which are not stabilized by attachment to actin? Not yet, it seems. Occasional actin filaments are truncated or missing in *Lethocerus* and where they are absent the regularly projecting rigor crossbridges are absent too, the periodic chevrons in a myac layer replaced by irregular, occasional bumps along the sides of the actin-free corridor (Fig. 10) or the arms of a flared X reduced decisively to 3 with the absence of one of the 4 actins (Fig.11c and see Fig.10 in Reedy, 1971). Perhaps during specimen preparation the unengaged bridges have become plastered against the myosin shaft, or thick filament backbone, as it is often called. Squire's (1972) suggestion to this effect offers a convenient mechanism for the concealment of the missing myosin heads, the heads we consider as missing when we compare the myosin count of 6 molecules (and thus 12 S-1 heads) per crown with the flared X count of 4 bridges per crown (Reedy, Bahr and Fischman, 1973). At the moment we don't know

---

Fig.10 Rigor crossbridges don't show up without an actin to hang on to. These two micrographs show myac layers containing incomplete actin filaments. The sides of the myosin filaments facing the missing length of actin fail to show the definite regular projections seen as double chevrons everywhere else, including on the opposite sides and remaining length of the same myosin filaments. The pertinent region of the myac layer in (b) is slightly less centered in the section than in (a); this tendency toward a split myosin region (striping 7:30 o'clock to 1:30 o'clock) probably accounts for the greater bumpiness and more frequent hints of faint periodic projections along the empty myosin sides in (b) as compared to in (a). Some bereaved bridges may be directed toward the two actin filaments underlying each empty corridor. Scale bar 100 nm.

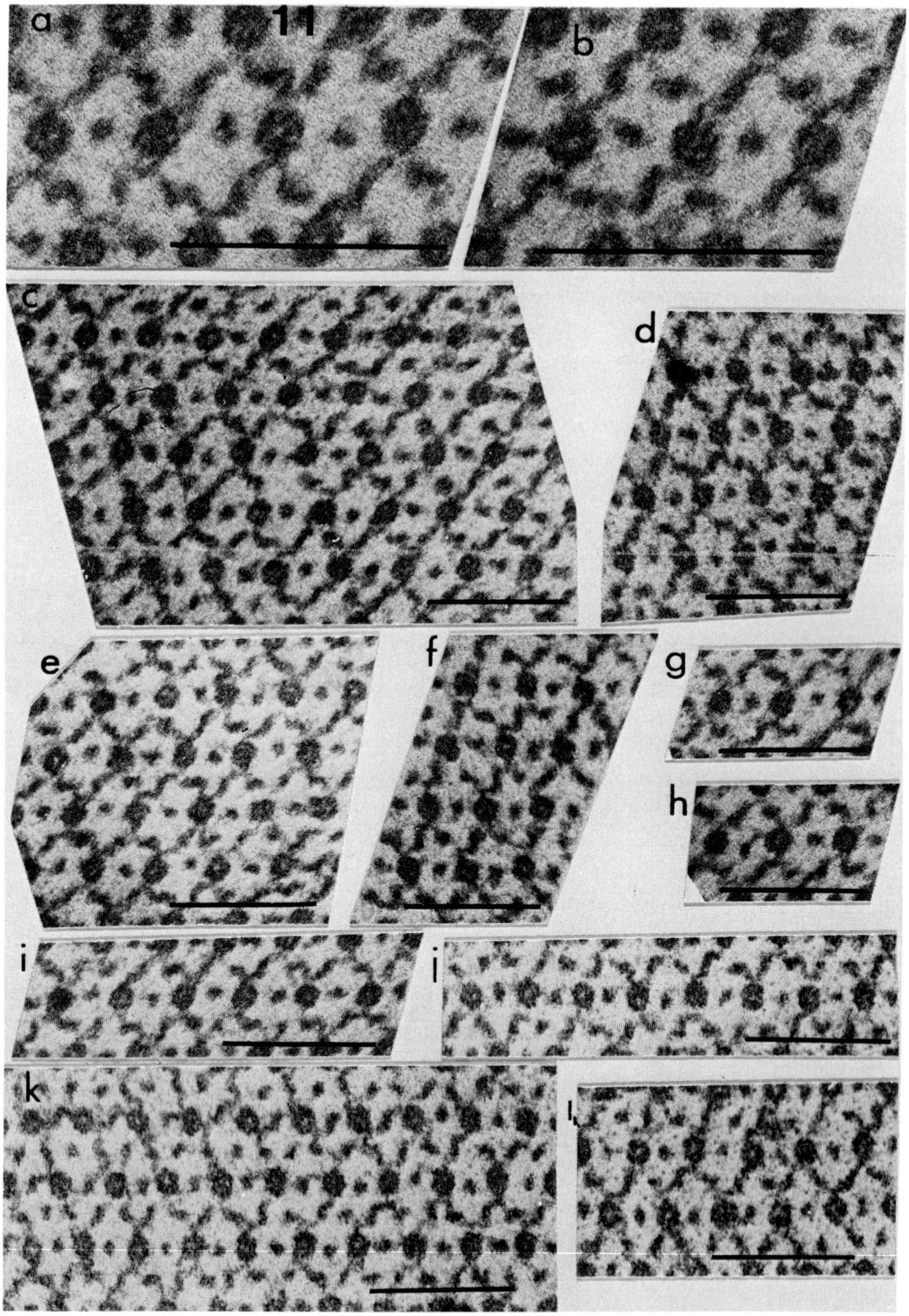

how many of these dozen heads are missing and how many (if more than 4) are included in the 4 preserved bridges.

Squire's (1972) diagrams of the flared X, which reproduce with some care the Moore, Huxley and De Rosier (1970) configuration of attached S-1 heads, prompted us to reconsider the actual images, and to analyze a previously ignored asymmetry which typifies the flared X. The point of junction between bridges and actin is usually a dense oblong region darker than the rest of the bridge, and specifically oriented in each arm of an X so that the overall flared X has a definite sense of twist (Fig.11). Clockwise and anti-clockwise forms can be designated and recognized (Fig.12). Only one of these two enantomorphs is present in any given fibril section. The anti-clockwise form turns out to occur when looking from Z-line towards M-line; the clockwise Xs occur when looking from M-region toward Z-region (research performed on Plate V, in Reedy, 1968). The chirality thus depends on whether the chevrons or arrowheads are pointing towards or away from the viewer. In Fig.12 we have derived end-on views of chevrons from the model photograph in Plate XIb of Moore, Huxley and De Rosier (1970), showing each bridge as a single S-1 head to make clear how the direction of arrowhead pointing fits into the clockwise and counter-clockwise forms of the flared X. Thus the presence of chirality in the flared X forms indicates reasonably good fidelity in preservation of the Moore, Huxley and De Rosier acto-S-1 conformation in a fair proportion of rigor crossbridges in thin sections of *Lethocerus* muscle. On the other hand, the diffraction patterns (X-ray and optical) from such fixed, embedded specimens show loss of layer lines beyond 100 Å; the loss of the moderately strong, smooth outer layer lines (59 Å etc.) from rigor patterns indicates some perturbation or disruption of order in the overall actin-crossbridge complex, since this complex is believed to give rise to the outer layer line system (Barrington-Leigh *et al.*, this volume).

Fig.11 These flared Xs are among the clearest from a very thin (less than 200 Å) transverse section of one *Lethocerus* rigor myofibril. Note waviness and kinking of bridge-actin complex between two myosins. The typical flared X does not have mirror plane symmetry but shows the following chirality. In these images, the dense oblong formed where bridges join actin is rotated anti-clockwise 20°-60° from the direct line of connection between 2 myosins. In some other fibrils, these oblongs were all rotated clockwise instead. These observations were partly stimulated by similar findings called to my attention by Roger Craig (1975) in his thick transverse sections of rigor lobster muscle, by similar appearances in high voltage electron micrographs of very thick cross sections of Drosophila flight muscle sent me by Hans Ris in January 1977, as well as by John Squire's (1972; his Fig.4) diagrams of the flared Xs. Scale bars 100 nm.

Fig. 12 Diagram on left interpretes the flared Xs of Figure 11 in terms of specific and uniform attachment of myosin heads to actin filaments according to Moore *et al.* (1970). The oblongs are composed of an actin filament with at least one S-1 head attached on either side. Differently oriented oblongs depend on 60° differences in orientation of successive actins around the actin envelope (Reedy, 1968, and Figure 13 here). The two different views, M-wards and Z-wards, were sorted out by identifying the Z-wards form in Plate V (c and b) of Reedy, 1968, where the viewing direction had been established by serial sectioning. The most inportant feature is that the myosin head curvature appropriate for the Z-wards view turns out to coincide with the curvature seen end-on in the model of Moore *et al.* when the arrowheads or chevrons are pointing directly toward the viewer. The opposite curvature occurs in the M-wards views of Figure 11. We estimated the azimuth of the actin filaments needed to fit with curved S-1 heads in order to approximate the flared X forms of Figure 11. The hinge point of the S-1 heads is best indicated in upper right and lower left bridges of Figure 11 flared Xs; the hinge is not usually bent in the upper left and lower right arms of the M-wards view. The S-1 head itself is rather uniformly bent in all views, corresponding to the rigor configuration discovered by Moore *et al.*

The versimilitude of the diagram in Fig. 12 is satisfying, although not meant at this time to urge a model of one S-1 head for each crossbridge, with one myosin shared out to two actins. Gerald Offer and Arthur Elliott (personal communications) deserve all credit for seriously exploring such a model in *Lethocerus* rigor. Speaking for one of us (M. K. R.), I had some difficulty facing up to the loss of two out of every six myosins at the time of our work for the Cold Spring Harbor Symposium (Reedy *et al.*, 1973; Bullard and Reedy, 1973). I need a bit more time or evidence to fortify my spirit before I can face the possible loss of two *more* myosins in accord with the suggestion of Offer and

Elliott.

One way to model the actin-myosin interaction which produces the *Lethocerus* rigor crossbridge lattice is to display the actin target lattice and the myosin surface lattice separately and then superimposed, as if they occurred at the same radius. This has been done in Fig. 13. If the actin target lattice were shifted axially about 110 Å toward the M region, then Figure 13c would be true to the rigor crossbridge angle of 45° (for an interfilament gap of 155 Å) but this multi-purpose diagram is already confusing enough without this modification so it has been omitted. The upper half of each diagram represents the way the actin and myosin lattices were first modelled by Reedy (1967, 1968) on the basis of electron microscopy of sectioned rigor muscle. The dotted ellipses represent the myosins acknowledged as missing from rigor in 1973 (Reedy *et al.*). The lower half of the diagrams has been modified to present features suggested by further study, particularly the studies of Garrett (1976). The actin target, originally suggested as a periodic single segment of helically recurring optimum azimuth for crossbridge attachment (Reedy, 1968) may be a split segment, 2 x 55 Å to 4 x 55 Å in length, presenting each myosin filament with two high-probability attachment sites flanking a lower-probability site. This is suggested by the extreme regularity of the filtered images of rigor actin layers, and especially by the precise spacing of the bead,rung doublets in such images. It is clear that this doublet spacing precision can be simply modelled by splitting each actin target into a bi-lobed target with centers 128 Å apart, but we have yet to agree on a physical basis for such splitting. On the other hand, it is clear that unsplit target segments could recruit bridges from two successive crowns (spaced 145 Å) which approximately straddle the target center. This would produce doublets easily part of the time, but not so easily or uniformly where the matching of actin and myosin filaments hits the awkward patch. The computer filtering program used here averaged all bridges. What is not clear at this stage is how far the averaging process might go in concealing awkward irregularities of crossbridge attachment to actin and in superimposing a precise doublet spacing on every 385 Å repeat of the filtered image.

The modified myosin surface lattice in the lower half of Fig. 13b is basically intended to offer a model in which the only certain parameter is the 145 Å axial period. Given 12 heads per crown, for a shaft of 90 Å radius, each head is allotted a 30° sector of 47 Å circumference at the surface. We have drawn this in Fig. 13c by deriving the S-1 heads from Squire's proposed surface lattice for insect thick filaments. However, any arbitrary lattice might do equally well, because the 40-50 Å heads could so nearly fill this circumference as to present an azimuthally almost undifferentiated "shelf" from which some S-1 heads would always be relatively convenient to actin targets,

13

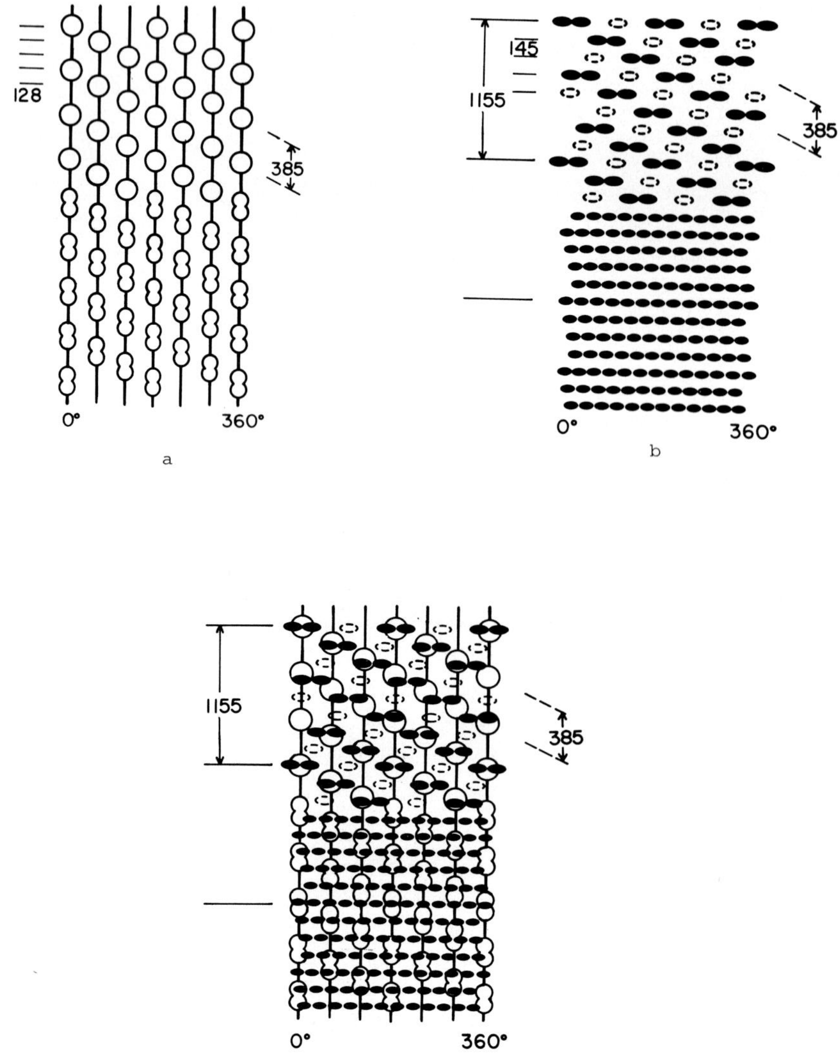

regardless of the precise azimuthal origin of the heads. If we decline to specify any particular azimuthal shift from one crown to the next, then what would account for the 385 Å and 233 Å layer lines in relaxed muscle upon which previous models base their specifications of myosin surface lattice (Reedy, 1967, 1968; Squire, 1973 and this volume)? What would suggest that even in relaxed muscle the actin target lattice might attract and stabilize the nearest of the available heads from the azimuthally crowded crowns, and that this selection of heads could approximate the surface lattice needed to give rise to weak 233 Å and 385 Å layer lines, regardless of the precise surface lattice of all heads. We do not share equal fondness for this model, but we do share the belief that

---

Fig. 13.a shows a radial projection of the actin target lattice along the six thin filaments surrounding one myosin filament. The unrolled actin envelope is viewed from outside. The segments marked by circles or double circles indicate 385 Å helical recurrence of actin subunits which are optimally directed to receive bridges from the myosin filament. The double circles in the lower half are to suggest a possible split target within the generally optimum region. This might help to account for the regularity of doublet bridge groupings in rigor.

b shows the myosin surface lattice of crossbridge origins from outside in radial projection, stretched to the same radius as the actin envelope projection in 13 a. Of the six myosin molecules which occur every 145 Å, at least two are believed missing (dotted ellipses) from the rigor crossbridge array on the basis of the four-bridge flared X groupings. The upper half shows two myosin molecules for each lattice point from which completed rigor crossbridges originate. There is no doubt that the rigor crossbridges mark a two-stranded helix pitched 2 x 385 Å on the thick filament surface. There is, on the other hand, no certainty that the relaxed crossbridges mark anything but a strong 145 Å period on the thick filament (See text). The lower half of 13 b portrays 12 myosin S-1 heads per crown, spaced evenly every 47 Å around the 565 Å circumference. Such a myosin filament might exert minimum azimuthal constraints on the selection of myosin heads for rigor bridges, and the constraints would not be expected to vary much regardless of the exact surface lattice of points or exact azimuthal specifications of each crown. This is meant to suggest that if each insect rigor bridge involves 2 S-1 heads, they might equally well come from two different myosin molecules as from a single myosin molecule.

c In this diagram the myosin surface lattice has been superimposed on the actin target lattice as if both were at the same radius. This exhibits recurrent matching of 145 Å and 385 Å periods every 1155 Å, and shows the spectrum of axial and azimuthal variations encountered in matching the two lattices. Such a diagram affords a basis for modelling the selection of myosin molecules or S-1 heads which form rigor bridges. We do not suggest any necessary conjunction of the doublet actin targets with the 12-head crown; the 12-head crown could equally well be modelled on the unsplit actin targets.

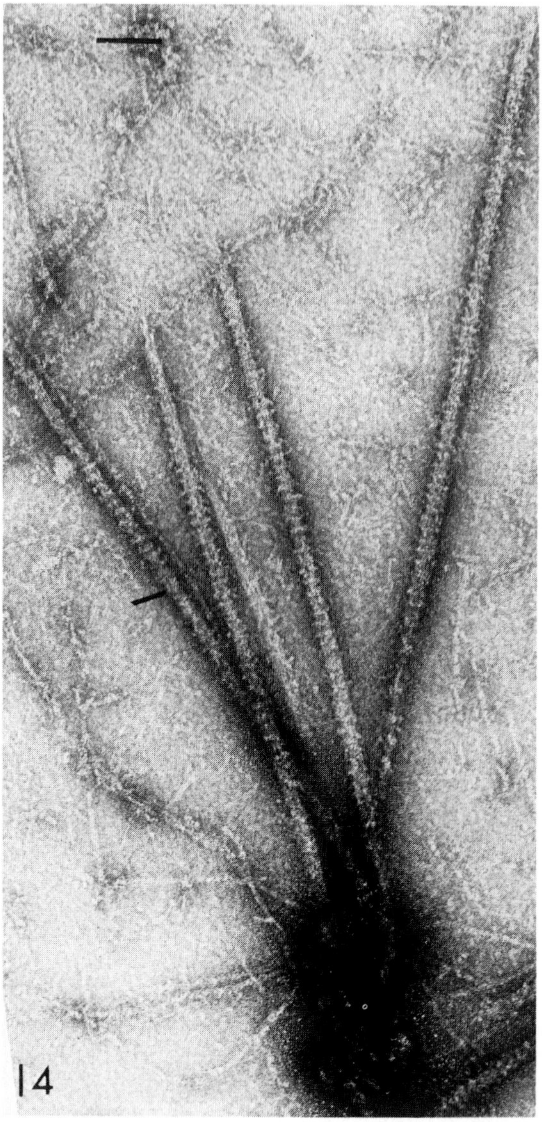
14

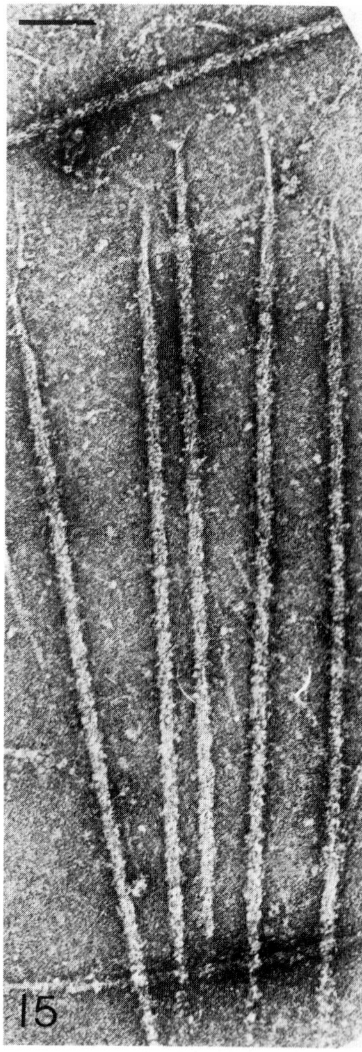
15

present evidence cannot exclude it. Among the evidence which permits this idea, we would mention the tendency for crossbridge attachment in relaxed muscle (Moss *et al.*, 1976) and the fact that insect actomyosin has been found rather resistant to complete dissociation under relaxing conditions (Maruyama, 1959, 1965).

An obvious and attractive source of some relief to the problem of the awkward patch in matching myosin and actin periods is provided by remembering that filaments can slide. It is not necessary even in insect muscle to suppose that all bridges attach simultaneously, either in rigor or during contractile activity. The bridges in the easy patches of the 1155 Å repeats would be expected to attach first and to produce some sliding which would then relieve the axial mismatch of the awkward patches by up to 150 Å. This would shift the regions of maximum mismatch by up to 450 Å, but the region newly made most awkward by sliding would already have recruited and finalized rigor bridge attachments before the sliding began.

Some light is thrown on the question of the myosin surface lattice by computer-filtered images of relaxed *Lethocerus* myac layers (Fig. 4d) and actin layers (Fig. 4e). In myac layers, the 145 Å period is clearly associated with thick filaments as a periodic dense shelf or stripe which attenuates markedly as it projects near the thin filaments. The 385 Å period is identifiable as a faint background wave of light gray between filaments in the myac layer, and as a dense segment which recurs every 385 Å along the thin filaments of both myac layers and actin layers. The 233 Å layer line has so far survived only marginally or not at all in sectioned relaxed muscle, and so neither the

---

Fig. 14 Thick *Lethocerus* flight muscle myofilaments, isolated by sonication of myofibrils in relaxing solution and negatively stained by uranyl acetate, are seen in this micrograph. They are clearly striped with a 145 Å repeat, a projecting shalf or ridge which betrays no substructure or azimuthal differentiation to inspection or optical diffraction. ATP has no effect on the periodic structure of isolated thick filaments. Scale bar 100 nm.

Fig. 15 Brief papain digestion, under conditions which suffice to liberate HMM S-1 from *Lethocerus* myosin or myofibrils, removes the 145 Å striping completely from *Lethocerus* thick filaments, as seen here, suggesting that striping was no more and no less than the periodic repeat of myosin heads. Scale bar 100 nm.

385 Å nor the 233 Å period has yet been traced to the relaxed myosin filament by electron microscopy and image processing. Optical diffraction of relaxed myac image transparencies where everything but the myosin shafts was painted opaque also failed to show 385 Å and 233 Å layer lines, although these layer lines did remain when the same selective ablation procedure was carried out with a rigor myac image. Now, even if later experiments were to preserve more strength in these layer lines, and even if they were to be identified in relaxed muscle with crossbridge mass close to the thick filament surface, it would remain possible that these bridges were expressive of actin influence as discussed in the previous paragraph, rather than of intrinsic thick filament structure.

Isolated and negatively-stained *Lethocerus* thick filaments can be placed beyond any possible influence of the actin lattice, but they have not yielded any more information; they look virtually identical to the thick filaments in relaxed myac layers, showing only the 145 Å striping but no hint of surface lattice (Fig. 14). Since papain removes the striping and all sign of the 145 Å period from isolated filaments (Fig 15), there is no reason to consider that the striping represents any feature other than myosin S-1 heads (Garrett, 1976). And since ATP has no influence on the striped appearance, negative staining studies give no reason to argue that ATP influences crossbridge angle apart from its effect on crossbridge attachment to actin.

The filtered image of a relaxed actin layer shows the same basic stagger of 385/3 Å between adjacent actin filaments that we saw in rigor actin layers. The density is smeared out rather than bead-like, actually resembling the F-actin helix somewhat, but we believe it largely represents labelling of the actins by relaxed bridges which become attached or trapped on the actins during the primary fixation by glutaraldehyde. During fixation of relaxed *Lethocerus* muscle, there occurs marked intensification of X-ray reflections on the 385 Å and 193 Å layer lines along the rowline over the equatorial 1,0 reflection, suggesting increased labelling of the actin target lattice (unpublished experiments, H.E. Huxley and Reedy and Tregear, 1966; Garrett and Reedy, 1976; Reedy, Goody, Rosenbaum and Hofmann, 1977). The stagger of dense segments in Fig. 4e would fit with this suggestion.

It seems rather important to consider the meaning of the observation that the increased labelling follows the actin target lattice. If glutaraldehyde is considered to promote crosslinking of relaxed bridges to actin with absolutely no regard for stereospecific matching of natural contact sites on myosin head and actin subunits, then the unattached crossbridges could become attached wherever the thick filament surface lattice presented them to the sixfold envelope of actins. If the surface lattice were a six-stranded (Squire) or twelve-stranded (Fig. 13b bottom) helix, this process could only attach bridges uniformly to all 6 actins every 145 Å, which would not cause the observed X-ray changes or the actin target lattice visualization in Fig. 4e. If the myosin surface lattice were a two-stranded helix, as originally proposed by Reedy (1967, 1968), then such a process *could* cause the observed labelling. However, in the Cold Spring Harbor update of the two-stranded model the "missing myosins" (dotted ellipses in Fig. 13b) should fill in the azimuth and obscure any actin lattice labelling, for this model is much like the six- and twelve-stranded models in this respect.

Now, the two-stranded helix model is true and appropriate for the rigor insect lattice of attached bridges. But it is very low down on the list of myosin models one might consider for relaxed insect muscle. The other models mentioned above all come higher on the list. So we do not believe that glutaraldehyde is betraying the myosin surface lattice when it produces labelling of the actin target lattice. Instead, we believe that this labelling could only arise if some degree of physiological and stereospecific relaxed bridge attachment were being trapped by glutaraldehyde fixation. The X-ray changes suggest that glutaraldehyde actually stimulates an increased number of such attachments to azimuthally appropriate sites in the actin target lattice. If not, if glutaraldehyde only traps previously existing contacts of resting bridges with actin, then some conformational change during fixation of these bridges would be required to explain the X-ray changes.

REFERENCES

Benzer, S. (1966). *Phage and The Origins of Molecular Biology*. Ed. Cairns, J., Stent, G.S. and Watson, J.D. *Cold Spring Harb. Lab. Quant. Biol.*, p. 157.

Garrett, W.E. Jr. (1976). Ph.D. thesis, Duke University.

Maruyama, K. (1959). *J. Ins. Physiol.*, 3:271.

Maruyama, K. (1965). *The Physiology of Insecta*, Ed. M. Rockstein. 2, 451.

Moore, P.B., Huxley, H.E. and DeRosier, D.J. (1970). *J. Mol. Biol.* 50:279.

Moss, R.L., Sollins, M.R. and Julian, F.J. (1976). *Nature*, 260, 619.

Reedy, M.K., Holmes, K.C. and Tregear, R.T. (1965). *Nature*, 207:1276.

Reedy, M.K. (1967). *Amer. Zool.*, 7, 465

Reedy, M.K. (1968). *J. Mol. Biol.*, 31:155.

Reedy, M.K. (1971). *Contractility of Muscle Cells and Related Processes*, Ed. R.J. Podolsky. *Prentice Hall*, p. 229.

Reedy, M.K., Bahr, G.F. and Fischman, D.A. (1973). *Cold Spring Harb. Symp. Quant. Biol.* 37, 397.

Squire, J.M. (1971). *Nature*, 233, 457.

Squire, J.M. (1972). *J. Mol. Biol.* 72, 125.

THE INTERPRETATION OF X-RAY DIFFRACTION FROM GLYCERINATED FLIGHT MUSCLE FIBRE BUNDLES: NEW THEORETICAL AND EXPERIMENTAL APPROACHES

John Barrington-Leigh, Roger S. Goody, W. Hofmann, Ken Holmes, H.G. Mannherz, G. Rosenbaum,* and Richard T. Tregear.†

Max-Planck Institut für Medizinische Forschung, 6900 Heidelberg, W.Germany.

* *E.M.B.L., Hamburg, West Germany.*

† *A.R.C. Unit, Department of Zoology, Oxford, U.K.*

Following the discovery that ATP changed the structure of glycerinated insect flight muscle (Reedy, Holmes and Tregear, 1965) it gradually became clear that the hypothesis which had been advanced to explain the X-ray and electron microscope observations, the swinging bridge hypothesis, did not uniquely explain the experimental data. Two problems contributed to the slowness of the progress made: the considerable difficulty of recording the low angle X-ray fibre diagram and the absence of any ready-made theory to interpret it.

In the last few years these problems have been partially overcome by:

1. The use of the synchrotron X-ray source
2. The production of modified diffraction patterns
3. A new theoretical approach

The synchrotron X-ray source has been discussed in a number of places (e.g. Barrington-Leigh and Rosenbaum, 1976) and Figs. 1-4 bear adequate testimony to its effectiveness. A satisfactory qualitative description of the rigor and β,γ-imido ATP patterns has now been obtained on the basis of the new theory. The experiments with subfragment-1 give support for the hypothesis underlying the theory. We are now in a position to carry out a quantitative comparison of the X-ray fibre diagrams with specific models and on the basis of a preliminary study it appears improbable that the data fit to a simple swinging bridge hypothesis.

THE FIBRE DIAGRAM

For ease of reference we will index all the layer lines of the fibre diagram (Fig. 1) on a repeat of c = 2326 Å. The strong low order layer lines with a 388 Å spacing typical of rigor become the 6th,12th,18th orders. The high order "actin" layer lines become the 27th, 33rd, 39th, 45th order. The 145 Å meridional is the 16th, and the 235 Å layer line becomes the 10th order (Fig. 4). The low order layer lines are lattice sampled whereas the outer layer lines are continuous. The Bragg reflexions on the inner layer lines index on a hexagonal lattice (a = 530-590 Å). The 6th, 12th and 18th layer lines arise from a pseudo-unit cell with $P3_1$ space group symmetry (see Holmes et al., 1976).

The elements of the new theory

In analysing the rigor pattern Miller and Tregear (1972) compared the diffraction which would be obtained from a lattice of decorated actin molecules of the Moore, Huxley and DeRosier (1970) type with the rigor pattern of insect muscle. They showed that a reasonable fit could be obtained for the outer layer lines, (27, 33, 39, 45) and pointed out that the inner part of the fibre diagram was not easy to interpret. In particular one can show that the innermost reflexion on the 6th layer line is inexplicably strong. This reflexion corresponds to a set of Bragg planes of spacing ca. 300 Å which run diagonally through the structure. Holmes *et al* (1976) realised that these planes would be strong if cross bridges occurred only every 388 Å along the filament axis in the way which had been proposed by Reedy (1968) on the basis of the electron micrographs. Reedy calls this periodicity of marking of the actin monomers by cross bridges "chevrons" or "double chevrons". The approach used by Holmes *et al*. was, however, unable to analyse or account for those aspects of the pattern arising from its decorated actin character. There remained the technical problem of fusing the two approaches. This has recently been accomplished (Holmes, Tregear, and Barrington-Leigh; manuscript in preparation) by considering the probability of attachment of cross bridges to actin to be a periodic function of position (the Z coordinate of the actin fibre). This description allows the whole of the diffraction pattern to be accounted for. The longitudinal concentration (or modulation function) is a phenomenological description which itself makes no statement concerning the molecular origin of the process. Model-fitting shows that the low-angle X-ray diffraction contains the effects of this and other "concentration waves".

It is easy to understand qualitatively the effects which such modulation functions would have on the diffraction pattern. Assume for the moment that the probability of attachment of a cross bridge to actin varies in a sinusoidal way with a period of 388 Å. The structure may now formally be composed by multiplying the decorated actin structure by the modulation function expressing the probability of attachment. From Fourier transform theory we know that the transform of the product of two functions is just the convolution of the transforms of the two functions. In this case the transform of decorated actin is the structure factor of a helix having 28 subunits in 13 turns giving, for example, J_o terms on the zero layer line, J_2 terms on layer line 6, J_4 terms on layer line 12 etc. The transform of the sinusoidal modulating function is simply a pair of points on the meridian at ± 1/388 Å (i.e. in the middle of the 6th layer line). On convoluting this pair of delta functions with the actin transform we produce J_o terms on the 6th layer line, J_2 on the 12th layer line, etc. in addition to the normal helix terms. In fact the modulation function in rigor has a more complex form than sinusoidal and we have to include 3 components with periods 388, 194, and 127 Å. Each of these would allow a meridional reflexion but the space group relationship between neighbouring actins leads to a cancellation on the meridian except for every third layer line of the $P3_1$ pseudo cell (i.e. the 18th layer line, 1/127 Å) which does show a meridional reflexion. The fact that the ghost reflexions obey the space group symmetry demonstrates that each decorated actin complete with its modulating functions is related to its neighbours by the $P3_1$ symmetry operators. In other words the modulating function (Q) is stepped by 1/3 of 388 Å between neighbouring actin molecules. The $P3_1$ lattice is in any case not very perfect and each of the layer lines is actually composed of a lattice sampled part and a continuous part added together. The continuous part arises from each decorated actin filament scattering as an independent unit. As the effect of the disorder factor weakens the lattice sampled scattering the continuous scattering gets stronger. By the 12th layer line the effect of the disorder is strong enough to reveal the meridional nature of the scattering on the layer line, proving the presence of a weak second order component of $Q(q_2)$.

The high order layer lines (27, 33, 39, 45) show no detectable lattice sampling and arise from the decorated actin molecules scattering independently. However, even here Q leaves its mark. The effect of convoluting the decorated actin pattern with the components of Q is to displace the diffraction pattern by 6 or 12 layer lines. Thus normally the main Bessel functions contributing to layer lines 27, 33, 39 are -5, -3 and -1 respectively. The first order components of Q (q_{+1}, q_{-1}) produce in addition -3, -7 on 27, -1, -5 on 33 and 1, -3 on 39. Since we are dealing with cylindrically averaged structures these terms add in the square in the normal way (Franklin and Klug, 1955). We also refer to these "ghost" terms as "leaky Bessel functions" in order to demonstrate the origin of the new terms. The Q function essentially breaks the actin symmetry and by so doing allows otherwise forbidden Bessel function terms, each representing a different kind of helical symmetry, to "leak"out. As in the low order pattern, the leaky Bessel functions appear as intensity nearer to the meridian than otherwise expected. Inclusion of the modulation factor allows a better fit to the rigor outer layer lines since they are much more radially extensive than the helix cross on the 27, 33, 39, 45 layer lines to be expected from the allocation of Bessel functions -5, -3, -1, 1 respectively. However it also considerably complicates the interpretation since one has to know the size of the components of Q before one can begin to model the outer layer lines. Only through a unified interpretation of inner and outer layer lines will this be possible.

Subfragment 1; an experimental check on the theory

We have proposed that a considerable part of the low angle X-ray fibre diagram arises from the unequal distribution of cross bridges along the length of the decorated actin filament. This is equivalent to saying that a lot of the diffraction pattern arises from holes. If this were so, then if we were able to fill up the holes, these new features in the diffraction should vanish. This suggested to us that the experiment with subfragment -1 loading performed on rabbit muscle by Rome (1973) should be repeated on insect flight muscle. Bundles of glycerinated fibres in rigor were immersed in rabbit S1 solutions. The uptake of S1 was monitored by following the changes in the 1.0 and 2,0 equatorial reflexions. There was a considerable strengthening of the 2,0 reflexion beyond the normal rigor value. The uptake of S1 is accompanied by a swelling of the hexagonal cell of the muscle fibre. An X-ray photograph obtained from 24h-soaked muscle is shown in Fig. 2. Compared with rigor (Fig. 1) we see that the 1,0 on the 6th and 12th layer lines have become weaker. These are parts of the diagram which should arise from components of Q. As a result of this experiment, there seems little doubt that the class of solution we have adopted is broadly correct.

Among the obvious features of Fig. 2 are the strong continuous outer layer lines (27, 33, 39, 45). They are approximately what one would expect as the diffraction patterns of decorated actin. They form the southern half of a helix cross, the minimum reciprocal space radius becoming larger as we descend the cross because the Bessel function order gets steadily greater. Preliminary calculations show that this pattern agrees well with the structure factor which can be calculated from the electron micrograph reconstruction of a cross bridge made by Moore, Huxley and DeRosier (1970). If we compare these outer layer lines (Fig. 2) with those of the rigor pattern (Fig. 1) we see considerable differences. These could arise because the added rabbit S1 has a different shape from the insect S1. This assumption, however, is not necessary since the observed rigor pattern can be approximately generated from the Moore, Huxley and DeRosier cross bridge taking due account of the components of Q. In other words the rigor pattern may be considered as a case of leaky Bessel functions, the broken symmetry allowing low order Bessel functions to appear on several layer lines of the helical series. Thus the S1 experiments are helpful in the interpretation of the fibre diagram.

We have checked the distribution of added S1 by electron miscropy and there is no doubt that the added S1 is contributing to the actomyosin matrix. One can also see additional bands of heavy uptake on the M line and near to the Z line but this additional loading does not invalidate our general conclusion. Loading with HMM produces similar results.

AMPPNP

Equipped with the new theory we turn our attention anew to the AMPPNP state (Fig. 3). The salient characteristics of this state have been elaborated in a number of publications (e.g. Goody et al., 1975; Marston, Rodger and Tregear, 1976). Essentially, the majority of the cross bridges are still attached to actin but the cross bridge conformation appears to be different from rigor. We first thought it to be like the relaxed state but the arguments presented below will show that this is improbable. The main differences from rigor are that the 145 Å meridional reflexion and the associated layer line become strong; the 1,0 on the 6th layer line becomes very much weaker, and the rest of the actin-based pattern all becomes somewhat weaker; the 1,0/2,0 equatorial ratio moves in the direction of relaxed muscle.

The strong 145 Å meridional reflexion presents us with a conceptual problem. If, as the mechanical measurements indicate (Marston *et al*, 1976), the majority of cross bridges are bound to actin, how can an incommensurate spacing such as 145 Å be generated? An appealing answer is to apply again the methodology we developed for the Q function. It is known from studies of relaxed muscle that the cross bridges originate every 145 Å along the myosin filament. Apparently, the 145 Å periodicity of cross bridges origins has a much more powerful influence in the presence of AMPPNP than in rigor. This leads to the decorated actin structure being modulated with a periodic function of Z which repeats every 145 Å. We call this modulating function M and its components m_1, m_2 etc. The primary effect of the modulation will be to produce a pair of points ± 1/145 Å on the meridian which acts as new origins for the original pattern generated by the $P3_1$ lattice of decorated actin. Thus a pair of strong spots appear at ± 1/145 Å which are essentially ghosts of the $F(00\bar{0}0)$ term produced by the M function modulation. The reduction of the 1,0 on the 6th layer line appears to reflect a considerable weakening of 388 Å component of $Q(q_1)$; this explanation requires that the bridges redistribute among actin sites on binding AMPPNP.

Having invoked the presence of the M function we can return to rigor and ask why the 145 Å meridional is so weak in rigor. The classical answer would have been that the shape of the cross bridge leads to a chance absence of the meridional 145 Å. With the help of the theory outlined above we can now test such contentions quantitatively and it seems unlikely on the basis of the Moore,

Huxley and DeRosier cross bridge that it could be so weak. An alternative is that the magnitude of the M function is reduced in rigor.

From the present qualitative analysis it seems likely that the M function is strong in the presence of AMPPNP and the Q function is weak whereas in rigor the opposite is true. There is at present no evidence that that part of the cross bridge which binds to actin with the symmetry of actin (i.e. is stereospecifically related to actin) has a different conformation in rigor or on binding of AMPPNP. However, the strong effect of AMPPNP on the Q and M function implies a change of stereo affinity between actin and myosin in the presence or absence of nucleotide. In rigor the cross bridges bind preferentially when the actin faces myosin with little regard for the effective Z registration between actin and myosin. In the presence of nucleotide this azimuthal specificity becomes much weaker but is replaced by a tight preference for the correct 145 Å level. This change of specificity is accompanied by some transfer of mass back to the myosin filaments, which requires a sharp azimuthal distortion in the bridges, unless they detach.

ATP

The ATP pattern (Fig. 4) is dominated by the 145 Å meridional reflexion and orders thereof. This reflexion arises mostly from rings of bridges projecting from myosin every 145 Å. Whether the bridges project at right angles or not is still undetermined. The 3-4 orders of the 145 Å spacing which are visible indicate that the Z-distribution of density is sharp which would favour an angle close to 90°. The other features of the ATP pattern are remnants of the rigor pattern, including a relatively strong 1,0 on the 6th layer line; these features may represent residual crossbridge attachment in the "relaxed" muscle, which in the insect case is relatively stiff.

Another interesting feature of the relaxed pattern is the 10th layer line (235 Å). It has been proposed by Reedy (1968) and Miller and Tregear (1972) that this arises from the helical symmetry of the myosin filament. An alternative explanation arises on the basis of the M function. If the bound bridges are strongly modulated by the M function then one may see not only the $F(00\bar{0}0)$ term imaged in the meridional points ($\pm$ 1/145 Å) but also other features of the diagram such as the second layer line. According to this explanation the 10th layer line would be a ghost of the second layer line produced by the first order of the M perturbation on the meridian of the 16th layer line. In fact the two explanations correspond to similar structures.

Conclusions

We have developed a unified theory for interpreting the various diffraction patterns from insect flight muscles which have been obtained. We have been able to verify the basis of the theory experimentally in the case of rigor by adding

S1 and HMM to the fibres. The theory shows that the main effect of binding AMPPNP is to cause a redistribution of cross bridges amongst available actin sites so that the 388 Å period, which is strong in rigor, is considerably weakened and that the 145 Å period is considerably strengthened. On this basis we deduce that in rigor the probability of crossbridge formation is determined by the azimuthal orientation of the actin relative to the thick filament and the crossbridge conformation itself is determined entirely by the actin. On the other hand in AMPPNP the probability of crossbridge formation is greatly influenced by the axial periodicity of the thick filament and the crossbridge conformation is probably also influenced azimuthally by the thick filament. It is fair to add that these changes of stereo affinity must reflect profound changes in the S1 myosin-backbone relationship which we presume are of direct relevance to the cross bridge cycle but which we at present can only detect through their effect on the Q and M functions.

References

Barrington-Leigh, J. and Rosenbaum, G. (1976). *Ann. Rev. Biophys. Bioeng.* 5, 239.

Franklin, R. and Klug, A. (1955). *Acta Cryst.* 8, 777.

Goody, R.S., Holmes, K.C., Mannherz, H.G., Barrington-Leigh, J. and Rosenbaum, G. (1975). *Biophys. J.* 15, 687.

Holmes, K.C., Goody, R.S., Mannherz, H.G., Barrington-Leigh, J. and Rosenbaum, G. (1976). *26 Colloquium Mosbach 1975*, eds. Heilmeyer, Rüegg and Weiland, Springer Verlag, Heidelberg.

Marston, S.B., Rodger, C.D. and Tregear, R.T. (1976). *J. Mol. Biol.* 104, 263.

Miller, A. and Tregear, R.T. (1972). *J. Mol. Biol.* 70, 85.

Moore, P.B., Huxley, H.E. and DeRosier, D.J. (1970). *J. Mol. Biol.* 50, 279.

Reedy, M.K. (1968). *J. Mol. Biol.* 31, 155.

Reedy, K.M., Holmes, K.C. and Tregear, R.T. (1965). *Nature(London)*, 207, 1276.

Rome, E. (1972). *Cold Spring Harbor Symp. Quant. Biol.* 37, 331.

FIGURE LEGENDS

Fig. 1. X-ray diffraction pattern obtained from glycerinated Lethocerus flight muscle in the absence of nucleotides. Conditions: 10 mM $MgCl_2$, 10 mM NaN_3, 4 mM EGTA, 20 mM imidazole. HCl, pH 6.8, 4°C. Scale on upper photographs is 2x that on lower ones. Horizontal arrows identify layer-lines, vertical arrows row-lines.

Fig. 2. X-ray diffraction pattern obtained after incubation of glycerinated Lethocerus flight muscle 24h. in 20 mg/ml myosin subfragment 1 at 4°C. Conditions otherwise as in Fig. 1.

Fig. 3. X-ray diffraction pattern obtained from glycerinated Lethocerus flight muscle in the presence of 1 mM AMPPNP. Conditions otherwise as in Fig. 1.

Fig. 4. X-ray diffraction pattern obtained from glycerinated Lethocerus flight muscle in the presence of 15 mM Mg ATP. Conditions otherwise as in Fig. 1. Horizontal arrows identify layer lines.

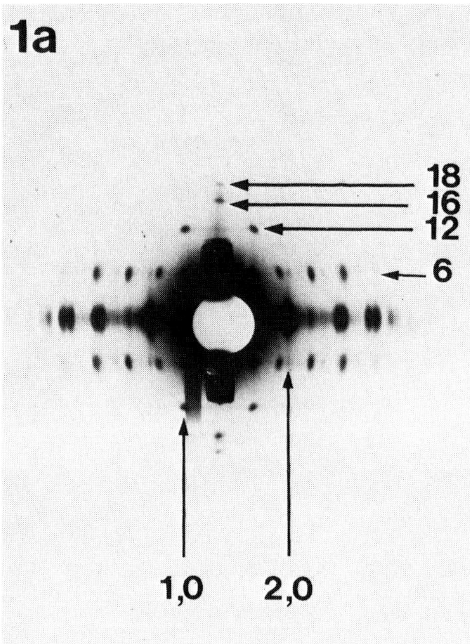
1a
18
16
12
6
1,0
2,0

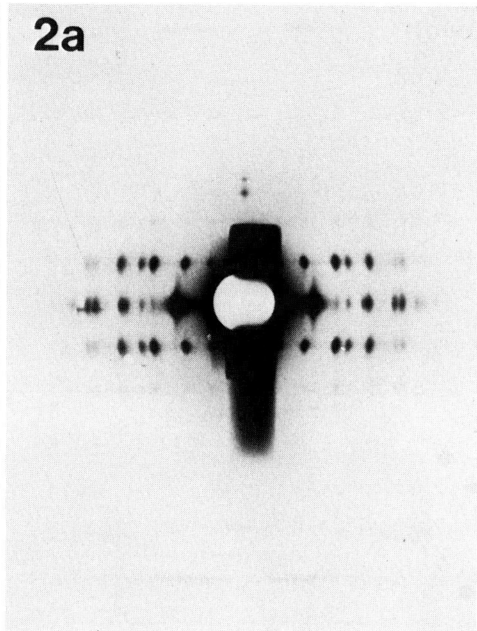
2a

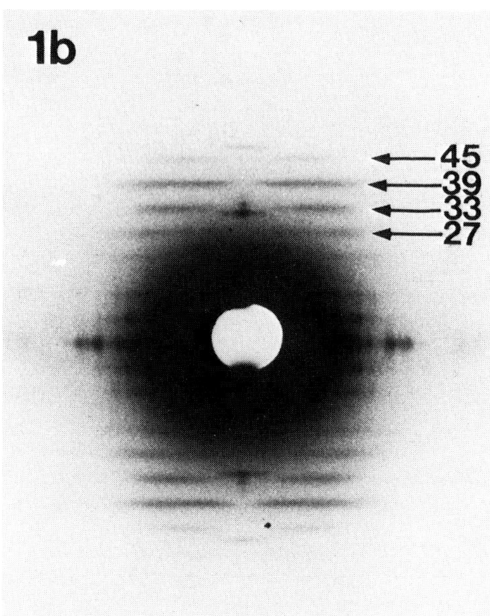
1b
45
39
33
27

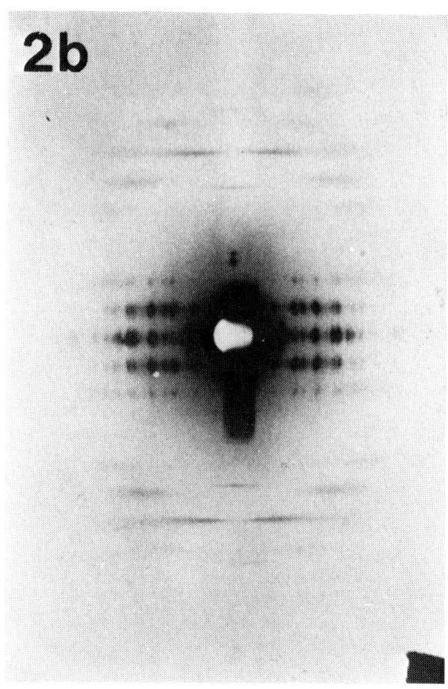
2b

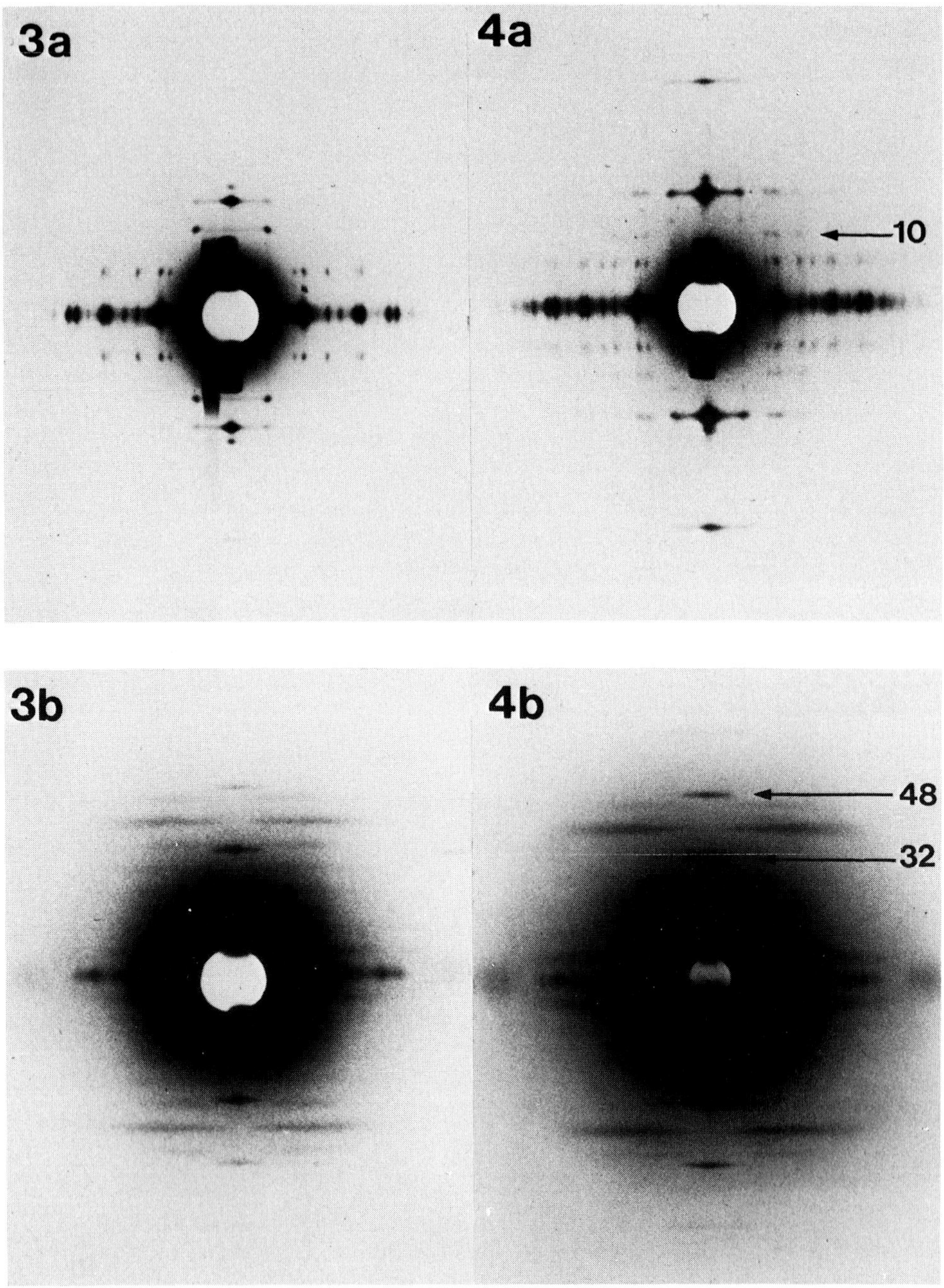
3a
4a
10
3b
4b
48
32

ELECTRON MICROGRAPHY OF *LETHOCERUS* MUSCLE IN THE PRESENCE OF NUCLEOTIDES

G. Beinbrech

Universität Münster, Zoologisches Institut, Arbeitsgruppe Muskelphysiologie, Hindenburgplatz 55, D-4400 Münster, BRD, West Germany.

INTRODUCTION

The molecular mechanism of muscle shortening is thought to consist of cyclic movements of crossbridges. The results of X-ray diffraction, electron microscope, mechanical and biochemical studies, (Reedy, *et al.*, 1965; Huxley, 1969; Huxley and Simmons, 1971; Lymn and Taylor, 1971), suggested a four state model of this cycle correlating the power producing step with the transition of cross-bridges from a perpendicular to an angled position. Two states of the model (1 and 3 in Fig. 1) could be identified as the relaxed state (state 1) and the rigor state (state 3). The relaxed state is stabilized by the absence of Ca^{2+} and the presence of ATP, the rigor state by the absence of nucleotides (Reedy, Holmes and Tregear, 1965). For this reason it was obvious to look for conditions where the position of state 2 is stabilized and preferably taken up by the myosin heads.

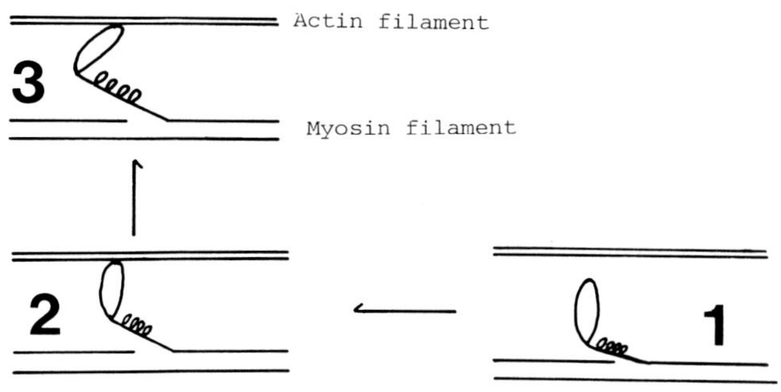

Fig.1. Model of three crossbridge states. State 1 is thought to represent crossbridge configuration in relaxed, state 3 in rigor muscle. State 2 is an intermediate state necessary for the explanation of force generation.

This model of crossbridge cycle has been supported by the possibility of demonstrating structural changes of the myosin molecules of insect flight muscle by electron microscopy and of correlating them with distinct functional states of muscle. In spite of the possibility of directly demonstrating different crossbridge configurations we have to be careful in interpreting the results of this method: comparison of X-ray patterns before and during fixation with electron micrographs and their optical diffraction patterns (Reedy, *et al.*, 1965, 1967 and 1971) showed different artefacts happening during fixation, embedding and cutting. Thus I had to take into account these artefacts and besides that to make optical diffraction pictures (a method I was introduced to in Oxford by Drs. Louise Johnson and Peter Eagle) and to compare them with X-ray patterns made by other groups (Lymn and Huxley, 1973; Barrington, *et al.*, 1973; Goody, *et al.*, 1975; Lymn, 1975; Marston, *et al.*, 1976) in order to exclude errors as far as possible.

MATERIALS AND METHODS

Bundles of glycerol-extracted dorsal longitudinal flight muscle of the water bug *Lethocerus* were first relaxed in relaxing solution (ATP 15 mM, $MgCl_2$ 15 mM, EGTA 5 mM, NaN_3 5 mM, imidazole 20 mM, pH 6.7) and then incubated in rigor solution (KCl 85 mM, $MgCl_2$ 5mM, EGTA 5 mM, NaN_3 5 mM, imidazole 10 mM, pH 6.7) to remove ATP. After the generation of a rigor tension various levels of relaxation were induced by adding the nucleotides β,γ-imido - ATP (AMP.PNP) or ADP* and equimolar amounts of $MgCl_2$ to the rigor solution. The ionic strength of all solutions was kept at μ=0.125 by adjusting the KCl concentration. In a second series of experiments the imidazole buffer was replaced by a 10 mM phosphate (P_i) buffer at pH 6.7.

The muscle fibres were fixed in one of these solutions by adding glutaraldehyde to the incubation medium and then postfixed with OsO_4. Sections were stained with uranyl acetate and lead citrate and examined by a Philips Em 300, a Siemens Elmiskop 101 or a Zeiss EM9. Optical diffraction patterns of electron micrographs were photographed in a Polaron diffractometer. The experiments were performed in collaboration with Prof.Dr. J.C. Rüegg, Dr. J.W. Herzig and Dr. H.J. Kuhn. (Further methods see Beinbrech, *et al.*, 1976).

---

* 0.2 mM P^1,P^5-di(adenosine-5') pentaphosphate, 0.5 mg/ml hexokinase and 2 mM glucose were added.

RESULTS AND DISCUSSION

About 20 years ago mechanical (Bozler, 1953 and 1956) and biochemical studies (Martonosi, *et al.*, 1960; Graenicher and Portzehl, 1964) demonstrated that pyrophosphate (PP) is able to imitate in some ways the plasticizing action of ATP, e.g. to relax glycerinated rigor fibres or to dissociate the actomyosin complex in solutions of high ionic strength. The interpretation of these findings however was complicated in the light of mechanical experiments: the stiffness of PP-relaxed fibres is comparable to that of rigor muscle if the fibres are stretched by applying sinusoidal length changes of small amplitude and at frequencies above 1 Hz (White, 1970). In correspondence with these findings the attachment of crossbridges to the actin filaments seems to be influenced by the action of PP in a characteristic way: Reedy (1967, 1968) could show that in rigor muscle the actin ends of the myosin heads are attached to the actin filaments in periods of 388 A* as demonstrated in Plate 1a: this micrograph of an actin filament layer of a rigor fibre shows lines vertical to the longitudinal axis of the myofibril which consist of attached myosin heads cut off from their filament backbone. If the fibre is relaxed by PP (Kuhn, *et al.*, 1972) these lines seem to lose much of their intensity (Plate 1b) but they re-appear correlated with an increase of fibre tension if the PP is removed by immersing the fibre in rigor solution again (Plate 1c). What might have happened during PP-relaxation? Regarding myosin-actin filament layers cross-bridges seem to be in different positions: in some areas most of the cross-bridges seem to be in the rigor position, in other areas most seem to be relaxed. In agreement with these findings optical diffraction patterns (Beinbrech, *et al.* 1972) and X-ray pictures (Lymn and Huxley, 1973) seem to combine the patterns of rigor and relaxed muscle. We could conclude that PP-treated fibres are somewhere between rigor and ATP relaxation.

The diffraction patterns change dramatically if we add to the rigor solution instead of PP similar concentrations of AMP.PNP (5 or 10 mM), a non-splittable analog of ATP. Optical transforms (Plate 4c) show an intense 145 A

---

* The periodicities on the electron micrographs presented here show deviations in the longitudinal axis of myofibrils of up to 4 per cent compared with those measured by X-ray diffraction (Reedy, *et al.*, 1965). For the sake of clarity I shall neglect these deviations and use the X-ray diffraction values of Reedy, *et al.*, (1965).

reflection similar to that of X-ray patterns (Barrington-Leigh, *et al.*, 1973; Lymn, 1975) and weak rigor lines of 194 and 388 A, whereas the 129A reflection seems to be unaltered. Dr. Kuhn has described in detail the decrease of tension and stiffness of these fibres suggesting that this pattern might be caused by a mixture of attached and detached crossbridges.

These conclusions fit in with other results of electron microscopy, H.E. Huxley (1963) demonstrated that shearing forces disintegrate myofibrils into separate actin and myosin filaments if relaxing conditions (presence of ATP and absence of Ca^{2+}) are maintained during homogenization. Likewise it should be possible to isolate single filaments if actin-myosin linkages are dissociated by the action of an analog. For this reason myofibrils of the flight muscle of *Locusta migratoria* have been homogenized in the presence of ATP, AMP.PNP or PP. After negative staining the grids were covered with single filaments if ATP or AMP.PNP in concentrations of 5 or 30 mM were used. Mg-PP relaxed myofibrils however seem to disintegrate first into bundles of myofilaments which may become smaller by further homogenization but are not broken up completely (Winkelhahn and Beinbrech, 1974). It is evident that these results do not permit any quantitative conclusions. They suggest however that there are more crossbridges attached to actin filaments in PP solution than in solutions with high concentrations of AMP.PNP.

The relaxing effect of AMP.PNP seems to depend on two parameters: on the AMP.PNP concentration and on the absence of phosphate. Let us consider first the effect of concentration. If we immerse fibres relaxed by 10 mM AMP.PNP in a pure rigor solution in order to wash out the analog, rigor tension and stiffness redevelop. Electron micrographs of actin filament layers show regular 388 A lines indicating that the crossbridges have re-attached to the actin filaments (Plate 2c). Myosin-actin layers on the other hand (Plate 2b) show the crossbridge pattern typical for rigor muscle. Thus we should not wonder that the optical transforms of these micrographs show the rigor pattern too, perhaps with a stronger 145 A layer line (Plate 4b).

Low concentrations of AMP.PNP (e.g. 0.5mM) show little effect on the rigor pattern (Plate 4c) of optical transforms except that we can see a more intense 145 A reflection. This again is in agreement with X-ray diffraction results (Lymn and Huxley, 1973; Goody, *et al.*, 1975; Lymn 1975; Marston, *et al.*, 1976) of other groups. Only small differences compared with rigor sarcomeres (Plate 3a) can be recognised in electron micrographs of fibres immersed in 0.5 mM AMP.PNP: as in the rigor-like PP-relaxed sarcomeres the myosin heads seem to be attached to the actin filaments in the rigor periodicity of 388 A (Plate 3c). In this respect the appearance of the crossbridge pattern does not seem to depend on the use of imidazole or phosphate buffer (Plate 3b).

However this is not the case if high concentrations of AMP.PNP are used. Fibres fixed in the presence of 10mM AMP.PNP and 10 mM phosphate buffer instead of imidazole give quite different patterns. Optical transforms of electron micrographs seem to combine in this case characteristics of the rigor pattern and those of ATP relaxed muscles (Plate 4d). They are comparable to those obtained from fibres relaxed in 5mM pyrophosphate solutions (Beinbrech, *et al.*, 1972).

Two or three questions arise regarding these results: in what way does a muscle fibre relax if the crossbridges seem to remain attached, in what way is the crossbridge cycle influenced by phosphate, one of the two products of ATP hydrolysis and last, but not least, what will happen if ADP, the other product, is added to rigor fibres? The problem in the latter case is to be sure that no ATP has been formed. Therefore we performed all experiments in the presence of the myokinase inhibitor AP_5A, hexokinase and glucose. In spite of that there remains a residual doubt which should always be kept in mind.

Electron micrographs of fibres relaxed in solutions with 0.5 mM ADP again show periodically attached crossbridges along the actin filaments (Plate 5). They are comparable to those of fibres relaxed by PP or 0.5 mM AMP.PNP. The optical transforms (Plate 4f) confirm this similarity: as in 0.5 mM AMP.PNP relaxed muscle the 145 A reflection is more intense, the rigor lines a little weaker. These results indicate that ADP too is capable of influencing the actin-myosin interaction. They seem to coincide with the mechanical measurements described by Dr. Kuhn, suggesting that the binding of Mg-ADP to the crossbridges changes their rigor state conformation.

Regarding these results all we can say is that we have good arguments for the hypothesis that analogs and ADP influence the interaction of actin and myosin and generate a state between rigor and relaxation. Discussing their X-ray diffraction patterns Goody, *et al.*, (1975) suggested that in solutions containing 0.5 mM AMP.PNP crossbridges are possibly attached to the actin filaments in a perpendicular position. To test this possibility, we compared the positions of crossbridges from fibres fixed in different solutions. For this purpose 1000-2000 crossbridges per experiment have been classified according to their position into three categories: rigor, perpendicular and antirigor. The results have been made up in Table I. No position appears to be preferred if the fibres have been fixed in ATP relaxing solution. In the rigor muscle about two-thirds of the crossbridges are in the rigor position, the rest appear to be equally distributed at perpendicular and antirigor angles (the fraction of antirigor bridges seems to be high compared with the observations of Reedy, 1971). It did not change in the other experiments, so that we may concentrate in the following on the distribution of crossbridges

TABLE I

Percentage of crossbridges in rigor (angle approximately 30-65°), perpendicular (70-110°) and antirigor position 115-150°) of glycerinated muscle fibres in rigor and different states of "relaxation".

	POSITION OF CROSSBRIDGES		
	Rigor	Perpendicular	Anti-rigor
Relaxation in 5 mM Mg-ATP	36	31	33
Rigor after ATP-relaxation	61	20	19
Relaxation in 3 mM PP	44	41	15
Relaxation in 0.5 mM AMP.PNP (imidazole buffer	40	44	16
Relaxation in 0.5 mM AMP.PNP (phosphate buffer)	48	35	17
Relaxation in 10 mM AMP.PNP (imidazole buffer)	26	57	17
Rigor after 10 mM AMP.PNP relaxation	51	35	14
Relaxation in 0.5 mM ADP	42	40	18

in rigor and perpendicular positions. It appears that just as many crossbridges have been shifted from rigor to the perpendicular position if 3 mM PP, 0.5 mM AMP.PNP in imidazole buffer or 0.5 mM ADP have been used for relaxation of the fibres. Phosphate seems to reduce this amount, high AMP.PNP concentrations together with imidazole buffer seem to increase it.

CONCLUSIONS

Summarizing these results and taking into account the mechanical results and their interpretation by Dr. Kuhn, one may propose the following model of PP, AMP.PNP and ADP effects (Fig.2.): in the presence of AMP.PNP or ADP in low concentrations or of PP,crossbridges in the rigor position move into the perpendicular position (thereby discharging the elastic elements located in the crossbridges) but remain attached to the actin filaments. If the concentration of AMP.PNP is increased the percentage of crossbridges in the perpendicular position also increases and at the same time a certain

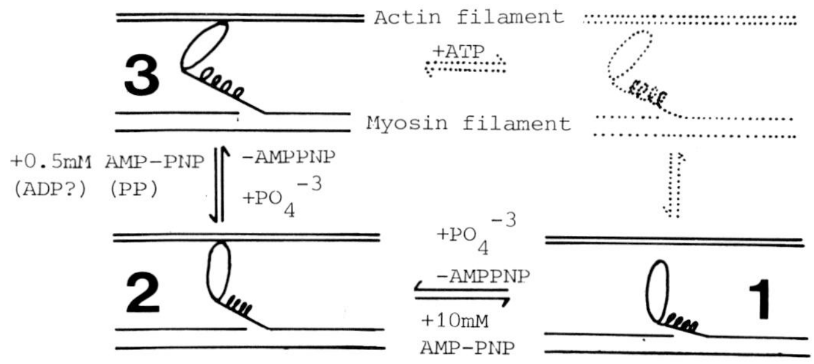

Fig.2. Model summarizing the conclusions concerning the influence of the investigated analogs on the structure of the myosin molecule and its interaction with actin. The dotted crossbridge of state 4 represents the hypothetical intermediate state postulated by Lymn and Taylor (1971) in order to close the crossbridge cycle.

fraction detach from the actin filaments. At least the first step (transition from state 3 to state 2) seems to be inhibited by the presence of phosphate. On the other hand, if the analog concentration is decreased, the myosin heads re-attach to the actin filaments in the perpendicular position and bend from a vertical to an acute angle (state 2 to state 3), thereby stretching the elastic components of the crossbridges again.

ACKNOWLEDGEMENT

I thank Prof. Dr. H.C. Lüttgau, Lehrstuhl für Zellphysiologie der Ruhr-Universität Bochum for the permission to use the optical diffractometer and to Mrs. Renate Schmidt-Tophoff for excellent technical assistance.

REFERENCES

Barrington-Leigh J., Holmes K.C., Mannherz H.C., Rosenbaum G., Eckstein F., and Goody R. (1973) *Cold Spring Harbor Symp. Quant.Biol.*, 37, 443.

Beinbrech G., Kuhn H.J., Herzig J.W. and Rüegg J.C., (1976) *Cytobiologie*, 12, 385.

Beinbrech G., Kuhn H.J., and Rüegg J.C. (1972) *Experientia*, 28, 511.

Bozler E. (1953) *J.gen.Physiol.*, 37, 53.

Bozler E. (1956) *J.gen.Physiol.*, 39, 789.

Goody R.S., Holmes K.C., Mannherz H.G., Barrington-Leigh J., and Rosenbaum G., (1975) *Biophys.J.*, 15, 687.

Goody R.S. Barrington-Leigh J., Mannherz H.G., Tregear R.T. and Rosenbaum G., (1976) *Nature*, 262, 613.

Graenicher D. and Portzehl H. (1964) *Biochim.biophys.Acta.*, 86, 567.

Huxley F.G. and Simmons R.M. (1971) *Nature*, 233, 533.

Huxley H.E. (1963) *J.mol.Biol.*, 7, 281.

Huxley H.E. (1969) *Science*, 164, 1356.

Kuhn H.J., Schroeder H. and Rüegg J.C. (1972) *Experientia*, 28, 510.

Lymn R.W. (1975) *J.Mol.Biol.*, 99, 567.

Lymn R.W. and Huxley H.E. (1973) *Cold Spring Harbor Symp. Quant.Biol.*, 37, 449.

Lymn R.W. and Taylor E.W. (1971) *Biochemistry*, 10, 4617.

Marston S.B., Rodger C.D. and Tregear R.T. (1976) *J.mol.Biol.*, 104, 263.

Martonosi A., Gouvea M.A. and Gergely (1960) *J.biol.Chem.*, 235, 3169.

Reedy M.K. (1967) *Amer.Zool.*, 7, 465.

Reedy M.K. (1971) *Contractility of muscle cells and related processes*, Ed. by R.J. Podolsky. Prentice -Hall Inc. p.229-246.

Reedy M.K. (1968) *J.mol.Biol.*, 31, 155.

Reedy M.K., Holmes K.C. and Tregear R.T. (1965) *Nature* 207, 1276.

White D.C.S. (1970) *J.Physiol.*, 208, 583.

Winkelhahn J.M. and Beinbrech G., (1974) *Experientia*, 30, 350.

PLATES

PLATE 1. Thin longitudinal sections through fibres of *Lethocerus* flight muscle which had developed tension in a rigor solution (a), have been relaxed by 5 mM pyrophosphate (b) or have been immersed in rigor solution again in order to wash out the pyrophosphate (c) (rigor after PP-relaxation). The micrographs show actin filament layers with myosin heads attached to the actin filaments. In (a) and (c) these heads are arranged at a regular spacing forming lines (indicated by arrows) crossing the longitudinal axis of the myofibrils. Z = Z-disc.

PLATE 2. Thin longitudinal sections through fibres of *Lethocerus*flight muscle fixed in rigor solution with 10 mM Mg-AMP.PNP, imidazole buffer (1), or in rigor solution where they had developed tension after relaxation in 10 mM AMP.PNP solution (b and c). In (c) the myosin heads are attached in regular arrangement to the actin filaments forming lines (indicated by arrows) crossing the longitudinal axis of the myofibril.

PLATE 3. Longitudinal sections through *Lethocerus* flight muscle fixed in the following solutions: Rigor solution (a); rigor solution with 0.5 mM AMP.PNP, 10 mM phosphate buffer (b); rigor solution with 0.5 mM AMP.PNP, 10 mM imidazole buffer (c).

PLATE 4. Optical transforms of electron micrographs of sarcomeres from myofibrils of *Lethocerus* flight muscle fixed in rigor solution (a), rigor solution after relaxation with 10mM AMP.PNP imidazole buffer (b), relaxing solution with 10 mM AMP.PNP imidazole buffer (c), solution with 10 mM AMP.PNP phosphate buffer (d), solution with 0.5 mM AMP.PNP imidazole buffer (e) and solution with 0.5 mM ADP imidazole buffer (f). The values of the layer lines correspond to those measured by X-ray diffraction (Reedy, 1967). Explanation - see text.

PLATE 5. Longitudinal section through fibres of *Lethocerus* flight muscle which had developed tension in rigor solution before they were immersed in 0.5 mM ADP solution at 12°C.

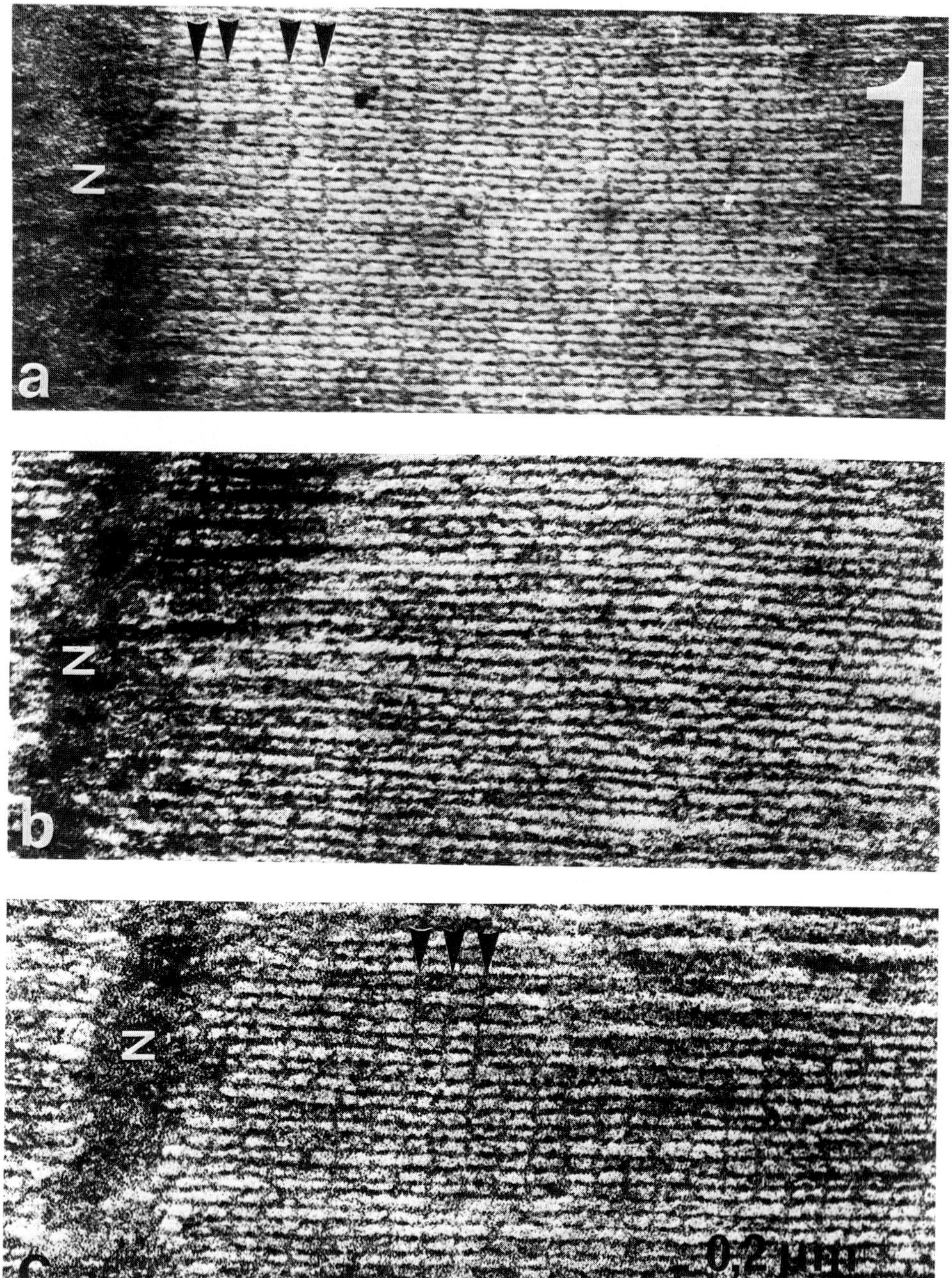
1
a
Z
b
Z
c
Z
0.2 μm

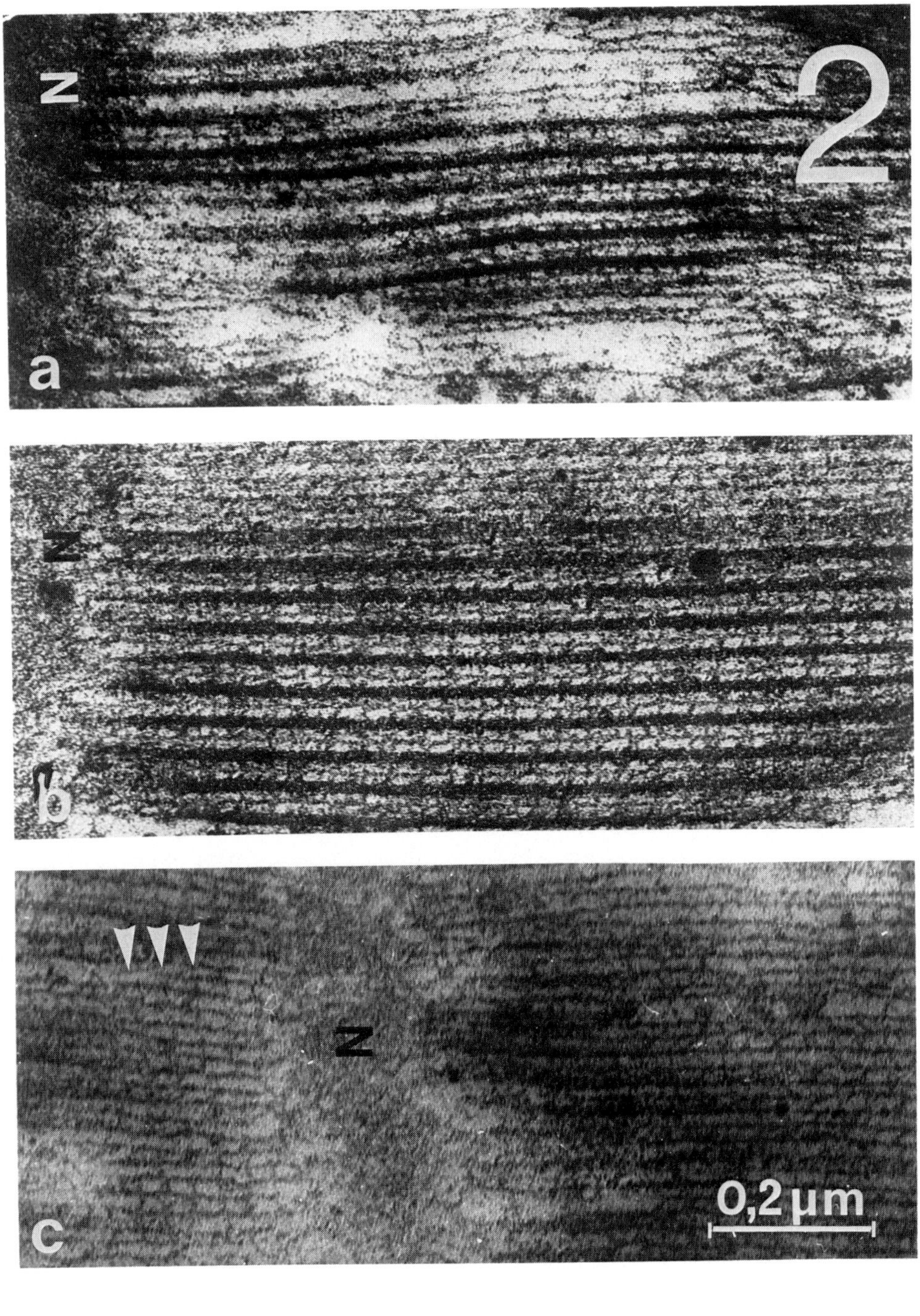
2
Z
a
Z
b
Z
c
0,2 µm

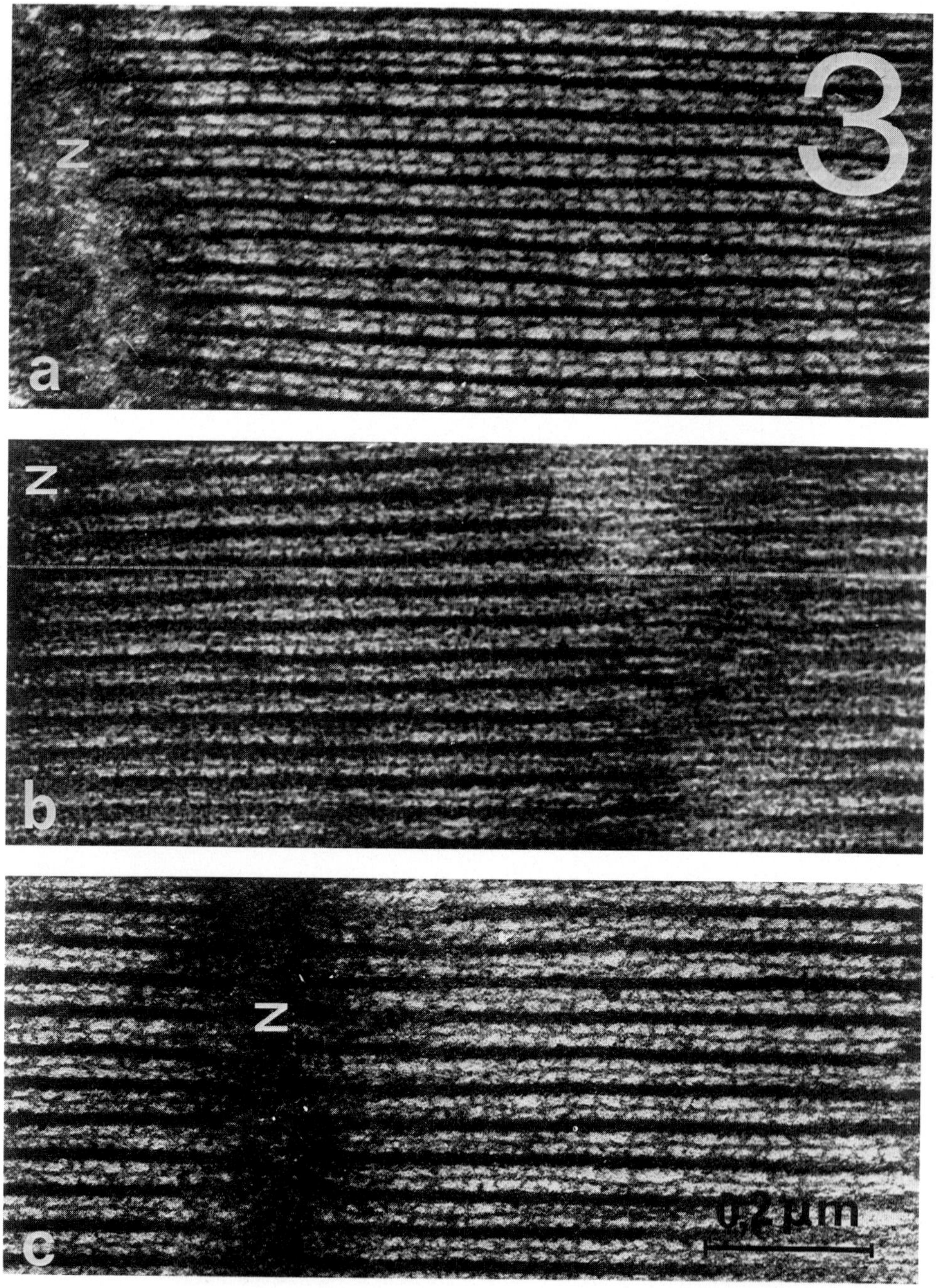
3
Z
a
Z
b
Z
c
0.2 µm

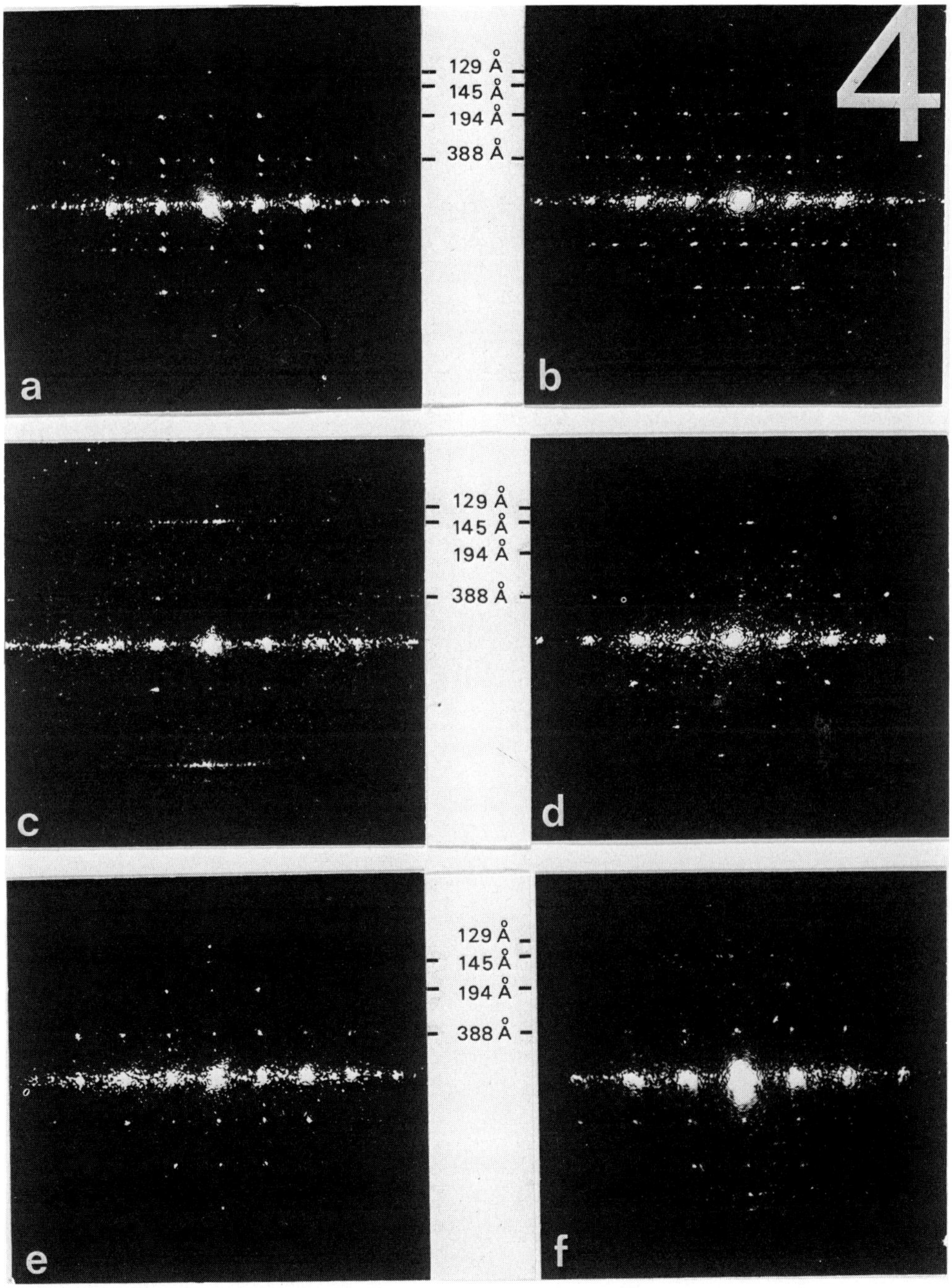
4
129 Å
145 Å
194 Å
388 Å
a
b
c
d
e
f

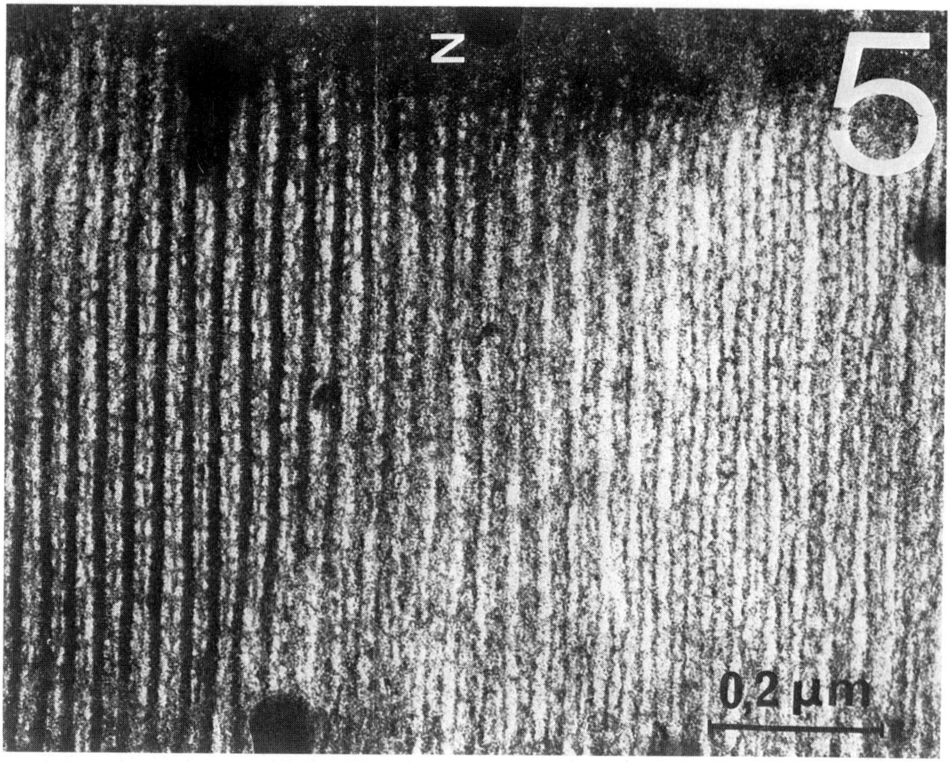
Z
5
0,2 µm

NEUTRON DIFFRACTION OF *LETHOCERUS* FLIGHT MUSCLE

C.D. Rodger [1]

D.L. Worcester [2]

A. Miller [3]

[1] *Laboratory of Molecular Biophysics, Department of Zoology, Oxford University, Oxford, OX1 3PS, England.*

[2] *Materials Physics Division, Atomic Energy Research Establishment, Harwell, Oxon, OX11 ORA, England.*

[3] *European Molecular Biology Laboratory, Grenoble Outstation, c/o L.M.A., C.E.N.G., Grenoble, France.*

INTRODUCTION

Neutron Diffraction Technology

The molecular arrangements of actin and myosin filaments in *Lethocerus* flight muscle have been a subject of intense research in recent years (Reedy, 1968; Miller and Tregear,1972, and Marston, Rodger and Tregear 1976). However, despite a variety of research methods employed, many basic and important details of the structural geometry have still to be elucidated. In this paper we present preliminary results of neutron diffraction from *Lethocerus* flight muscle. Neutron diffraction has become a well established biophysical research technique and is here applied to insect muscle for the first time.

The method of small-angle diffraction using neutron beams has developed rapidly in recent years by the establishment of an international facility for neutron beam research at the Institut Laue-Langevin, in Grenoble, France. This facility is based on the provision of a high flux reactor, equipped with a liquid deuterium 'cold source' to enhance the production of low energy neutrons of wavelengths suitable for small angle diffraction experiments. The small angle camera D 11 (Schmatz, Springer, Schelten and Ibel 1974, and Ibel 1976), which was used for these experiments is provided with interchangeable wave-guides before the specimen to maintain a high neutron flux at some distance away from the reactor shell. The neutron flux produced from the cold source of the reactor is further monochromated before the specimen by a mechanical velocity selector. Adjustment of the rotational speed of this selector allows selection of neutron wavelengths down to 0.4 nm with a maximum intensity between 0.6 nm and 1.2 nm. Detection of the diffracted information is afforded by an area detection system with a 64 x 64 matrix of 1 cm² individual detector elements. The digital output of the counting elements is handled by an on-line computer, before transfer to a Dec-system PDP 10,which is programmed to produce contour plots of the recorded intensities such as are reproduced in this paper.

The main difference between this small angle neutron facility and the more conventional X-ray facility is one of size. Despite the provision of the high flux reactor at Grenoble, the inherent flux is still low compared with X-ray sources; maximum flux at the specimen on the D 11 camera is between 10^5 and 10^8 neutrons per second per cm². In order to minimise this problem, one must use a large incident beam cross section and consequently, a large sample. The bulk of the experiments reported here employed samples of approximately 0.5 cm² useful area, wavelengths of incident radiation between 0.6 nm and 1.0 nm and a specimen detector distance of some 5 metres.

Contrast Variation with Neutrons

Neutrons do not replace X-rays in the analysis of molecular structure because of the low flux and poorer resolution presently available. Their use in this branch of research is based largely on the very different nature of the scattering of neutrons, which is primarily by the nuclei of atoms compared with X-rays which are scattered by the extranuclear electrons. There is, for example, no large linear increase in scattering amplitude with increasing atomic number of the scattering elements as is the situation with X-ray diffraction (Bacon, 1962). However, the most important feature is the relatively large difference in scattering amplitude of the hydrogen nucleus and the deuterium nucleus (Table I).

TABLE I

COMPARISON OF NEUTRON AND X-RAY SCATTERING LENGTHS FOR COMMON ELEMENTS IN UNITS OF 10^{-12}cm

	H	D	C	O	N
Neutrons	-0.374	0.667	0.665	0.580	0.940
X-ravs	0.28	0.28	1.70	2.26	1.98

If the ratio of hydrogen and deuterium in the aqueous environment of the structural proteins in the muscle is changed, the 'contrast' between these proteins and the solvent will also be changed, as judged by the neutron diffraction pattern. With different H_2O/D_2O mixtures it is possible to measure different intensities for the reflections in the diffraction pattern and these changes can give information about the variation of neutron scattering density between different proteins and also within different regions of the same protein in the muscle.

A convenient representation is to plot $\sqrt{I}$ against the molar fraction of D_2O, in order to obtain the 'zero-intensity intercept' for a particular reflection. If the neutron scattering density of the protein is evenly distributed at the resolution of the reflection the plot of $\sqrt{I}$ versus D_2O molar fraction is linear and the zero-intensity intercept reveals the average scattering density of the protein. If the neutron scattering density varies from lower to higher values over distances comparable with the spacing of the reflection, this will result in different intercepts for the different reflections, the intercept having been shifted from the mean scattering density of the protein. The plot of $\sqrt{I}$ versus D_2O molar fraction in this case is not necessarily linear, but will be so if the variation of scattering density is centrosymmetric.

It is possible to obtain values for the mean scattering density of individual proteins for comparison with the values of the zero-intensity intercepts obtained by contrast variation of the reflections in the muscle diffraction pattern. One can directly obtain these mean scattering densities from the contrast variation behaviour of zero angle scattering from the proteins in solution, but data for muscle proteins are not yet available. One can also make theoretical predictions of these mean scattering densities using the amino acid composition of the proteins together with the relationship:

$$\rho = \frac{\sum_i \sum_j b_{ij}}{\sum_i \nu_i}$$

where ρ = scattering density, ν_i = the partial specific volume of the ith amino acid residue, the summation over j is for the coherent scattering amplitudes of the atoms in a particular residue and the summation over i is for the amino acid residues in the protein. In previous work on neutron diffraction from vertebrate striated muscle Worcester *et al.*,(1975) made calculations of the mean scattering densities of several vertebrate muscle proteins in this way. The surprising feature was that the mean scattering amplitudes of the myosin S1 and S2 subunits, light meromyosin, actin, troponin, tropomyosin and C-protein were all in the very small range 42-44% D_2O concentration. This meant that it would be very difficult to match the aqueous scattering density to some of the muscle proteins whil leaving some contrast to others. However it left open the possibility of using neutrons to examine any departures in the observed zero-intensity intercepts from the theoretical values, thus providing information about changes in neutron scattering density over distances comparable with the spacing of the reflections concerned. In the work reported here the same approach is applied to insect flight muscle.

MATERIALS AND METHODS

The main problem with specimen preparation was to produce oriented bundles of glycerinated *Lethocerus maximus* and *cordofanus* muscle fibres of approximately 1 cm length, in a band about 1 cm wide. This was achieved by laying individual bundles alongside each other and glueing them with cellulose glue at each end to stainless steel wires. The fibres were supported during the glueing process by a glass plate between the wires to prevent the effects of the considerable surface tension forces disrupting the oriented array. Then the fibres were irrigated with a suitable bathing medium and stretched slightly to improve orientation. This muscle mounting assembly was fixed into a perspex cell with quartz windows, 1-2 mm thick, to allow transmission of the neutron beam.

During the experiments between 25 and 100 ml of the various bathing media were continuously circulated through the cell by a peristaltic pump, the temperature being maintained at about 8°C (D_2O freezes at +4°C). The solution could be changed between exposures without altering the position of the specimen in the neutron beam, to enable accurate comparisons of diffracted intensities in solutions of different H_2O/D_2O composition. The D_2O used in the experiments was obtained from Harwell and redistilled to avoid possible contamination from heavy metal phosphate sometimes found in commercial D_2O (Huxtable and Bressler 1974). The experiments attempted to measure contrast variation behaviour of some of the strongest reflections in the *Lethocerus* diffraction pattern from rigor muscle and muscle relaxed by the addition of MgATP. A few trial experiments were also attempted, to obtain data from the altered acto-myosin configuration in the muscle structure when the fibres are bathed in solutions containing β-γ adenylylimidodiphosphate (AMP.PNP) (Marston, Rodger and Tregear, 1976). All solutions were buffered at near-normal pH/pD by phosphate buffer or PIPES[*], contained 5mMEGTA[†] and the ionic strength was about 0.1.

The diffraction pattern could be monitored during the course of the exposure by a visual display on a computer-lined C.R.T. which displayed the counts collected in the various detector elements. Accurate intensity measurements were made later by summing the counts in the individual detector elements and subtracting the background. Allowance was also made for exposure time and

---

[*] PIPES - Piperazine-N,N'-bis-2-ethane sulphonic acid.

[†] EGTA - ethylene clycol-bis (β-aminoethylether)-N,N'-tetracetic acid.

a measurement of the absorption by the specimen (being significant in predominantly H_2O media) was facilitated by a semi-transparent beam stop.

RESULTS

Equilibration

When the D_2O content of the bathing media was changed it was, of course, necessary to wait until rapid exchange of labile hydrogens was completed and the interfilament space was completely equilibrated. The provision of the on-line data handling system enables the equilibration process to be followed by the automatic collection of consecutive short spectra. A typical result of the analysis of such spectra is shown in Fig.1. Data collection was commenced as soon as a 100% D_2O solution was pumped into the specimen cell to replace the H_2O solution, and the intensity of the 2.0 equatorial reflection was monitored in a series of 3 minute measurements. Figure 1 shows that equilibration is substantially complete in about 10 minutes. This contrasts markedly with the findings for vertebrate muscle, which required about 1 to 2 hours for equilibration to be completed. This difference seems to be the result of the thinner specimens, and of the fact that the insect muscle was glycerinated and the membranes therefore disrupted.

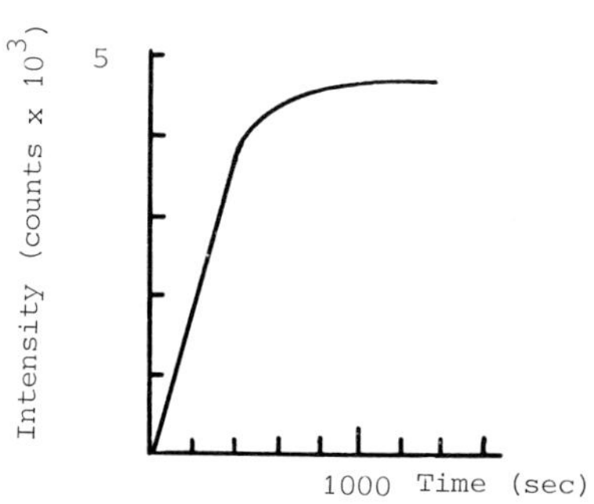

Fig.1. Graph showing the rate of equilibration of the myofilament system after the introduction of a rigor solution made up with 100% D_2O. The intensity measurement is of the rigor 2.0 equatorial reflection (corrected for background) and the data was automatically collected by computer in a series of 171 second measurements.

Background Scatter from Membranes

Considerable difficulty was encountered in the analysis of the patterns obtained in D_2O solutions because of very high background scattering at low angles. This was similar to the situation already encountered with vertebrate muscle but the background was not so obviously equatorially oriented and it was much stronger relative to the Bragg reflections from the contractile proteins. This background is almost certainly from material of low scattering density, such as membranes, and a higher scatter might be expected in *Lethocerus* muscle compared to that from vertebrate because of the greater mitochondrial concentration. It is interesting to note that despite glycerination a large amount of lipid material apparently remains in the matrix of the insect muscle fibre.

Diffraction Patterns

As shown in the contour plots of Plates 1 - 5 a basic result is that it is relatively easy to obtain a low-angle neutron diffraction pattern from *Lethocerus* flight muscle on a realistic time scale. As with X-rays, the low-angle diffraction pattern from muscle in the rigor state shows two prominent equatorial reflections from the hexagonal lattice of the myofilaments, the 1,0 and 2,0. The ratio of their intensities in H_2O solutions is much the same as has already been described for X-ray diffraction patterns. In some exposures the 1,1 equatorial can just be made out as a shoulder on the inner edge of the 2,0 but it is not properly resolved. With a shorter wavelength (0.6nm) and a specimen-detector distance of 5 metres the 2,2 and 3,1 equatorial reflections can be clearly seen as an unresolved complex (Plate 1), but thus far the 2,0 and 3,0, which are weaker than these other equatorials in X-ray patterns, have not been detected. Reflections from the 38.5 nm layer line in rigor have also been recorded, notably the row lines at the 1,0; 1,1; and 2,1 positions (Plates 3,4,5). In D_2O solutions, a very strong pattern is seen, but as already described, the low angle reflections from the myofilament array are superimposed on a very strong background (Plates 2-3). As the contour plots of the diffraction data from muscles in D_2O solutions show, this background particularly affects the resolution of the 1,0 equatorial, which lies on a very steep background that has prevented accurate intensity measurement of this reflection in several spectra. The 2,0 reflection is not so badly affected and no problem has been encountered with intensity measurement of any other equatorial reflection.

As with X-rays, the relaxed diffraction pattern is radically altered by the presence of a strong meridional reflection at a spacing of 14.5 nm (Plate 1). This reflection is sharply defined and well clear of the low-angle background scatter in D_2O solutions (Plate 2). The relative intensities of this meridional and the 1,0 and 2,0 equatorials in H_2O relaxing solutions are similar to those found for X-rays. Strong exposures have also revealed weak sampling reflections on a layer line at the same spacing as this meridional reflection which are also seen with X-rays (Plate 1). The 2,2/3,1 unresolved equatorial reflections are also seen, but no low angle layer line reflections have been observed from relaxed muscles. Attempts have also been made to obtain diffraction patterns from muscles in the presence of the nucleotide analogue, AMP.PNP. The rigor pattern is certainly changed in this condition and a meridional reflection is observed at about 14.5 nm. However, so far the exposures have not been strong enough to reveal if the characteristic alterations in the sampling pattern that have been observed with X-rays for this structure,(Marston, Rodger and Tregear 1976, and Goody, Holmes, Mannherz, Barrington-Leigh and Rosenbaum, 1975) also occur with neutrons. At this stage, therefore, it is not possible to be certain that the pattern is not from truly relaxed muscle.

Contrast Variation

The diffraction from rigor and relaxed specimens was recorded in various H_2O / D_2O mixtures and the intensities of the strongest reflections were subjected to contrast variation analysis. After corrections had been applied for background and absorption in the specimen, the square root of the diffracted intensity was plotted as a function of the H_2O - D_2O composition.

Data from the experiments revealed zero intensity intercepts for the 1,0 and 2,0 rigor reflections at the same value of 42% (Fig.2.). For measurements of relaxed muscle, the wavelength was set to include the 14.5 nm reflection on the detector and consequently the 1,0 equatorial was not sufficiently resolved for accurate calculation of the zero intensity intercept. The 2,0 equatorial from relaxed muscle gave a value similar to that obtained for rigor muscle, but the 14.5 nm meridional had an intercept some 1-2% lower (Fig.3.). However, in view of the counting statistics, this difference is only marginally significant. It is apparent from the data obtained so far, that the relatively large difference in intercept of the equatorial reflections and the 14.3 nm meridional reported for vertebrate striated muscle (42% and 37%) is probably not a feature of *Lethocerus* protein structure.

Theoretical Neutron Scattering Density

Neutron scattering densities of some of the major proteins found in insect flight muscle were calculated from their known amino acid compositions and the partial specific volumes of their constituent residues. These calculations assume complete exchange of labile hydrogen atoms, and this assumption may lead to some degree of uncertainty regarding the scattering amplitude density calculated for the different amino acids. The amino acid composition of rabbit actin (Worcester, Gillis and O'Brien, 1975; Elzinga, Collins, Keuhl and Adelstein, 1973), *Lethocerus* myosin (Hammond and Goll, 1975) and insect paramyosin were used in the calculation. Rabbit actin is probably similar to insect actin and the amino acid composition of paramyosin from the dung beetle *Heliocopris japetus* (Bullard, Luke and Winkelman, 1973) is probably very similar to that of *Lethocerus*.

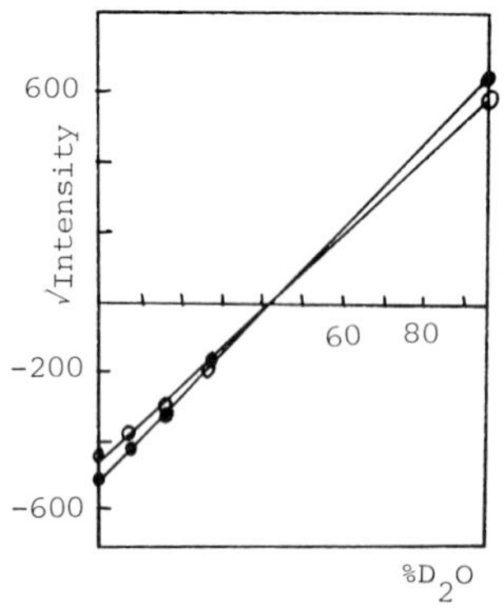

Fig.2. Contrast variation of the 1,0 and 2,0 equatorial reflections from *Lethocerus* flight muscle in rigor. ● - 2,0; o - 1,0.

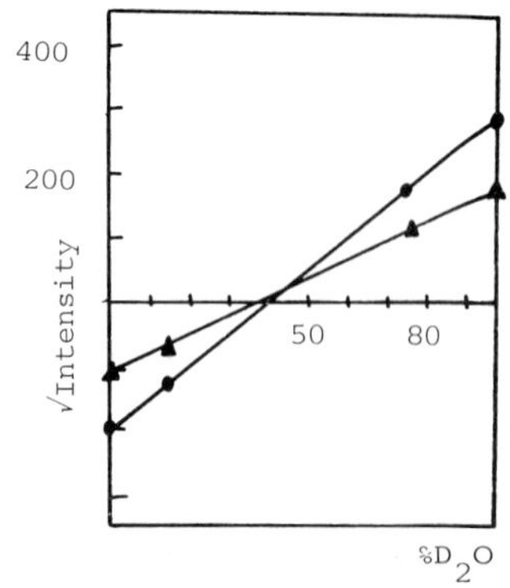

Fig.3. Contrast variation of 2,0 equatorial and 14.5 nm meridional reflections from *Lethocerus* flight muscle relaxed by the addition of MgATP. ● - 2,0; ▲ - 14.5 nm meridional.

Although not conclusively proved, it appears very likely that the paramyosin of *Lethocerus* muscle is a core protein of the myosin filaments (Squire, 1972; Squire, 1973, and Bullard, Hammond and Luke, 1977), and it forms about 11% of the myosin/paramyosin fraction from the muscle (Bullard, *et al.*, 1973). The values calculated, expressed in terms of their zero intensity intercepts, are shown in Table II.

TABLE II

NEUTRON SCATTERING AMPLITUDE DENSITIES OF MUSCLE PROTEINS
(ASSUMING COMPLETE EXCHANGE OF LABILE HYDROGENS)

Actin	(rabbit)	42.6%
Myosin	(*Lethocerus*)	43.4%
Paramyosin	(*Heliocopris*)	43.3%

(expressed as % D_2O composition)

DISCUSSION

It is apparent from the calculation of neutron scattering density for insect muscle proteins that there is very little predicted difference between them in terms of their average scattering density. This is very like the situation already encountered with vertebrate striated muscle. The theoretical zero intensity intercept for insect myosin is admittedly for whole myosin and is an average value obtained over the globular enzymic subunits and the coiled-coil rod; however, with vertebrate myosin, no significant difference was seen with either region, e.g. Myosin rod-43.3%; HMM S 1-41.8%; HMM S 2-42.8%, (Worcester, *et al.*, 1975). A particular disappointment is the zero intensity intercept derived for insect paramyosin. Despite the different amino acid composition of paramyosin, the mean scattering density is very similar to that for myosin and so contrast variation studies on the low angle equatorials, for instance, are unlikely to prove useful in structural analysis of the thick filaments at this resolution, (Fig.4.).

Examination of the contrast variation behaviour of the 14.5 nm reflection which arises from the axial repeat of the crossbridges was a high priority in these studies, in view of the findings reported for vertebrate muscle (Worcester, *et al.*, 1975). The vertebrate relaxed 14.3 nm meridional reflection was shown to have a lower intercept (37%) than either of the two main equatorials (about 42%), but so far we have not been able to detect such a large difference for the equivalent reflections in insect flight muscle. One possible cause for such a departure from the predicted intercept is a variation in neutron scattering density over a distance comparable with the spacing of the reflection. However, it is also possible that the calculated mean scattering density of vertebrate HMM is lower than the experimental value because of errors in estimates of the partial specific volume of the protein. The only way of resolving this uncertainty is an experimental determination of the mean scattering density of the crossbridge subunits by zero angle neutron scattering methods.

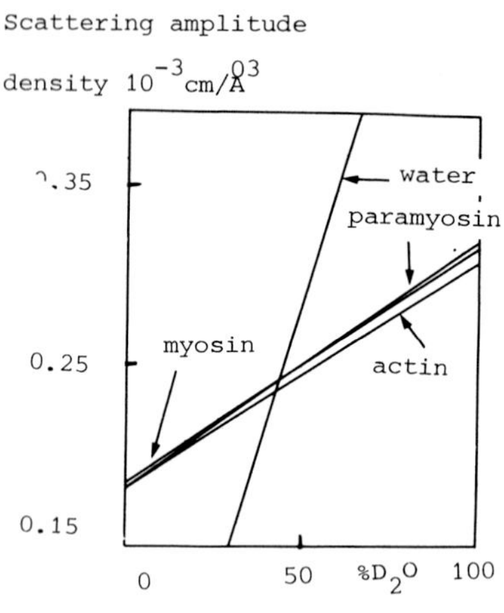

Fig.4. Neutron scattering amplitude densities of muscle proteins. The point where the water scattering density intersects the protein density gives the zero intensity intercept; the point where the H_2O/D_2O composition has the same mean scattering as the protein.

Investigation of the 14.5 nm reflection may also prove useful for studies on the structural changes produced by AMP.PNP. If the AMP.PNP 14.5 nm reflection is derived from bridges which are distally attached to actin filaments, as the evidence suggests, (Marston, *et al.*, 1976; Goody, *et al.*, 1975), then contrast variation studies may detect differences between the scattering density of the protein contributing to this reflection and that which gives rise to the reflection at the same spacing in ATP-relaxed muscle fibres. The data on the neutron diffraction from the analogue state is far from complete at the moment, however, and difficulty has been experienced in maintaining the AMP.PNP-induced state in so many fibres during repeated changes of different H_2O/D_2O solutions.

It is hoped to extend this work further by investigating the structure at a higher resolution. In particular, the well sampled transform makes investigation of the outer equatorials (such as the 2,2/3,1 complex) in terms of contrast variation of interesting possibility, and the advent of area detectors with up to 4 times the number of elements previously available should make resolution of discrete reflections simpler.

In the longer term it will be interesting to pursue the possibility of enhancing differences in the muscle protein scattering densities by selective deuterium labelling of the muscle and then undertaking direct structural analysis by contrast variation.

ACKNOWLEDGEMENTS

We should like to thank the S.R.C. for financial support, the Institut Laue-Langevin, the European Molecular Biology Laboratory Outstation at Grenoble and A.E.R.E. Harwell for technical assistance. In particular we should like to thank Dr. K. Ibel of the I.L.L. for generous support and Mr. T. Greenway of the Nuclear Physics Laboratory, Oxford University for assistance in preparing contour plots with its P.D.P. 10 computer system.

REFERENCES

Bacon B.E. (1962) *Neutron Diffraction,* Oxford University Press.

Bullard B., Luke B., and Winkelman L. (1973) *J. Mol. Biol.,* 75, 359.

Bullard B., Hammond K.S. and Luke B., (1977) - Paper in preparation.

Elzinga M., Collins J.H., Kuehl W.M. and Adelstein R.S. (1963) *Proc. Nat. Acad. Sci.,* 70, No.9, 2689.

Goody R.S., Holmes K.C., Mannherz H.G., Barrington-Leigh J. and Rosenbaum G. (1975) *Biophys. J.,* 15, 687.

Hammond K.S. and Goll D.E. (1975) *Biochem. J.* 151, 189.

Huxtable R. and Bressler R. (1974) *J. Membr. Biology,* 17, 189.

Ibel K. (1976) *J. Appl. Crystallogr.,* 9, 296.

Marston S. Rodger C.D. and Tregear R.T. (1976) *J. Mol. Biol.,* 104, 263.

Miller A. and Tregear R.T. (1972) *J. Mol. Biol.,* 70, 85.

Reedy M. (1968) *J. Mol. Biol.* 31, 155.

Schmatz W., Springer T., Schelten J. and Ibel K. (1974) *J. Appl. Crystallogr.* 7, 96.

Squire J. (1972) *J. Mol. Biol.,* 72, 125.

Squire J. (1973) *J. Mol. Biol.,* 77, 291.

Worcester D.L., Gillis J.M. and O'Brien E.J. (1975) *Brookhaven Symp. in Biology,* 27, 101.

Zamyatnin A.A. (1972) *Progr. Biophys. Mol. Biol.,* 24, 107.

PLATES

Contour maps of neutron diffraction data from glycerinated *Lethocerus* fibres bathed in various media. In each case the area of muscle exposed in about ½cm^2, which corresponds to the size of the elements in the detector.

KEY:

A-14.5nm meridional reflection.

B-1,0 equatorial reflection.

C-2,0 " "

D-2,2/3,1 unresolved equatorial reflections.

E-1.0 row line reflection on the 38.5 nm rigor layer line

F-1,1 " " " " " " " " "

G-2,1 " " " " " " " " "

PLATE 1

Diffraction pattern from A.T.P.-relaxed *Lethocerus* fibres in H_2O solution. Specimen-detector distance 5 Metres, neutron wavelength 0.59 nm. Exposure time 40 minutes.

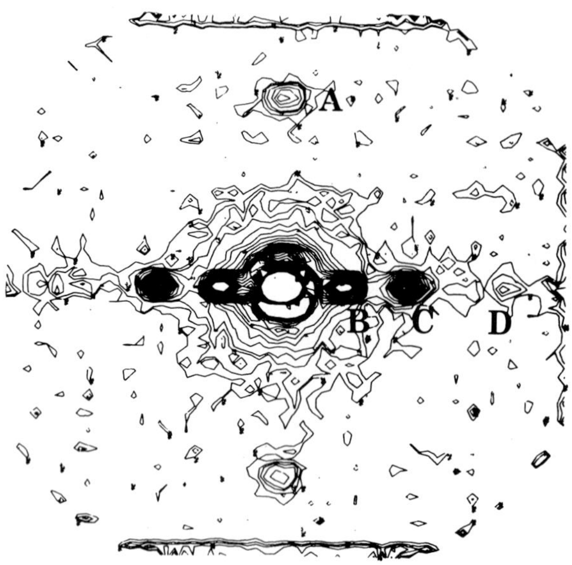

PLATE 2

Muscle fibres in 100% D_2O relaxing solution. Exposure time 20 minutes.

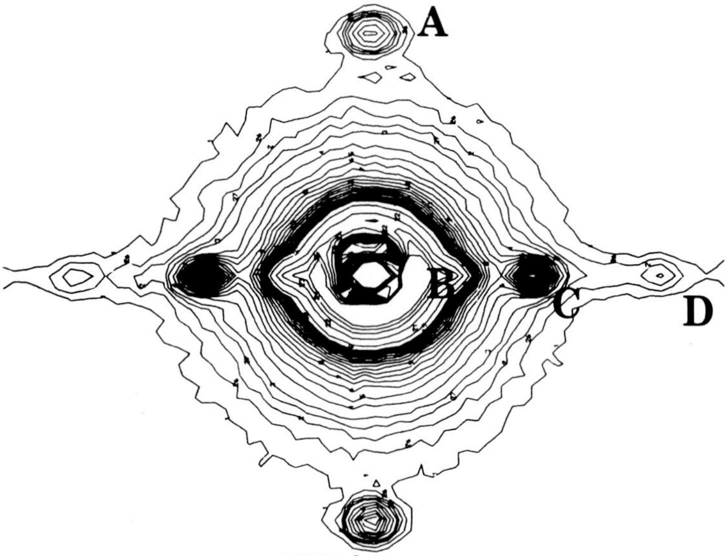

PLATE 3

As above but in 100% D_2O rigor solution. Exposure time 20 minutes.

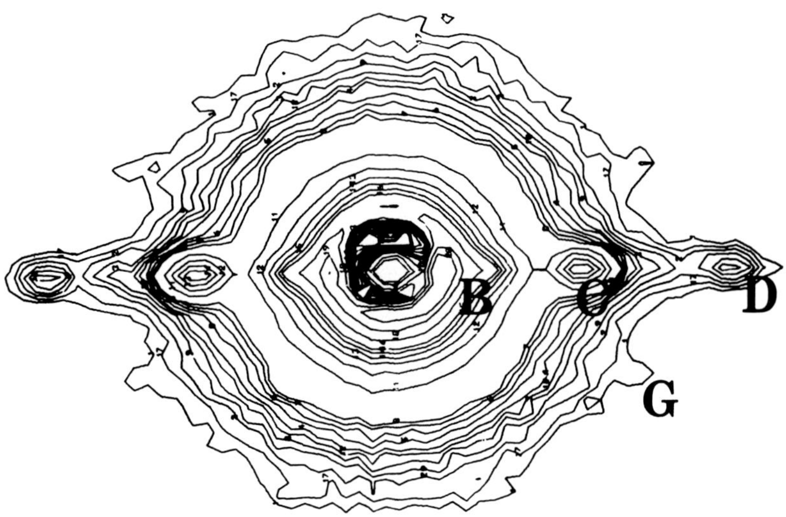

PLATE 4

Diffraction pattern from fibres bathed in H_2O rigor solution. Neutron wavelength 0.7 nm, exposure time 4 hours.

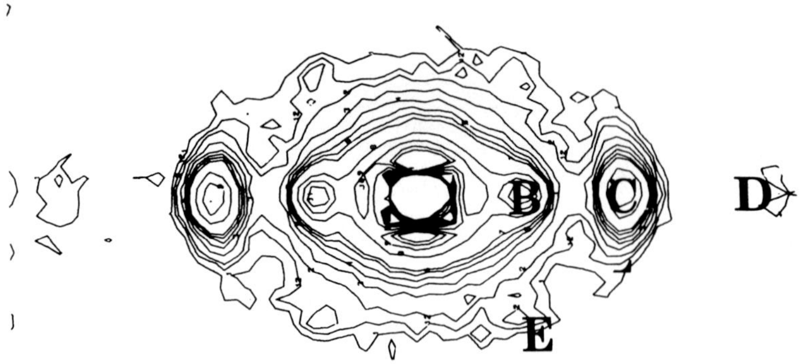

PLATE 5

As above but wavelength 1 nm, exposure time 3.3 hours.

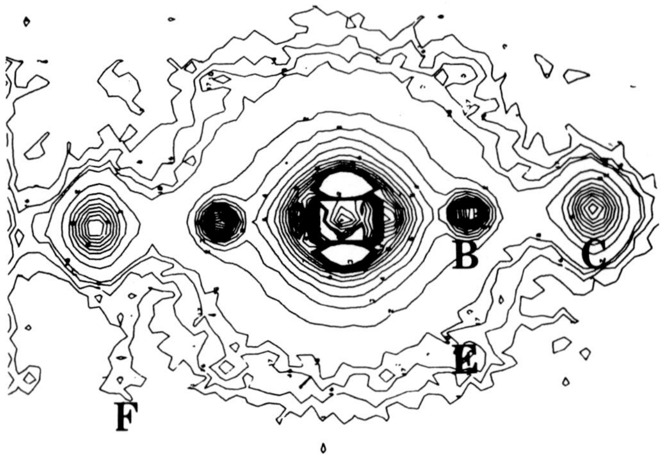

MECHANICAL KINETICS

THE MECHANICAL CHARACTERISTICS OF INSECT FIBRILLAR MUSCLE

J.W.S. Pringle

A.R.C. Unit, Department of Zoology, South Parks Road, Oxford OX1 3PS, U.K.

INTRODUCTION

The existence of an unusual type of striated muscle in the flight system of certain higher insects was first recognised by von Siebold (1848) though it was Kolliker (1888) who first used the name "fibrillar muscle" by which it is identified in this article. The term recognises the ease with which the fresh material can be teased into its constituent myofibrils. In the great days of light microscopy, much work was done on this type of tissue; Jordan (1933) summarises the information it provided on the histological changes accompanying contraction.

Renewed interest in insect fibrillar muscle was aroused by the discovery (Pringle, 1949) that, in addition to its structural peculiarities, it also has an unusual method of activation. In the living insect the mechanical activity which determines the rhythm of the wing beats is not accompanied by synchronous activity in the motor nerve fibres. The taxonomic distribution of this property was surveyed by Roeder (1951) and has been completed by Cullen (1974). Table 1 lists the main Orders of insects in which the two types of power-producing flight muscles are found and shows also some other correlated properties.

In Type I flight muscles, each muscular contraction is initiated by a volley of impulses in the motor nerves (Fig.1A). The rhythm of the wing beats is therefore neurogenic in origin and is maintained by oscillators in the thoracic ganglia (Wilson, 1968). In Type II flight muscles, rhythmic mechanical activity results from the fact that the pterothorax possesses mechanical resonance (inertia of the wings and elasticity of the muscles and skeleton) and that the muscles show delayed stretch activation (Machin and Pringle, 1959; Pringle 1967). The power output remains under the control of the motor nerves, but nerve impulses do not determine the timing of contractions (Fig.1B).

The internal membrane systems of the muscle fibres show a clear correlation with this difference in mode of activation (Fig.2). Both types of flight muscle have a well developed T-system, but the sarcoplasmic reticulum of Type I flight muscles is very well developed while that of Type II muscles is sparse (Smith 1966). Type I muscles require a large surface area of sarcoplasmic reticulum in order to change the myofibrillar calcium concentration at each twitch. In Type II muscles the oscillatory mechanical activity takes place at a constant intracellular calcium concentration. Rapid control of flight and

TABLE I

CHARACTERISTICS OF INSECT POWER-PRODUCING FLIGHT MUSCLES

PROPERTY	TYPE I	TYPE II
Nervous control	Synchronous	Asynchronous
T-system	Regular	Irregular
Sarcoplasmic reticulum	Rich	Poor
Myofibril structure	Non-fibrillar	Fibrillar
Myofibril diameter	About 0.5 μm	1-3 μm
I-band	Short	Very short
Passive elasticity	Moderate	Strong
Stretch activation	Transient ?	Maintained

Taxonomy		
	TYPE I	Cockroaches (Dictyoptera) Locusts (Orthopera) Dragonflies (Odonata) Mayflies (Ephemeroptera) Butterflies (Lepidoptera) Lacewings (Neuroptera) Caddis flies (Trichoptera) Some bugs (most Homoptera)*
	TYPE II	Flies (Diptera) Bees & Wasps (Hymenoptera) Beetles (Coleoptera) Some bugs (Heteroptera) Booklice (Psocoptera) Thrips (Thysanoptera)

* Jassidae, Psyllidae and Aphidae have Type II flight muscles (Cullen, 1974)

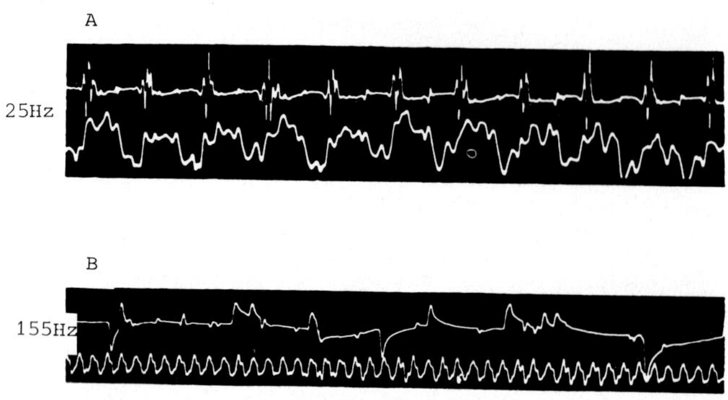

Fig.1 Electrical (upper trace) and mechanical (lower trace) records from the thorax of tethered flying insects. (A) Cockroach (synchronous). (B) fly (asynchronous). The frequency of the mechanical rhythm (wing beats) is indicated (Pringle, 1957).

the initiation and termination of wing movements are all achieved through auxiliary mechanisms involving small, twitching muscles (Pringle, 1966; Barber & Pringle, 1966; Nachtigall, 1968).

In both types of flight muscle the construction of the thorax and wing hinge provides a large amplification of movement, so that the extent of muscle shortening is small. Locust and dragon-fly flight myofibrils (Type I) nevertheless have a distinct I-band. Fibrillar (Type II) muscles function almost isometrically in the intact insect and the I-band is very short in unstretched fibres (Fig.2).

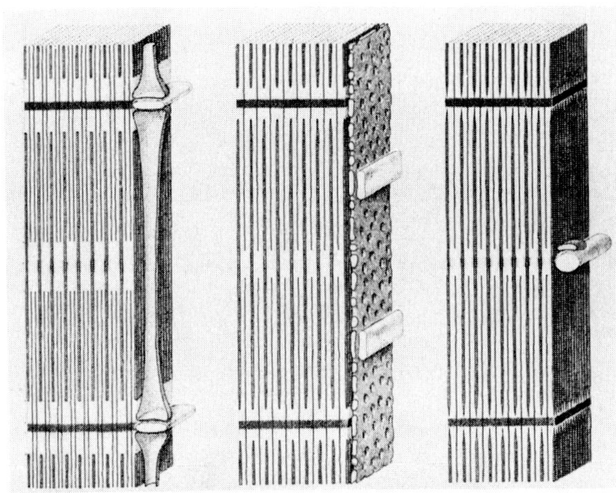

a b c

Fig.2 Diagrams of the structure of the sarcomere, T-system and sarcoplasmic reticulum in (a) vertebrate muscle; (b) synchronous flight muscle (dragon-fly); (c) asynchronous fibrillar flight muscle (wasp). (Smith, 1965).

The structure of a tissue is best understood by reference to its ontogeny; the embryology of fibrillar flight muscles has been studied by Tiegs (1958), Beinbrech (1968) and Auber (1969). In Diptera, the myofibrillar rudiments are formed at an early stage after pupation by extensive aggregation of myoblast cells. The myofibrils then enlarge by the incorporation of more filaments and elongate by longitudinal division of the dense bodies which will form the Z-lines. It is presumably this feature of the ontogeny which gives rise to the discrete myofibrils of the mature tissue.

General reviews of insect fibrillar muscle have been written by Pringle (1967, 1972) and Tregear (1975).

PASSIVE ELASTICITY OF THE FLIGHT MUSCLES

In all insects, the intact pterothorax has a considerable degree of mechanical resonance and part of the kinetic energy of the wing movements is stored in the elasticity of the skeleton and muscles. This elasticity is functionally important, since the up-stroke of the wings contributes little to the aerodynamic output; much of the mechanical energy of the up-stroke muscles is stored in elastic structures and released during the down-stroke. In the locust *Schistocerca*, one third of the kinetic energy is stored in the passive elastic system of the flight muscles (Weis-Fogh, 1965); the fraction may be even higher in insects with fibrillar muscles.

The location of the structural elements responsible for this elastic behaviour of flight muscles was discussed by Buchtal and Weis-Fogh (1956), who made comparative measurements on isolated insect and vertebrate muscles. In frog muscle, most of the resistance to stretch of relaxed fibres resides in the connective tissue sheath. In locust flight muscle the sarcolemma contributed only one third of the passive elasticity while in the fibrillar flight muscle of a beetle its contribution was negligible. Fig.3 summarises the measurements that have been made on intact relaxed muscles.

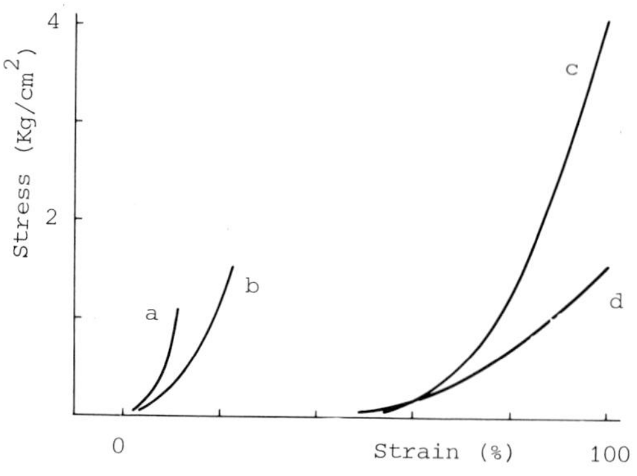

Fig.3 Stress-strain curves for living relaxed muscles. (a) insect fibrillar muscle (beetle); (b) non-fibrillar flight muscle (locust); (c) crustacean leg muscle (crayfish); (d) vertebrate leg muscle (frog). (Pringle, 1972).

Final proof that the elasticity of fibrillar muscles resides in the myofibrillar contractile mechanism came from measurements on glycerol-extracted fibres, in which the myofibrils are the only remaining continuous longitudinal elements (White, 1967, discussed by Pringle, 1974). Tension-length curves of glycerol-extracted fibres from *Lethocerus* flight muscle in relaxing solution show a slope similar to that of relaxed beetle muscle (Fig.4).

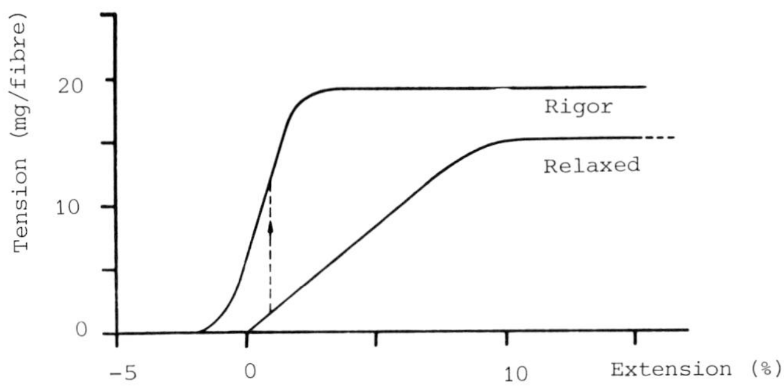

Fig.4 Tension-length curves for *Lethocerus* fibres in relaxing and rigor solutions, after 4 min. of stress relaxation. (White, 1967).

Before discussing the possible explanation of this property in terms of the fine structure and composition of the myofibrils, it is important to have a clear picture of the nature of the phenomenon. If relaxed fibres are subjected to a quick stretch of a few per cent, the increased tension decays slowly owing to stress relaxation (Fig.5). There is no truly maintained tension and over a time period of three decades, the tension decay curve is of the form:-

$$T = A\ t^{-k}$$

where A is a constant depending on the amplitude of stretch and k is a small fraction.

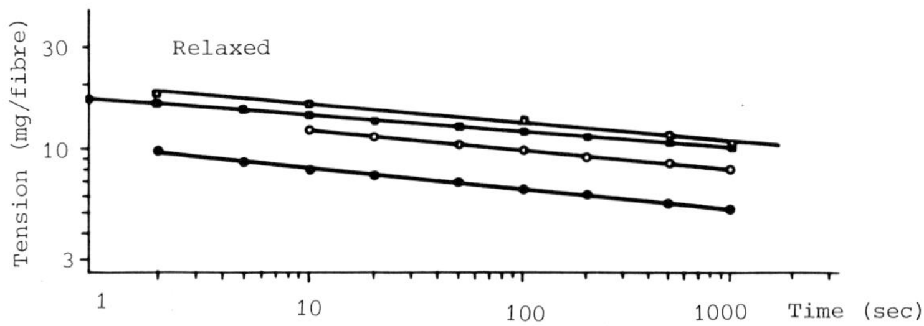

Fig.5 Stress relaxation of relaxed *Lethocerus* fibres, for quick stretches of 3%, 5%, 7% and 9% (White, 1967).

Because of stress relaxation, tension-length or stress-strain curves of the type of Fig.4 depend critically on the speed of stretch and the extent of stress relaxation before the tension measurement is made. Sinusoidal modulation of length, which can also be used to study the characteristics of the passive elasticity, gives a tension change that depends on the frequency of the modulation. For many purposes it is convenient to define a parameter "stiffness" as the complex modulus at 5 Hz oscillation. Fig.6 shows measurements of tension and stiffness for relaxed *Lethocerus* fibres stretched under two different mechanical regimes; the curves are of similar form but the position of the maximum of stiffness and the break in the tension curve occur at different degrees of extension depending on the mechanical conditions.

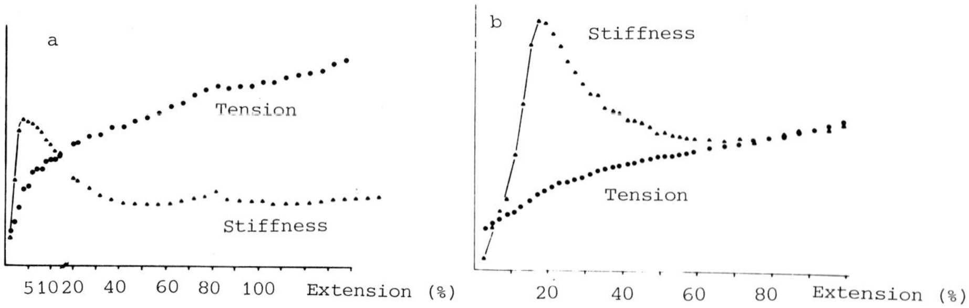

Fig.6 Tension-extension and 5 Hz stiffness-extension curves for relaxed *Lethocerus* fibres. (a) Extension by 1% steps in 1 sec, followed by 24 sec stress relaxation; (b) extension by 2% steps in 21 sec, followed by 40.5 sec stress relaxation. (Pringle, 1974).

Stress relaxation is a reversible phenomenon but the initial conditions are only reached after a long period of time. With a slow cycle of stretch and release, the tension curve shows pronounced hysteresis; tension is less during release than during stretch. Only after a period of up to one hour is it possible again to obtain the initial relationship.

The high-frequency components of the stress-strain relationship have not been studied critically in glycerol-extracted fibres under relaxing conditions. Machin and Pringle (1964), who measured the mechanical properties of living beetle muscle with sinusoidal length modulation over the frequency range from 5 Hz to 200 Hz, suggested that it could be represented by a three-element viscoelastic model with a time constant of 1.2 msec (Fig.7). Machin *et al.*

(1962) showed that neither the tension/length curve nor the vector modulus plot between 5 Hz and 105 Hz were significantly altered over the temperature range from 15 to 40°C. These results have to be explained by any model of the structures responsible.

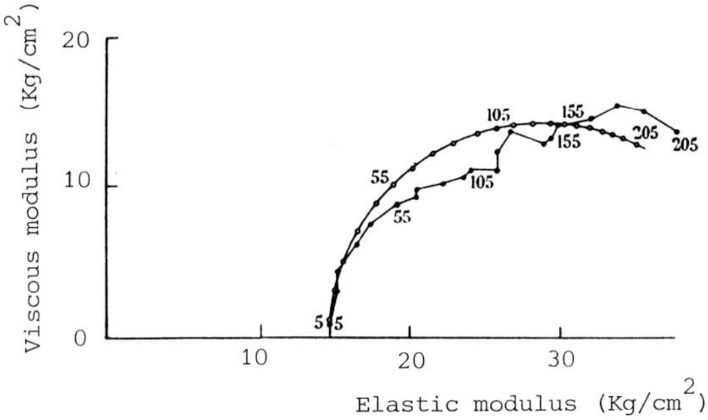

Fig.7 Vector modulus plot for relaxed living beetle muscle and theoretical curve for a three-element viscoelastic model. (Machin and Pringle, 1960).

STRUCTURAL EXPLANATION OF PASSIVE ELASTICITY

In the classical picture of the fine structure of the myofibril of a striated muscle, there is no continuous element stretching from one Z-line to another across the sarcomere. It has been suggested (Garamvölgyi, 1972; Garamvölgyi *et al.*, 1974) that a significant longitudinal force might be developed in insect flight muscles due to lateral interaction between the myofilaments. From electron micrographs of bee flight muscle fixed at various degrees of extension, it was stated that the inter-filament distance reduces more in stretched fibres than would be expected from a constant volume relationship in the overlap region. A theory is developed in which it is assumed that there is a lateral force between each point on the myosin filaments and the points on the actin filaments which are opposite to these in the unstretched fibre. On extension, this force acts diagonally and therefore has a longitudinal component which is supposed to account for the passive elasticity.

There are a variety of reasons why this theory is difficult to accept. First, importance is attached to the particular filament lattice found in insect muscles, which is different from that of vertebrate muscle. However, this lattice is characteristic of all arthropod muscles and is found not only in flight muscles with high passive elasticity but also in leg muscles, including the crustacean muscle whose stress-strain curve is shown in Fig.3c and whose elasticity is mainly due to the sarcolemma. Secondly, no controls were performed by Garamvölgyi (1972) to check the extent of shrinkage due to osmium fixation, which could have been different in differently stretched muscles. Finally, the known arrangement of myosin and actin monomers in the filaments lends no support to the idea that there is a lateral interaction between fixed points, as demanded by the theory. This type of explanation of the passive elasticity will therefore not be considered further and it will be assumed that there is some continuous longitudinal structure in the sarcomeres which is capable of bearing tension, which shows a power-law stress relaxation, and which has a maximum 5 Hz stiffness at a degree of extension between 3 and 20% depending on the speed at which the fibres have been stretched.

Four possible locations for the continuous structural element will be discussed in turn.

1. S-filaments connecting actin filaments in the middle of the sarcomere.

The term S-filament was used by Hanson and Huxley (1955) to describe a highly elastic structure which, they considered, must exist between the ends of the actin filaments in order to account for the fact that vertebrate myofibrils stretched to the point of no overlap return to their original length when released. No filamentous structure in this position in the sarcomere has ever been found by electron microscopy of any striated muscles, nor is there a higher density in the H-zone of fibrillar muscles as would be expected if additional non-filamentous material was present in this region. As in other striated muscles, double overlap of the actin filaments can occur in contracted fibres. In spite of the attraction of this hypothesis in connection with stretch activation (see later), it must be concluded that S-filaments are not responsible for the passive elasticity.

2. Filaments in parallel with the myosin and actin filaments

Very thin filaments, with a diameter of 2.5 to 4.0 nm, stretching from one Z-line to the next through the sarcomere,were described by McNeill and Hoyle (1967) from a variety of vertebrate and arthropod muscles including the non-fibrillar flight muscles of *Schistocerca*; the name T-filaments was suggested for these structures. Other authors have made similar suggestions. Toselli and Pepe (1968) described two distinct diameters of filament (modes 5.5 nm and 7.5 nm) attaching to dense bodies of an insect abdominal muscle;

the smaller of these is clearly not a normal actin filament. Initial filaments 5.5 nm in diameter have been described by most authors who have studied the embryology of arthropod muscles; these precede the appearance of actin and myosin filaments and persist longest in the region of the Z-lines (Auber, 1965).

While it cannot be excluded that T-filaments may indeed be present in striated muscles, there is nothing in published evidence to suggest they have significant mechanical strength or that they are present in greater number or density in flight muscles. Sufficient material to preserve the sarcomere structure remains after myosin and actin filaments have been dissolved from vertebrate muscle by KI, but removal of the myosin filaments alone from insect fibrillar muscle by 0.6M KCl completely eliminates the passive elasticity. This hypothesis must therefore also be discarded.

3. Fixed or very slowly-acting crossbridges

A continuous structural element would be present through the sarcomere if crossbridges joined the myosin and actin filaments in the overlap region or in any part of it. The evidence for and against this hypothesis must be considered carefully.

Fixed crossbridges evenly distributed in the overlap region cannot account for the observations. Fig.6 shows that in relaxed *Lethocerus* fibres tension continues to increase up to the point where the fibres break at 120-140% extension. If the amount of overlap was being progressively reduced by extension, as would be true for frog fibres, then this observation in itself would eliminate the hypothesis, but in fact stretching relaxed *Lethocerus* fibres destroys the regularity of the A-band; some myosin filaments move towards one Z-line and others towards the other, so that some overlap is present even at 120% total extension. This effect is, however, much reduced in fibres from the fibrillar flight muscle of bees. Fig.8 (overleaf) shows the appearance of longitudinal sections of fibrillar fibres from a tropical carpenter bee at two extensions. At 40% extension, overlap is markedly reduced and distortion of the A-band occurs only at extensions greater than this. Tension continues to increase steadily in stretched carpenter bee fibres, as in *Lethocerus*, showing that evenly distributed crossbridges cannot explain the phenomenon. The straight transverse boundary of the H-zone in Fig.8b is good evidence against the presence of S-filaments.

The appearance of Fig.8B and the steady rise of tension with extension could perhaps be explained by fixed crossbridges at the ends of the myosin filaments, but such bridges would have to be capable of an enormous degree of extension. It is also very difficult on this basis to explain the peaks of 5 Hz stiffness in Fig.6. Fixed crossbridges can therefore be excluded.

8
a
b
c

Crossbridges with a slow rate of attachment and detachment in relaxed muscle provide a much more attractive possibility and will be considered in the mechanism proposed.

4. Material connecting myosin filaments to the Z-line

The evidence from electron microscopy for material in this position in the sarcomeres of fibrillar muscle is considered by Ashhurst (this volume). The suggestion was first put forward by Auber and Couteaux (1963) from examination of sections of blow-fly flight muscle and since, under some circumstances, it can be filamentous in appearance, the name C-filaments ("connecting") has been used in the literature (Pringle, 1967; Garamvölgyi, 1969). Material in the correct position in the lattice to represent a connection between myosin filaments and the Z-line becomes most obvious when fibres are stretched in rigor; the actin filaments then break off from the Z-line, leaving "clear zones" of low optical density (Hanson, 1956). Reedy (1972) and White and Thorson (1973) have published electron micrographs showing regularly arranged filaments crossing these clear zones and have pointed out that the fact that the actin filaments break away evenly on both sides of the Z-line shows that the material possesses structural strength.

The nature of this material has been studied by Bullard(this volume), who has shown that it consists of a protein not previously identified from muscle tissue. A preliminary report that the protein was paramyosin (Pringle, 1974) was in error.

It seems clear from the work of Ashhurst (1967) that the connecting material is not in filamentous form in fibres that are completely unstretched, but that filaments can be formed from it. This fact and the mechanical properties which must be ascribed to it suggest that its physical state may be that of a microcrystalline polymer (Tobolsky, 1960). In particular, the power-law characteristic of stress relaxation resembles that of low-density polyethylene below the critical temperature. The molecular state can be crudely visualized as a hydrated chain with a tendency to aggregate laterally into filamentous crystallites, particularly when oriented by flow. In such material, low-frequency stiffness would be maximal when the molecules are maximally oriented but hydration would slowly destroy the crystallinity and produce longitudinal slip if overstrained. More stable binding to the ends of the myosin filaments and the Z-line material would lead to the creation of ordered crystallites opposite the ends of the myosin filaments, giving the appearance of filaments in this position in the lattice in stretched fibres.

Fig.8 Longitudinal sections of bee flight muscle (*Xylocopa*) fixed in relaxing solution. (a) unstretched; (b) 40% extension; (c) 90% extension. (Pringle, 1974). Scale bar 1 μm.

A MODEL FOR PASSIVE ELASTICITY

The two continuous structural elements which have not been exluded in the previous section are crossbridges attaching and detaching at a low frequency and material connecting the ends of the myosin filaments to the Z-line. The mechanical properties which have to be explained are a stress relaxation with power-law characteristics, a tension-length curve showing a break between 3% and 20% extension depending on the speed of stretch with a peak of 5 Hz stiffness at the same extension and a high-frequency response (above 5 Hz in living beetle muscle) which is simulated by a three-element viscoelastic model with a temperature-invariant time constant of about 1.2 msec.

It seems most plausible to attribute the low-frequency behaviour of relaxed fibres to the connecting material between myosin filaments and the Z-line and the high-frequency behaviour to slow-cycling crossbridges. These two structural elements are in parallel in the sarcomere and the whole system can therefore be represented by a Maxwell model of the type shown in Fig.9. Element A must have non-linear elastic properties and show stress relaxation. On the slow time scale at which stress relaxation is apparent, the turnover rate of residual crossbridges will be fast enough to make little contribution to the mechanical response.

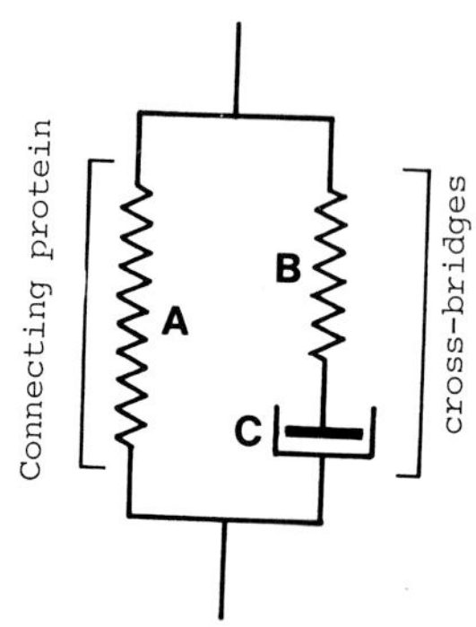

Fig.9 Three-element Maxwell model for passive elasticity, showing the suggested location of elements in the sarcomere of relaxed *Lethocerus* fibres.

At frequencies above about 5 Hz in living beetle muscle and above 1 Hz in glycerol-extracted *Lethocerus* fibres, element A behaves as an almost perfect elasticity and will therefore not contribute to the shape of the vector modulus plot (Fig.7) apart from moving its position along the X-axis. It is perhaps significant that the time constant of the system in beetle muscle is the same as the fastest time constant measured by Abbott and Steiger (1977) in activated *Lethocerus* fibres and attributed by them to the visco-elasticity of the bridges themselves. The small temperature coefficient is also a common feature of these two independent measurements. At high frequencies, therefore, one may suppose that these residual crossbridges are sufficiently numerous to contribute to the mechanical response.

Involvement of slow-cycling crossbridges in the passive elastic system would explain the reversibility of stress relaxation. Maintained stretch of the fibres will lead to sliding of actin filaments and reduction of overlap, yet the initial state can be restored by leaving the fibres unstretched in relaxing solution for a long period. If it is assumed that the residual crossbridges are able to do work, albeit at a slow rate, then the effect is explained. It is a fact that there is significant ATPase activity of *Lethocerus* fibres even in relaxing solution; Chaplain and Tregear (1966) estimated a turnover rate of myosin enzyme sites of 0.1 sec^{-1} which would be fast enough to produce the slow sliding needed.

The results of White (this volume) are not at variance with this model. He has found that in activated fibres there is a linear relationship between tension and ATPase activity under various conditions but that the tension line intercepts the axis at a value corresponding to the passive tension. In a model with a viscoelastic time constant of 1.2 msec, elements B and C (Fig.9) will not contribute to the tension measurements made by White and the passive elasticity measured by him will be that of element A; this is indeed in parallel with the crossbridges in activating conditions.

This model makes certain testable predictions. The amount of passive elasticity differs in different flight muscles; there should be a correlated difference in the concentration of connecting material between myosin filaments and the Z-line. Furthermore, when it is available in sufficient quantity as a pure solution, the connecting protein should show rheological properties similar to those of microcrystalline polymer. The high-frequency response of glycerol-extracted fibres in relaxing solution should be qualitatively similar to that of activated fibres and differ quantitatively from it in the manner expected from a difference in the properties of crossbridges.

THE PROBLEM OF STRETCH ACTIVATION

The oscillatory behaviour of fibrillar flight muscles in the intact insect derives from the fact that, when excited through their motor nerves, these muscles are activated by stretch and develop a large component of tension with the delay. Glycerol-extracted fibres show the same property (Jewell and Rüegg, 1966) and with small amplitudes of stretch and release the effect is symmetrical (Fig.10).

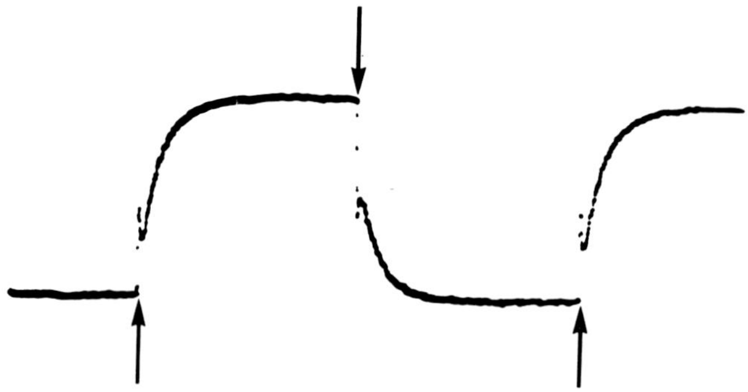

Fig.10. Stretch activation of activated *Lethocerus* fibres, showing symmetrical delayed tension changes on stretch and release when the amplitude of length change is 0.1%. (Tregear, 1975).

It is now known that stretch activation is shown by most types of muscle which have been investigated with this phenomenon in mind; other muscles are discussed by Steiger (this volume). In insect fibrillar muscle, the delayed tension component is much larger than in vertebrate muscles (Abbott and Steiger, 1977) but the time constant and temperature coefficient are similar in rabbit psoas muscle.

Models of the mechanism of muscular contraction including stretch activation have been proposed by Thorson and White (1969) and White and Thorson (1972, 1973). A four-state model is proposed by Abbott (this volume). In all these models a common feature is that stretch increases the rate constants of the attachment and/or detachment steps in the crossbridge cycle. The problem to be considered in this article is the location in the sarcomere of the element which senses the length change and produces changes in the rate constants of the cycle.

Insect muscles have a dual system of calcium activation (Lehman, this volume). Calcium binds both to the troponin of the actin filaments and to the myosin and *in vitro* studies using hybrid actomyosins suggest that full calcium control is achieved only when both components are activated. An obvious first hypothesis for the mechanism of stretch activation is that stretch increases the binding constant of one of these two systems for calcium. However, Marston and Tregear (1974) showed that there is no increase in calcium binding on stretch of *Lethocerus* fibres, although the amount bound suggested that both troponin and myosin sites were being occupied. Since the calcium concentration for half-activation is also unaffected by stretch (Schädler, 1967), since full ATPase activity cannot be achieved in unstretched fibres by any concentration of calcium and since stretch and calcium have different effects on the amplitude and rate constant of stretch activation (Abbott, 1973), it must be concluded that the two methods of control are independent. No delayed tension or increased ATPase activity are found when fibres are stretched in relaxing solution. Both calcium activation and stretch activation are required for full activity.

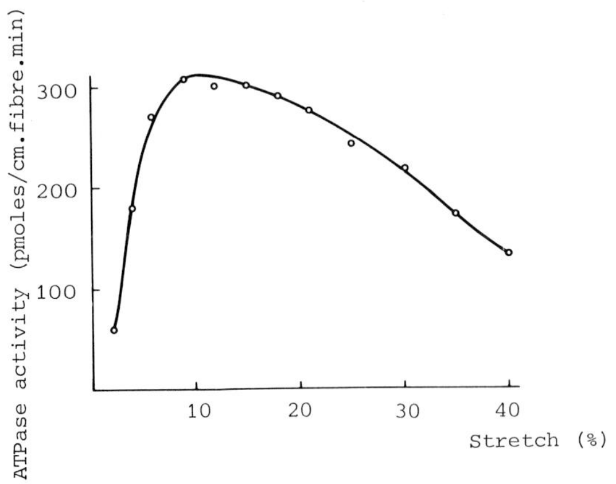

Fig.11 ATPase activity of *Lethocerus* fibres stretched in relaxing solution and subsequently activated by 1.5 x 10^{-8}M Ca^{2+} (Chaplain, 1969).

There are two important qualitative differences between the fibrillar muscle of *Lethocerus* and other muscles in the manifestation of stretch activation. In *Lethocerus* muscle, the fibres can be stretched by up to 10% under relaxing conditions and if they are then activated by calcium the ATPase activity is found to be increased (Fig.11) (Chaplain, 1969). In vertebrate muscles, stretch while relaxed merely reduces the amount of overlap between myosin and actin filaments and leads to a lower ATPase activity (Ward *et al.*, 1965); the same effect is seen in *Lethocerus* muscle fibres stretched by more than 10% (Fig.11). Since a stretch of about 10% is also required to reach the peak of passive 5 Hz stiffness (and the break in the tension-length curve) when stress relaxation has proceded to an extent intermediate between the two conditions shown in Fig.6 (as in Chaplain's experiments), it is very tempting to conclude that there is a causal connexion between passive elasticity and stretch activation in fibrillar muscles.

The second qualitative difference becomes apparent when calcium-activated fibres are subjected to quick stretch. In *Lethocerus* fibres the delayed tension is maintained for the duration of the stretch. In rabbit psoas fibres the delayed tension declines with a rate constant of about 5 sec^{-1} (20^oC) to a value only slightly different from that before the stretch was applied (Rüegg *et al.*, 1970). Unfortunately, there is at present insufficient information from other fibrillar muscles or from non-fibrillar insect flight muscles to decide whether this difference is a general one or whether the maintenance of delayed tension is merely quantitatively greater in *Lethocerus*. The early work of Boettiger (1957) on living bumble-bee muscle suggested that delayed tension might not be maintained in this insect. From a functional point of view, failure to maintain the higher level of tension would not matter in an insect with a high wing-beat frequency.

The mechanical conditions for stretch activation in glycerol-extracted fibres were studied most critically by Pringle and Tregear (1969). The solutions used in these experiments did not contain inorganic phosphate; the absence of this ion is now known to reduce the optimum frequency for the performance of oscillatory work but not otherwise to affect the performance (White, 1972). There was also no "ATP regenerating system" but problems with the diffusion of ATP into the tissue were avoided either by using single fibres or by limiting the periods of high work output. Sinusoidal length modulation was used and at large amplitudes of oscillation when the system is highly non-linear, work output was estimated by measurement from tension-length loops on the oscilloscope. The ionic strength of solutions was 0.1M and the pH 6.9-7.1.

The amplitude of length change required to elicit maximum work output at the optimum frequency varied between preparations from 1.8 to over 5%. In order to obtain this work output, the mean level of extension had to be adjusted. At a constant oscillation amplitude there was an optimum mean extension which varied from 2.8 to 4.8% if the fibres were extended in activating solution and could be up to 16% if the fibres were extended in relaxing solution before calcium activation. If the mean extension was too great, work output was reduced but returned to its maximum value in a few seconds. It was clear from these results that the significant parameter controlling stretch activation was not strain but stress and that stress relaxation is greater if the fibres are first extended under relaxing conditions and then calcium activated.

Taken together with the earlier conclusion that stress relaxation arises from the properties of the material connecting the myosin filaments to the Z-line, it can be strongly argued that this material plays an important role in stretch activation of fibrillar muscle. Its mechanical properties are the most variable element in different preparations of glycerol-extracted fibres and it seems likely that the variability of Pringle and Tregear's results might have resulted from this.

THE STRUCTURAL BASIS OF STRETCH ACTIVATION

Hypotheses about the location in the sarcomere of the stress-sensing element responsible for stretch activation must take account of the fact that the phenomenon is shown by most or all types of muscle. As already stated, fibrillar muscles differ only in that the effect is larger, that stretch can be applied before calcium activation and that, at least in *Lethocerus* muscle, delayed tension is maintained. The objective must be to find structural or biochemical differences between fibrillar and other muscles that can account for these differences in properties.

A priori, it would be attractive to suppose that the sensor for stress resides in the actin filaments. The accepted mechanism for calcium control in vertebrate muscles involved the movement of tropomyosin molecules in the groove of the actin helix under the influence of the calcium-troponin complex (Spudich *et al.*, 1972). Tropomyosin is an α-helical protein oriented longitudinally to the filament axis and it is not difficult to conceive how stress in this filament could promote such an azimuthal movement, exposing actin sites. Unfortunately there is no suggestion from either structural or biochemical studies on fibrillar muscle as to how such stress could reach the tropomyosin in the actin filaments in relaxed muscle. Both S-filaments and T-filaments have been definitely excluded as explanations of passive elasticity and the same arguments effectively exclude the thin filament as the location of the stress sensor in fibrillar muscle.

The presence of connecting protein is the only known biochemical peculiarity of fibrillar muscle and we have seen that this material provides a good explanation for the stress relaxation of passive elasticity. If the same material is responsible for maintained stretch activation, all the differences between fibrillar and other muscles would be explained, but the sensor must, in that case, reside in the myosin filaments. Transient stretch activation after calcium activation in non-fibrillar muscles would then be explained by the fact that stress can be applied to the myosin filaments through attached crossbridges. It would be expected that in such muscles stretch activation would be smaller and transient and that it would be absent when stretch is applied before calcium activation.

No suggestions can be put forward at the present time as to how stress on the myosin filaments might affect the rate constants of actin-myosin interaction. Recently Hazelgrove (1975) has concluded from experiments with vertebrate muscle that the myosin filaments can exist in two different states and that there is an increase from 143.4 Å to 144.8 Å in the axial spacing of myosin molecules on passing from the relaxed to the rigor or activated state. In *Lethocerus* fibres the axial spacing is 145 Å and Armitage *et al* (1973) state that the change on going from relaxation to rigor is less than 0.3%. Even in vertebrate muscle there is no exidence that longitudinal stress in the filament backbone can promote axial reorganisation. There is probably still too much ignorance of the molecular structure of the myosin filament to make it profitable to speculate further and resolution of the problem must be left to the future. The best hope of progress would seem to lie in the improvements which are taking place in the techniques of electron microscopy and X-ray diffraction and in the possibility of making measurements under rigorously controlled chemical and mechanical conditions in glycerol-extracted fibres from fibrillar muscle.

REFERENCES

Abbott, R.H. (1973) *J. Physiol.*,231, 195.

Abbott, R.H. and Steiger, G.J. (1977) *J. Physiol.*, 266, 13.

Armitage, P., Miller, A., Rodger, C.D. and Tregear, R.T. (1973) *Cold Spring Harb. Symp. Quant. Biol.*, 27, 379.

Ashhurst, D.E. (1967) *J. mol. Biol.*, 27, 385.

Auber, J. (1965) *C.R. Acad. Sci.*, Ser.D 260, 668.

Auber, J. (1969) *J. Microsc. (Paris)*, 8, 197.

Auber, J. and Couteaux, R. (1963) *J. Microsc. (Paris)*, 6, 309.

Barber, S.B. and Pringle, J.W.S. (1966) *Proc. Roy. Soc. B*, 164, 21.

Beinbrech, G. (1968) *Z. Zellforsch. mikrosk. Anat.*, 90, 463.

Boettiger, E.G. (1957) In *Physiological Triggers* (ed. Scheer, B.T.), 117. University of Oregon Publications.

Buchtal, F. and Weis-Fogh, T. (1956) *Acta physiol. scand.*, 35, 345.

Chaplain, R.A. (1969) *Pflügers Arch.*, 307, 120.

Chaplain, R.A. and Tregear, R.T. (1966) *J. mol. Biol.*, 21, 275.

Cullen, M.J. (1974) *J. Ent. (A)*, 49, 17.

Garamvölgyi, N. (1972) *Acta biochim. biophys. Acad. Sci. Hung.*, 7, 157.

Garamvölgyi, N. (1969) *J. Ultrastruct. Res.*, 27, 462.

Garamvölgyi, N. Biczo, G., Ladik, L. and Eöry, A. (1973) *Acta. biochim. biophys. Acad. Sci. Hung.*, 8, 57.

Garamvölgyi, N., Biczo, G., Eöry, A. and Suhai, S. (1974) *Acta biochim. biophys. Acad. Sci. Hung.*, 9, 233.

Hanson, J. (1956) *J. biophys. biochem. Cytol.*, 2, 691.

Hanson, J. and Huxley, H.E. (1955) *Symp. Soc. exp. Biol.*, 9, 228.

Hazelgrove, J.C. (1975) *J. mol. Biol.*, 92, 113.

Jewell, B.R. and Rüegg, J.C. (1966) *Proc. Roy. Soc. B*, 164, 428.

Jordan, H.E. (1933) *Physiol. Rev.*, 13, 301.

Kolliker, A. (1888) *Z. wiss. Zool.*, 16, 374.

McNeill, P.A. and Hoyle, G. (1967) *Amer. Zool.*, 7, 483.

Machin, K.E. and Pringle, J.W.S. (1959) *Proc. Roy. Soc. B*, 151, 204.

Machin, K.E. and Pringle, J.W.S. (1960) *Proc. Roy. Soc. B*, 152, 311.

Machin, K.E. Pringle, J.W.S. and Tamasige, M. (1962) *Proc. Roy. Soc. B*, 155, 493.

Marston, S. and Tregear, R.T. (1974) *Biochim. biophys. Acta*, 347, 311.

Nachtigall, W. (1968) *Z. vergl. Physiol.*, 61, 1.

Pringle, J.W.S. (1949) *J. Physiol.*, 108, 226.

Pringle J.W.S. (1957) *Insect Flight*, Cambridge University Press.

Pringle J.W.S. (1967) *Progr. Biophys. Mol. Biol.*, 17, 1.

Pringle J.W.S. (1968) *Advan. Insect Physiol.*, 5, 163.

Pringle J.W.S. (1972) In: *Structure and Function of Muscle* (ed. Bourne, G.) 491.

Pringle, J.W.S. (1973) In *The Physiology of Insecta* (ed. Rockstein, M.) III, 433.

Pringle, J.W.S. (1974) *Symp. Biol. Hung.*, 17, 67.

Pringle, J.W.S. and Tregear, R.T. (1969) *Proc. Roy. Soc. B*, 174, 33.

Reedy, M.K. (1972) In *Contractility of Muscle Cells and Related Processes* (ed. Podolsky, R.J.), 229.

Roeder, K.D. (1951) *Biol. Bull., Woods Hole*, 100, 95.

Rüegg, J.C. Steiger, G.J. and Schädler, M. (1970) *Pflügers Arch.*, 319, 139.

Schädler, M. (1967) *Pflügers Arch.*, 296, 70.

Siebold, C.T. von. (1848) *Lehrbuch der vergleichenden Anatomie der wirbellosen Thiere*. Berlin.

Smith, D.S. (1965) *Sci. Amer.*, 212, 76.

Smith, D.S. (1966) *Progr. Biophys. mol. Biol.*, 16, 107.

Spudich, J.A. Huxley, H.E. and Finch J.T. (1972) *J. mol. Biol.*, 72, 619.

Thorson, J. and White, D.C.S. (1969) *Biophys. J.*, 9, 360.

Tiegs, O.W. (1958) *Philos. Trans. Roy. Soc. B*, 238, 221.

Tobolski, A.V. (1960) *Properties and Structure of Polymers*. John Wiley, New York and London.

Toselli, P.A. and Pepe, F.A. (1968) *J. Cell Biol.*, 37, 462.

Tregear, R.T. (1975) In *Insect Muscle* (ed. Usherwood, P.N.R.), 357. Academic Press.

Ward, P.C., Edwards, J.C. and Benson, E.S. (1965) *Proc. Nat. Acad. Sci.*, 53, 1377

Weis-Fogh, T. (1965) *Proc. 13th Congr. Ent.*,p. 186.

White, D.C.S. (1972) *Cold Spring Harbor Symp. Quant. Biol.*, 27, 201.

White, D.C.S. and Thorson, J. (1972) *J. Gen. Physiol.*, 60, 307.

White, D.C.S. and Thorson, J. (1973) *Progr. Biophys. mol. Biol.*, 27, 173.

Wilson, D.M. (1968) *Adv. Insect Physiol.*, 5, 289.

THE RESTING ELASTICITY OF INSECT FIBRILLAR FLIGHT MUSCLE, AND PROPERTIES OF THE CROSS-BRIDGE CYCLE

D.C.S. White, Madelaine M.K. Donaldson, G.E. Pearce and M.G.A. Wilson

Department of Biology, University of York, Heslington, York, England.

INTRODUCTION (by D.C.S. White)

Resting Elasticity

When I started work in the Oxford Zoology Department as a D. Phil. student, Professor Pringle suggested that I might investigate the problem of the high stiffness of the relaxed fibrillar flight muscle. At the time there was a suggestion (Auber & Couteaux, 1963) that there was a structure linking the A-filament to the Z-line, but as I discussed (White, 1967; see also Ashhurst, 1971), the evidence was very weak because of the likelihood that the A-filaments themselves extended into the Z-line in unstretched muscle. Preparations of *Lethocerus* dorsal longitudinal muscle, stretched slightly, failed to show such structures. Their clearest direct visualisation was first made from fibres which had been stretched 10-20 % in rigor (well beyond the elastic limit of the rigor muscle) and then fixed. Such pictures (White & Thorson, 1973; Ashhurst, this volume) showed electron-dense structures linking the A-filament and the Z-line, and were named C-filaments ("C" for connecting - White, 1967). This nomenclature is perhaps unfortunate because the structure may well not be filamentous under physiological conditions (Pringle, this volume). In this article we will give evidence from mechanical measurements that a structure in parallel with the cross-bridges is responsible for the resting stiffness.

Mechanical Properties of the Cross-Bridge Cycle

My other area of contribution has been largely in collaboration with Dr. John Thorson, and originated with his realisation that the delayed tension changes characteristic of fibrillar muscle (Machin & Pringle, 1960; Jewell & Rüegg, 1966) could be well explained in terms of a 2-state (attached-detached) cycle of activity in which the probability of attachment per unit time was dependent upon muscle strain (Thorson & White, 1969). The modelling of that paper, which provided an explanation of the mechanical responses for small-amplitude length changes, predicted the form of the response to higher amplitude changes, and for our next paper (White & Thorson, 1972) we performed such experiments. We also had the idea that, if the mechanical cycle were linked to the hydrolysis of ATP, then we might observe changes in the mechanical response when we changed the concentrations of ATP, ADP and inorganic phosphate (P_i) in the solution bathing the fibres. It was with surprise that we found that the response of the fibres in the presence of millimolar concentrations of P_i differed so qualitatively from that in its absence. Although the original 2-state cycle, with the additional assumption that the torque exerted by a cross-bridge was linearly dependent upon its distortion (its displacement from its minimum free energy configuration), was able to account for the high-amplitude responses in the presence of P_i, it provided no comparable straightforward explanation of the form of the non-P_i response.

The 2-state model accounted quantitatively for the finding (Ruegg & Tregear, 1966) that the ATPase of the fibres increased markedly with maintained stretch (in rough terms a doubling of the ATPase activity with a 1% increase of fibre length). It failed however to provide any explanation for the observation (Steiger & Rüegg, 1969) that if the fibre length were changed sinusoidally then the ATPase activity was markedly higher for those frequencies at which the power output was greatest.

A 3-state model (White, 1972), in which just one of the states generates significant tension, is capable of accounting for both the P_i and the non-P_i responses if the effect of phosphate is to increase both the backward and forward rate constants of one of the transitions. Furthermore this model also showed an increased cycling rate (equivalent to ATPase activity if one ATP is hydrolysed per cycle) at frequencies at which the power output is greatest (White & Thorson, 1973).

A direct prediction of this model is that the number of attached bridges should increase throughout the rising phase of the delayed tension, and fall during the falling phase. It is possible to test this prediction experimentally by assuming that the stiffness of the muscle, if measured sufficiently rapidly, is a direct measure of cross-bridge attachment number, as shown for frog muscle (reviewed by Huxley, 1974). Accordingly in this paper the stiffness is measured by applying an initial length change to invoke the response, followed by a second step to measure the stiffness. These are described as double-step experiments. Both the interval between the steps, and the magnitude and direction of the second step, was varied in order to obtain the stiffness at different times during the response.

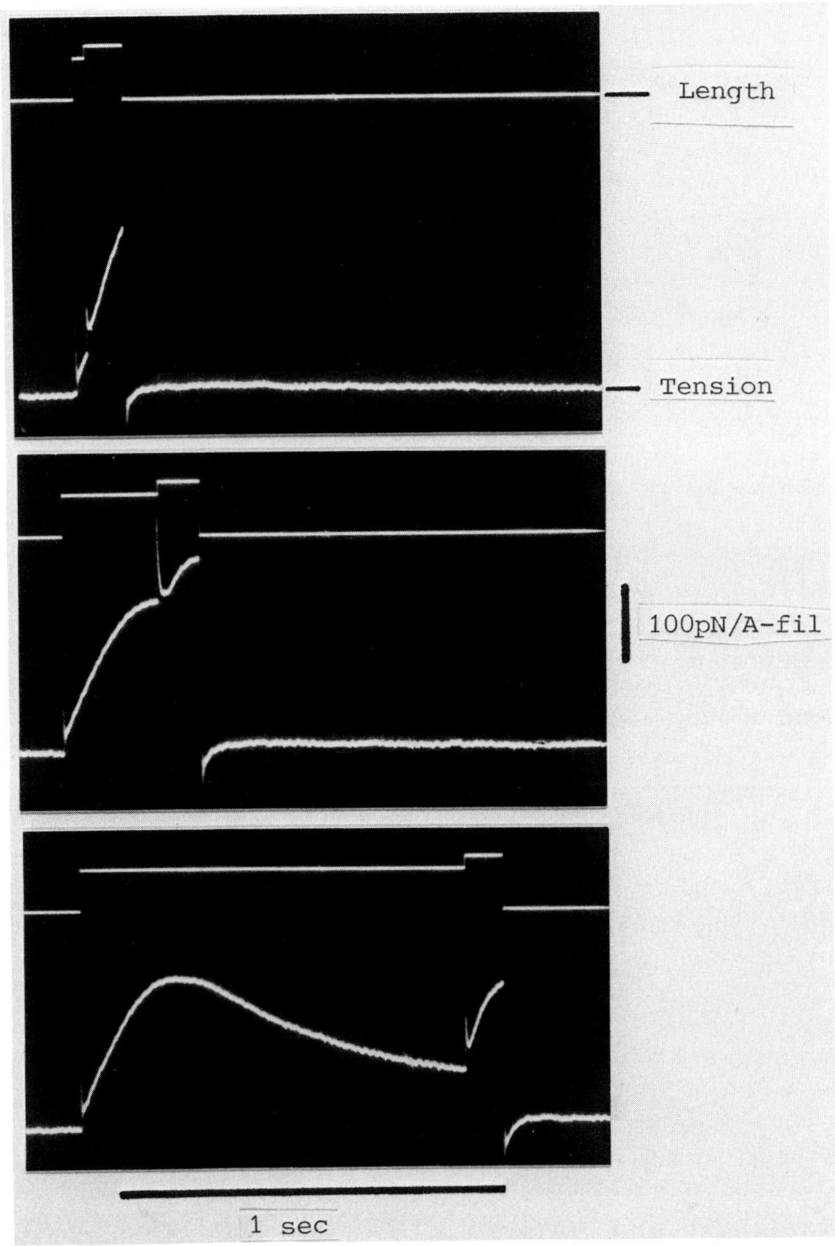

Fig. 1. Responses to a 1% step increase of length followed by a 0.3% increase at various times afterwards, in non-P_i activating solution - ($MgCl_2$, 10mM; Na_2ATP, 10mM; EGTA, 5mM; $CaCl_2$, 5mM; KCl, 45mM; HistidineHCL, 20mM; pH 7.0; Ionic strength. 117mM; estimated Ca^{++}, 0.32μM).

MATERIALS AND METHODS

The experiments were performed on fibres from the dorsal longitudinal muscle of the giant water bug *Lethocerus cordofanus*, glycerol extracted according to the procedures described by White & Thorson (1972). The source of the chemicals was Fisons Scientific, except for: EGTA (Sigma), Histidine (BDH), ATP, creatine phosphate and creatine phosphokinase (Boehringer, Lewes).

Bundles of 2-5 fibres were attached, using cellulose acetate dissolved in acetone as the glue, between two metal wires. One of these was attached to a tension transducer. For experiment A this was a metal-coated glass capacitance tranducer (Huxley & Simmons, 1968) with a resonant frequency of 2.2 kHz, and slightly underdamped. For experiment B the tension transducer was a semiconductor strain gauge (Akers type AE 803, from AME, Norway) with a resonant frequency of 2.7 kHz (measured when loaded with a relaxed fibre) and critically damped with a damping vane and oil. The other wire was attached to the moving arm of an electromagnetic vibrator (Ling Dynamics, type 101) whose movement was controlled by a servoloop which was in turn driven from the D/A output of a Hewlett-Packard minicomputer type 2114A. The rise time of the apparatus was slightly faster than 1ms.

RESULTS

The experiments described were all performed in the absence of inorganic orthophosphate. Just two experiments will be described, performed on fibres from different individuals.

Experiment A

In this experiment a 1% initial length step was applied to the muscle, and a 0.3% test step applied at varying times during the response. A brief account of these experiments was described by White & Donaldson (1975). Figure 1 shows three recordings with different inter-step intervals. The magnitude of the initial 'spike' obtained in response to the second step (magnitude of the elastic response) was used as a measure of muscle stiffness. In figure 2 the spike height is plotted as a function of the tension in the muscle at the instant the step was applied. Points from measurements made during the rising phase (marked up), at the peak (top) and during the falling phase (down) of the delayed response are distinguished.

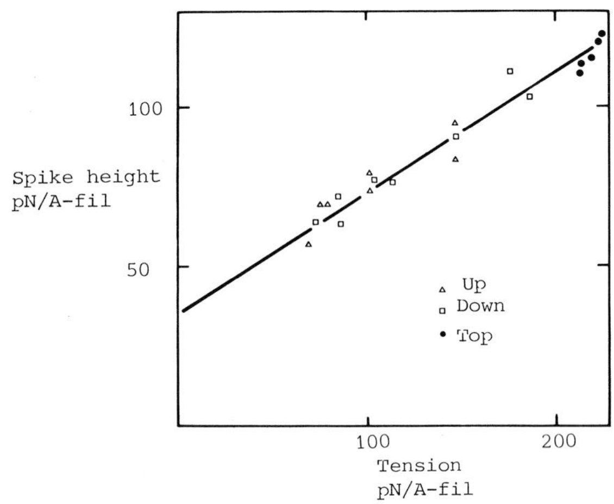

Fig. 2. Spike height of the initial tension response to the 0.3% length step in the experiment of which three records are presented in figure 1.

Experiment B

The experiment was performed in a similar way, except that only three intervals between the initial and test steps were used: one at the peak of the delayed tension, one on the rising and one on the falling phase. The intervals for these latter two responses were chosen so that the starting tension was approximately the same for both. At each time interval a series of responses was obtained, varying in the magnitude of the second step length change. The results are shown in figure 3. Also shown in the figure is a series of single steps of varying magnitude, used to obtain the stiffness of the muscle in the steady state in the activating solution.

The responses of figure 3 are analysed in figure 4 in the following way. For the single-step responses, the tension at the peak of the instantaneous response (i.e. the tension at the peak of the spike) is plotted as a function of the magnitude of the applied length change. For the double steps the tension at the peak of the instantaneous response to the second step is plotted as a function of the magnitude of the second step. Also plotted, as a "zero applied length-step" points on the tension axis, are the tensions in the muscle

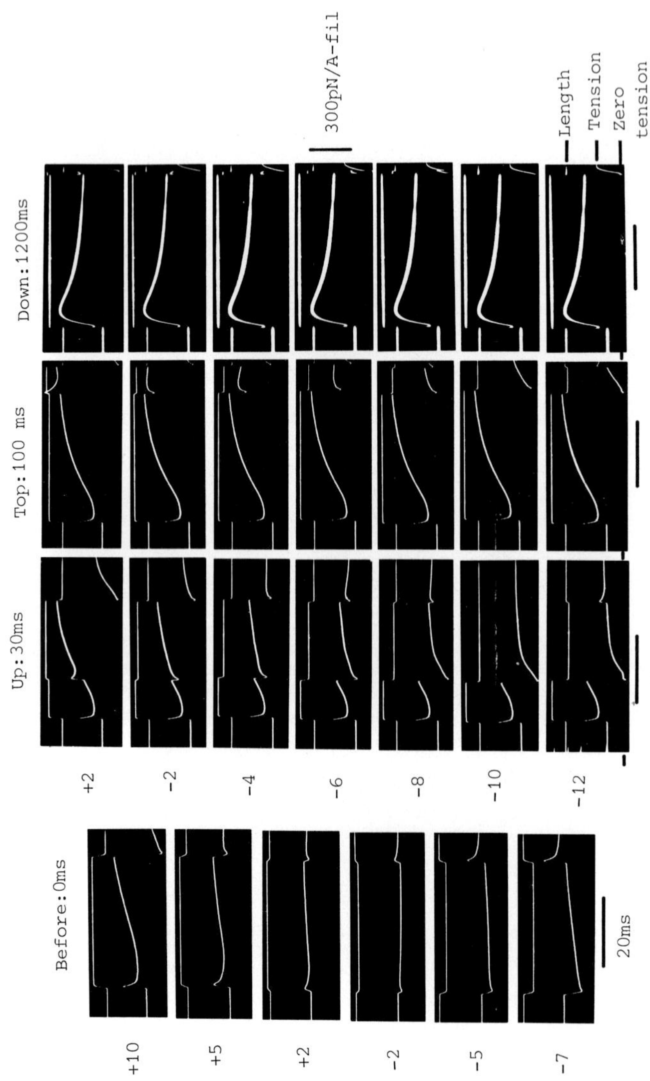

Fig. 3. Double step experiments similar to thoese described in figure 1, but in which varying magnitudes and directions of the second step are applied to the muscle. The left hand set of records is a series of single steps. The figures to the left of the records indicate the magnitude of the step in nm/½ sarcomere for the left hand column, and the magnitude of the second step for the three columns on the right hand side, in which the initial step was a 1% increase of length. Solution as for figure 1 with the addition of 2 mg/ml creatine phosphate and creatine phosphokinase.

at the instant the relevant steps were applied. This procedure is the same as that adopted by Huxley & Simmons (1971) to generate their T_1 curve. The symbols used in figures 2 and 4 correspond.

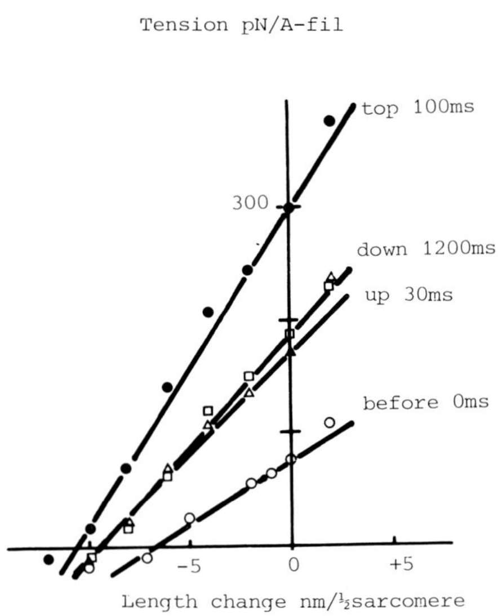

Fig. 4. Analysis of figure 3, as described in the text.

DISCUSSION

Resting Elasticity

The experiments described here indicate that the stiffness of the activated muscle is greater at greater tensions. This is to be expected if the cause of tension generation is cross-bridge activity, the idea being that the tension in the muscle is proportional to the number of attached cross-bridges (Gordon, Huxley & Julian, 1966). However, if no other structure were involved in contributing to the tension of the activated muscle, it would be expected that the stiffness should be directly proportional to the tension. This direct proportionality would show up as a linear relationship between stiffness and tension *passing through the origin* when plotted as in figure 2, and as a series of lines *passing through a common point* on the 'applied length change' axis when plotted as in figure 4. It is obvious from the results presented that this direct proportionality does not occur, although the relationship between tension and stiffness is linear.

By far the most straightforward interpretation of these results is that there is a structure in parallel to the active cross-bridges, but distinct from them, responsible for the stiffness at zero tension. Furthermore the stiffness of the muscle, obtained from the relationship between stiffness and tension extrapolated back to zero tension, is approximately equal to the stiffness of the relaxed muscle. From figure 2 this stiffness is about 116 pN/A-filament/% strain (the spike height on the axis is about 35 pN/A-filament, obtained from a 0.3% applied length change). The stiffness of the muscle in relaxing solution was 103 pN/A-filament/% strain, measured in the same way.

The conclusion from these experiments therefore is that the stiffness seen in relaxed insect flight muscle is caused by a structure in parallel to the cross-bridges, and not by the cross-bridges acting in a different way in relaxed muscle. It should be noted that in vertebrate striated muscle Hill (1968) attributed the extremely small viscoelasticity observed in the relaxed vertebrate muscle to activity of the cross-bridges. Ford, Huxley & Simmons (1976) showed that this viscoelasticity was not present in the active muscle, providing support for there being a different type of cross-bridge interaction in relaxed and active frog muscle. Our results indicate that for the insect muscle, in which the stiffness of the relaxed muscle is very much greater, the interpretation is different.

Where is the structure in parallel with the cross-bridges? One possibility is that there are two types of cross-bridge: 'relaxed' types and 'activatable' types. Note that if this is the case our results indicate that the 'relaxed' type cannot be converted to the 'activatable' type, unless an equal number are converted to the 'relaxed' type; the 'relaxed' type must (within the time scale of the experiments) be permanently maintained.

A more attractive possibility, of course, is that the C-filament discussed in the Introduction is the structure responsible for the elasticity. Our data in this paper, together with the structural data and the biochemical data identifying the probable protein of the C-filaments (Bullard & Winkleman, this volume) provide strong evidence that the C-filaments are the functional cause of the resting stiffness. (Well, since the time that my thesis came to this tentative conclusion (White, 1967), I've waited ten years before making this statement. It's a great relief to be making it now, and I look forward to someone proving it to be wrong. D.C.S.W.)

Mechanical Properties of the Cross-Bridge Cycle

The important experimental results reported here are (a) that the stiffness of the muscle increases linearly as the tension increases, and (b) that the tension on the rising phase of the delayed tension response to a step increase in length in the absence of phosphate ions is the same as that on the falling phase when measured at the same tension.

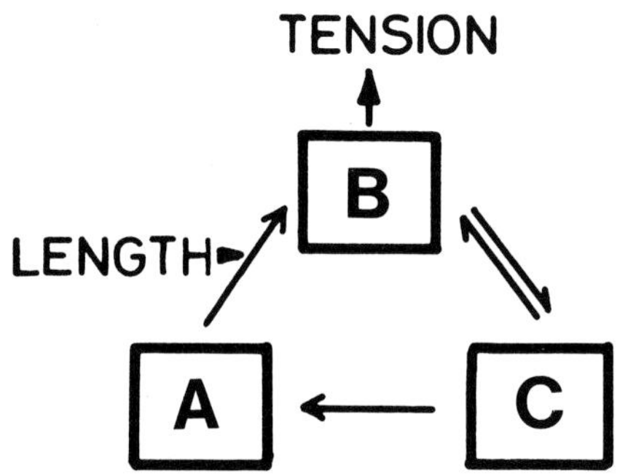

Fig. 5. 3-state cycle from White (1972). Tension is generated in state B. The effect of P_i is to speed up the equilibrium between states B and C.

We do not propose to discuss the details of the 3-state model in great detail. The form of the cycle is shown in figure 5. Details may be found in White (1972). The predicted form of the relationship between stiffness and tension during the response to a step increase in length for the non-P_i response, using the parameters of figure 11 of White (1972), is shown in figure 6. Three curves are shown, corresponding to the three amplitudes of stretch of figure 11 from the earlier paper. The arrow indicates the direction of time. To obtain this curve the stiffness of the muscle is assumed to be proportional to the total number of bridges in the attached tension-generating state. There is an immediate increase in tension, with no corresponding increase in stiffness, during the stretch (A to B). Immediately after the stretch the attached bridges will all have been distorted by the stretch, and so they exert a greater tension than before. These distorted bridges now detach, with a rate constant dependent upon the degree of distortion. The rate constants used for these predictions, and which give suitable fits to the tension responses of the experimental data, are such that all these distorted bridges detach during the delayed rising phase of tension. The detached bridges can attach, but it is assumed that the tension they exert once they have done so is not influenced by the stretch.

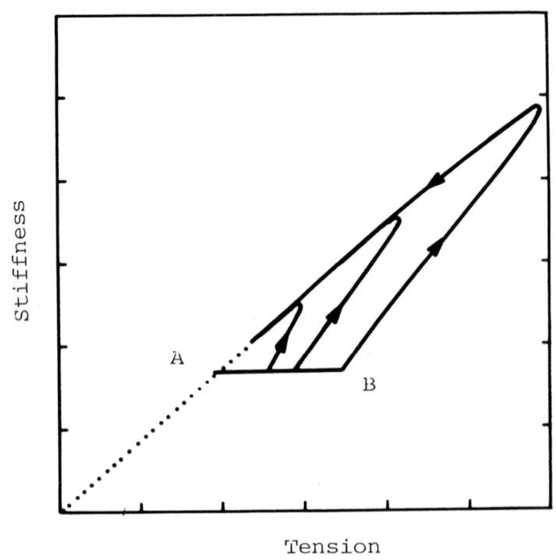

Fig. 6. Stiffness vs. tension curves predicted from the 3-state model of figure 5, using values of the parameters applicable to figure 11 of White (1972), which were chosen to give the best fit to the mechanical data. The three curves correspond to the responses of three differing magnitudes of length step. Further details in the text.

The stiffness is less on the rising phase of the response than on the falling phase because fewer bridges are required to generate a given tension on the rising phase (on the rising phase some of the tension originates from bridges which have been distorted by the stretch and so exert greater tension, but, it is assumed, not greater stiffness).

The experimental results (figure 2) do not agree with the predicted results, even if it assumed that the difference between the extrapolated intersect on the stiffness axis is accounted for in terms of the resting elasticity, in the sense that the experimental curve is closed, whereas the predicted curve is open. The difference is far too great to be accounted for in terms of experimental error.

The discrepancy can be accounted for in two ways. The first is if the stiffness of the bridges is nonlinear, so that the greater the bridge distortion the greater the stiffness. The second is in terms of the type of model for force generation proposed by Huxley and Simmons (1971). In their model they suggested that the attached cross-bridges could exist in a number of attached states, each exerting a greater tension than the preceeding one in the cycle. The energy stored in the cross-bridge during a stretch influenced the rate constants in such a way as to change the equilibrium towards the lower tension generating states (it thus provides an explanation of the 'spikes' observed experimentally immediately following the stretch in terms of the relaxation towards the new equilibrium). The effect of this on a model in which the tension generating state is a conglomeration of such states, but which in other respects is the same as the 3-state cycle of White (1972), is to tend to close the loop observed between stiffness and tension as illustrated in Fig. 6.

Further comments

1. Herzig (this volume) and Schädler, Steiger and Rüegg (1971) provide evidence that if the initial stretch given to the fibres is larger than the 1% used here, then the stiffness does not remain proportional to the tension, but reaches a plâteau value early during the delayed rise of tension. At low degrees of stretch Herzig reports a finding similar to that reported here. These results are not in conflict in that the form of the response appears from their work to change with the amplitude of the applied stretch. None of their results at higher degrees of stretch invalidates the conclusions reached in this paper.

2. Abbott (1972 & this volume) discusses a number of points not covered in this paper to show that the 2-state cycle provides an inadequate explanation of the muscle in terms of the response to small-amplitude length changes at different degrees of stretch and at different calcium ion concentrations. He proposes (this volume) a 4-state cycle, with two tension-generating states, to account for these responses. The reader is advised to be wary of trying to compare, say, the 3-state model of this paper and the 4-state cycle of Abbott's. At the present time they each account for only a (different) part of the total available experimental data.

REFERENCES

Abbott, R.H. (1972) *37th Cold Spring Harbor Symposium of Quantitative Biology* 647

Ashhurst, D.E. (1971) *J. molec. Biol.*, 55, 283

Auber, J. and Couteaux, R. (1963) *J. de Microscopie*, 2, 309

Ford, L.E., Huxley, A.F. and Simmons, R.M. (1976) *J. Physiol.*, 260, 29P

Gordon, A.M., Huxley, A.F. and Julian, F.J. (1966) *J. Physiol.*, 184, 170.

Hill, D.K. (1968) *J. Physiol.*, 19, 637

Huxley, A.F. (1974) *J. Physiol.*, 243, 1

Huxley, A.F. and Simmons, R.M. (1971) *Nature*, 233, 533

Jewell, B.R. and Rüegg, C. (1966) *Proc. Roy. Soc. B.*, 164, 428

Machin, K.E. and Pringle, J.W.S. (1960) *Proc. Roy. Soc. B.*, 152, 311

Rüegg, C. and Tregear, R.T. (1966) *Proc. Roy. Soc. B.*, 165, 497

Schädler, M., Steiger, G.J. and Rüegg, C. (1971) *Pflügers Archiv.*, 330, 217

Steiger, G.J. and Rüegg, C. (1969) *Pflügers Archiv.*, 307, 1

Thorson J. and White D.C.S. (1969) *Biophys. J.*, 9, 360.

White D.C.S., (1967) D. Phil. Thesis, Oxford.

White D.C.S. (1972) *37th Cold Spring Harbor Symp. Quant. Biol.*, 201.

White, D.C.S. and Donaldson, M.M.K. (1975) *Ciba Foundation Symposium*, 31, 341

White, D.C.S. and Thorson, J. (1972) *J. gen. Physiol.*, 60, 307

White, D.C.S. and Thorson, J. (1973) *Progr. Biophys. molec. Biol.*, 27, 173
Reprinted as a monograph in Pergamon Series in the Life Sciences.

Acknowledgement

This work was supported by the Science Research Council.

A MODEL OF STRETCH ACTIVATION BASED ON STIFFNESS MEASUREMENTS IN GLYCEROL EXTRACTED INSECT FIBRILLAR FLIGHT MUSCLE*

J.W. Herzig

Dept. Physiol. II, University of Heidelberg, Im Neuenheimer Feld 326, D-6900 Heidelberg, FRG.

* Supported by the Deutsche Forschungsgemeinschaft, Grant RU 154/9-12.

INTRODUCTION

Jewell and Rüegg (1966) showed that the myogenic oscillation of insect fibrillar flight muscle (Pringle, 1967) is independent of the functions of the sarcolemma and the SR system. Contractile structures, functionally isolated by glycerol extraction, can be activated by addition of Mg-ATP and Ca^{++}. Quick stretch of a Ca^{++}- activated fibre causes, after a rapid viscoelastic tension peak, a delayed transient development of tension (stretch activation). Schädler *et al* proposed in 1971 that the delayed tension development may be due to a synchronization of the motion of attached cross-bridges. On the other hand, White and Thorson (1973) found that the delayed tension rise may be due to recruitment of attached cross-bridges rather than synchronization.

In this paper, new stiffness measurements during Ca^{++}- activated contraction and stretch activation of glycerol extracted insect fibrillar flight muscle are described. As immediate stiffness is assumed to be a measure of the number of cross-bridges attached to actin at any one moment (Huxley and Simmons, 1971) the results suggest that Ca^{++}ions act by recruiting "activated" cross-bridges. Furthermore, evidence is presented that conformational changes of already attached cross-bridges may play an important role during the quick phase following quick stretch (cf. Huxley and Simmons, 1971). Under certain conditions, the delayed tension development may be due to a synchronous "power stroke" of attached cross-bridges. This problem has previously been briefly discussed (Herzig, 1976, Herzig, *et al.*, 1976).

MATERIALS AND METHODS

Dorso-longitudinal muscles were dissected from the water bugs *Lethocerus maximus* and *L. annulipes*. The muscles were then extracted for 2 x 24 h at $5^{\circ}C$ in extraction solution containing 50% glycerol, 50% water (v/v), 20 mM histidine, 10 mM NaN_3, pH 7.0. After that the preparations were transferred into fresh extraction solution and stored in the deep-freeze (cf. Jewell and Rüegg, 1966).

Bundles of about 10 glycerol extracted fibres, approximately 4 mm in length, were mounted with a fast setting glue between a feedback controlled vibrator (Ling Dynamics 101) and an RCA 5734 force tranducer. Rectangular changes in fibre length up to 200 μm stretch or release could be performed with a rise time of 1 msec. The force transducer had a resonance frequency of approximately 2.3 kHz.

At the beginning of an experiment, the glycerol was washed out of the fibre bundle with the washing solution. This consists of 20 mM KCl, 20 mM imidazole, 5 mM NaN_3, 15 mM $MgCl_2$ and 4 mM EGTA, pH 6.7. After washing the fibre bundle was relaxed in the relaxation solution and then contracted with contraction solution. The relaxation solution consists of 15 mM ATP, 15 mM $MgCl_2$, 20 mM imidazole, 4 mM EGTA and 10 mM NaN_3, pH 6.7, resulting in a free Ca^{++} concentration of $< 10^{-8}$M. The contraction solution is identical to the relaxing solution with the exception of 4 mM Ca-EGTA instead of EGTA, pH 6.7, resulting in a free Ca^{++} concentration of 2×10^{-5}M. The Ca^{++} concentration was varied by mixing relaxation and contraction solution.

In the state of relaxation and at various levels of Ca^{++}- activated contraction quick stretch and quick release experiments were performed at a temperature of 23°C.

The immediate elastic responses (ΔT) obtained in phase with the rectangular length change (ΔL) were related to stiffness according to the equation $S = \frac{\Delta T}{\Delta L} \times L_i$, where L_i is the initial fibre length before the length change. The elastic responses observed in relaxation were assumed to be due to passive parallel elasticity. Thus in order to determine purely series elastic properties of the contractile system the elastic responses obtained in relaxation were subtracted from those measured during Ca^{++}- activation and stretch activation.

RESULTS

Glycerol-extracted fibre bundles from *Lethocerus sp.* DLM were isometrically contracted by Ca^{++} ions in the presence of Mg-ATP (contraction solution). At different Ca^{++} concentrations and in the absence of Ca^{++} (relaxation) the fibres were then rapidly stretched or released rectangularly within 1 msec (Fig. 1). The immediate elastic responses occurring in phase with the length change were, after subtraction of the corresponding elastic responses obtained in relaxation (due to passive parallel elasticity), plotted against the length change (Fig. 2). This results in T_1 curves (Huxley and Simmons, 1971) which can be described as straight lines which deviate from the linear function in the range of releases

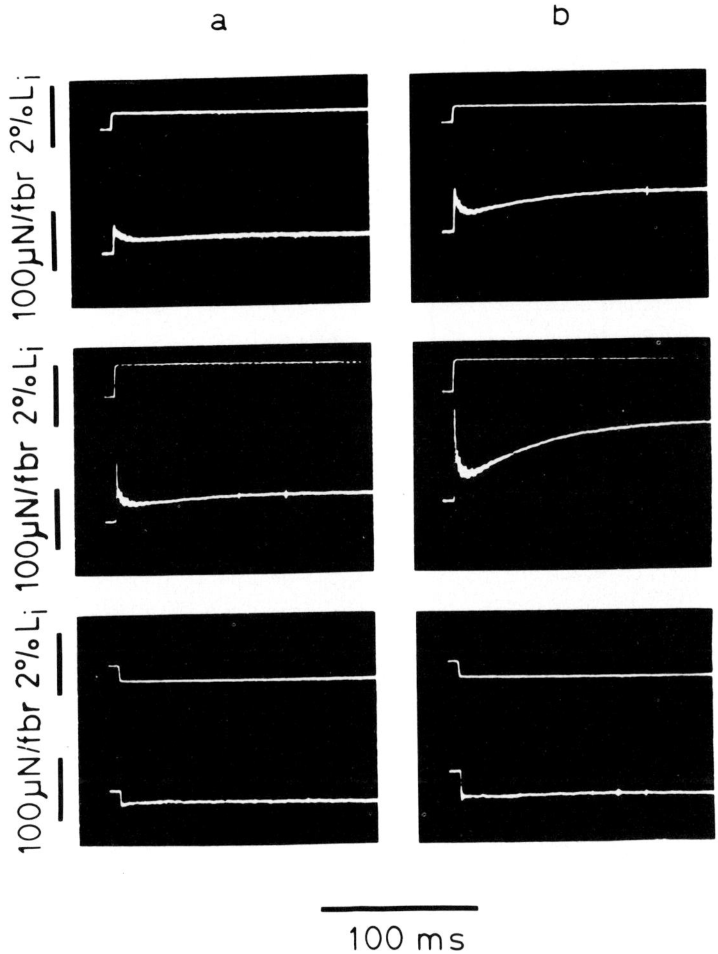

Fig. 1. Measurement of immediate stiffness by quick stretch and release at different Ca^{++} concentrations.
Glycerol extracted fibre bundle from *Lethocerus maximus* DLM in Mg-ATP salt solution, pH 6.7, 23°C.
a: 3×10^{-6} M Ca^{++}; b: 2.15×10^{-5} M Ca^{++};

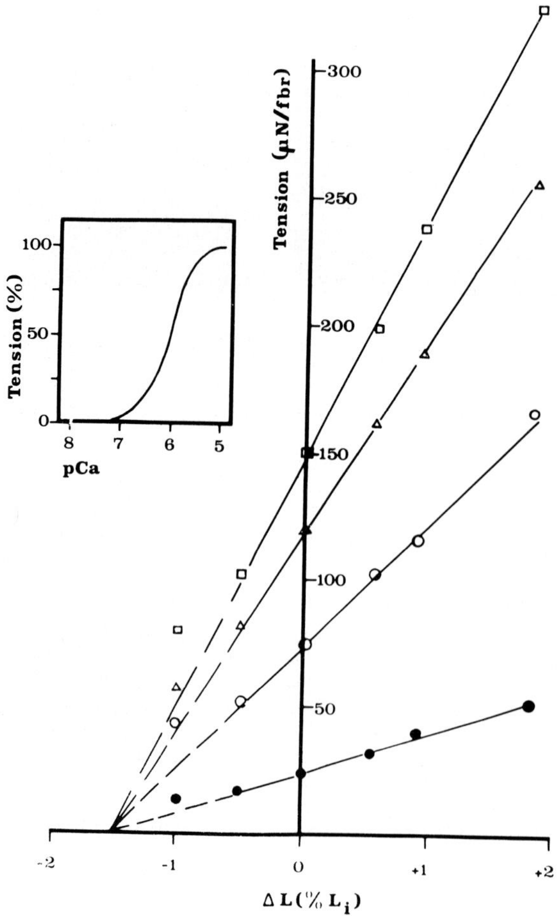

Fig. 2. T_1 curves obtained at various Ca^{++} concentrations.
Glycerol extracted fibre bundle from *Lethocerus maximus* DLM in Mg-ATP salt solution, pH 6.7, 23°C.

● : 3.4×10^{-7} M Ca^{++} Δ : 3×10^{-6} M Ca^{++}
○ : 1.2×10^{-6} M Ca^{++} □ : 2.15×10^{-5} M Ca^{++}

Insert shows Ca^{++} dependence of isometric tension.
T_i curves are corrected for passive parallel elasticity (*c.f.* methods).

greater than 0.6% L_i. With increasing Ca^{++} concentration the slope of the T_1 curve increases. The extrapolations of the linear portions of the T_1 curves obtained at various Ca^{++} concentrations reach the abscissa at a constant intercept of 1.5% L_i.

The slopes of the T_1 curves were used to calculate stiffness as a measure of the number of cross-bridges attached to actin at any one moment (cf. Huxley and Simmons, 1971, 1973). Plotting stiffness against Ca^{++}- activated isometric tension (Fig. 2, inset) results in a linear relationship (Fig. 3).

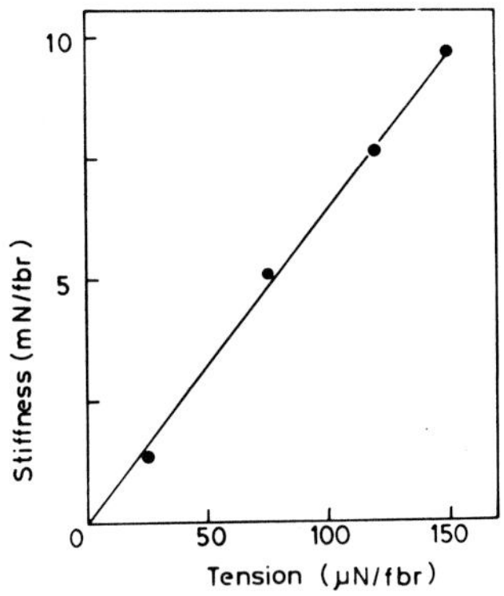

Fig. 3. Fibre stiffness as a function of Ca^{++}- activated isometric tension. Glycerol extracted fibre bundle from *Lethocerus maximus* DLM in Mg-ATP salt solution, pH 6.7, 23°C.
Tension was varied by variation of Ca^{++} concentration.
Stiffness was calculated from the slopes of the T_1 curves shown in Fig.2.

As shown in Figure 1 the immediate elastic tension increment in phase with a quick stretch is followed by a rapid tension drop (quick phase) and a subsequent quasi-exponential tension rise (stretch activation, delayed tension development). After a maximum tension declines slightly to a constant value. During the quick phase and the delayed tension development stiffness measurements as described above were performed by quick stretch and release (Fig. 4). After correction for parallel elasticity both isometric tension and stiffness were plotted against time (Fig. 5). It is shown that during the quick phase stiffness remains constant at the same value as under steady state isometric conditions (cf. Fig. 2). In some cases stiffness increases slightly during the quick phase (Fig. 6).

The time course of stiffness during the delayed tension rise depends on the amplitude of the conditioning step, i.e. the stretch inducing stretch activation. Figure 5 shows two examples of stretch activation, one evoked by a conditioning stretch of 0.5% L_i, the other by 1.5% L_i. In the first case, the delayed tension rise is paralleled by an increase in stiffness, both reaching a maximum approximately 60 ms after the conditioning step. If stretch activation is induced by a conditioning stretch of greater than 1% L_i, stiffness reaches a constant plateau value early after the onset of the delayed tension development. Thus a great portion of the stretch induced activation occurs without a corresponding increase in stiffness.

In Figure 6 T_1 curves obtained during quick phase and stretch activation (insets) in response to a conditioning stretch of 3% L_i are shown. The T_1 curves

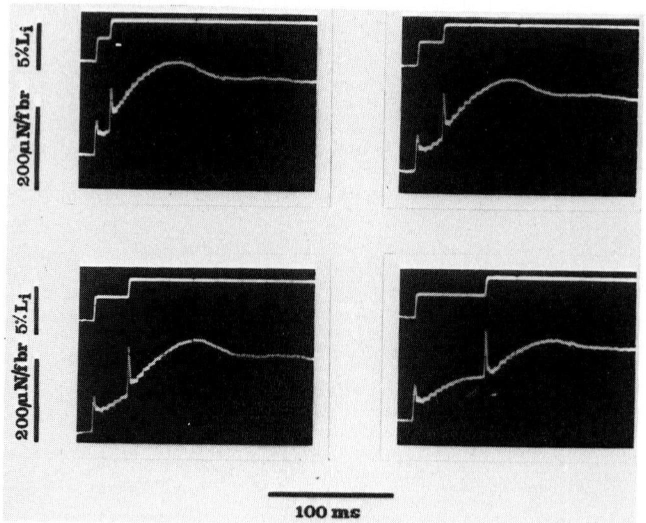

Fig. 4. Stiffness measurements at various moments during stretch activation (Double step experiment).
Glycerol extracted fibre bundle from *Lethocerus maximus* DLM in Mg-ATP salt solution, pH 6.7, 23°C.
3×10^{-6} M Ca^{++}
Amplitude of the conditioning stretch (cf. text): 3% L_i.
Note that the immediate elastic response in phase with the test stretch remains constant during the delayed tension development.

are corrected for passive parallel elasticity. As pointed out before, the slope of a T_1 curve is proportional to stiffness. Figure 6 A shows that stiffness is constant during the conditioning stretch. During the quick phase tension falls but stiffness increases. Consequently, with decreasing tension during the quick phase, the intercept at which the extrapolated T_1 curves reach the abscissa is diminished. During the subsequent tension rise (Fig. 6 B) stiffness remains constant, i.e. the T_1 curve is shifted in parallel to the left, so that the abscissa intercept increases with increasing delayed tension.

Quick length changes during the tension transients following quick release showed that the quick release recovery is always paralleled by a corresponding increase in stiffness.

DISCUSSION

The series elasticity of muscle is assumed to be linear and to be located within the cross-bridges. In consequence immediate stiffness is a measure of the number of cross-bridges attached to actin at any one moment (Huxley and Simmons,

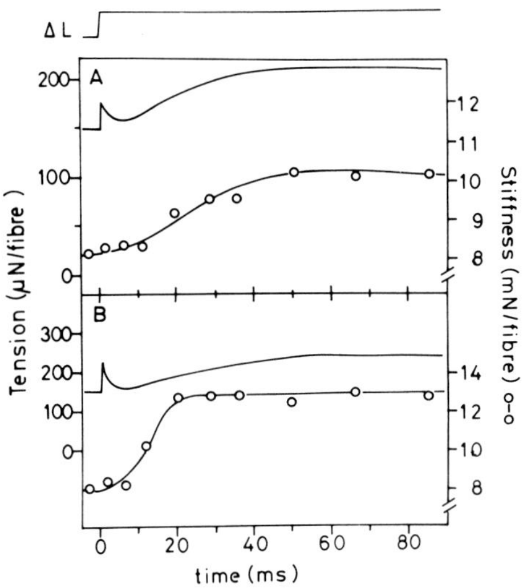

Fig. 5 Time course of tension and stiffness during stretch activation. Glycerol extracted fibre bundle from *Lethocerus maximus* DLM in Mg-ATP salt solution, pH 6.7, 23°C, 2.15 x 10^{-5} M Ca^{++}
The tension transients (solid lines in A and B) in response to a quick stretch ΔL (schematically shown in the uppest trace) and the stiffness (o — o) are corrected for passive parallel elasticity.
Amplitude of the conditioning stretch:
A = 0.5% L_i B = 1.5% L_i
Note that during stretch activation induced by a stretch with low amplitude (A) the delayed tension rise is paralleled by an increase in stiffness, while in response to a large conditioning stretch (B) stiffness reaches a plateau immediately after the onset of the delayed tension development. The tension transients are achieved by retracing the differences between the responses obtained in contraction and relaxation.

1971). The results may therefore be interpreted as follows:

1. The length-tension diagrams (T_1 curves) obtained during Ca^{++}- activated contraction and during the tension transients following quick rectangular changes in length can, after correction for passive parallel elasticity, be described as straight lines. It is probable that the deviation from the linear function in the range of releases greater than 0.6% L_i is due to truncation (Huxley and Simmons, 1971). Thus, the cross-bridges of insect flight muscle seem to behave like linear springs.

2. During Ca^{++}- activated contraction tension and stiffness increase in proportion (cf. Fig. 3). The same phenomenon has been described for glycerol extracted preparations of heart muscle (Herzig and Herzig, 1974, Herzig, 1975) and skeletal muscle (Dickhaus *et al.*, 1974, Rüegg *et al.*, 1975). This might mean that Ca^{++} ions act by recruitment of attached cross-bridges while the force per cross-bridge remains constant. This explanation is supported by the obser-

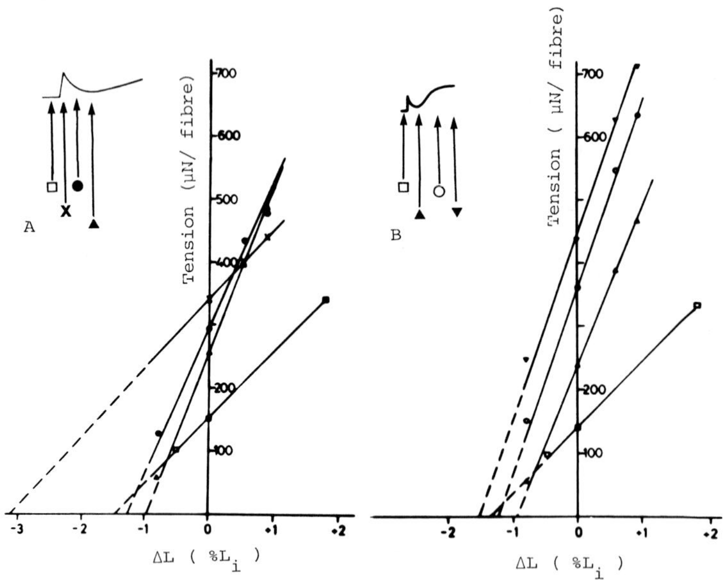

Fig. 6 T_1 curves obtained during the quick phase and the delayed tension development.
Glycerol extracted fibre bundle from *Lethocerus maximus* DLM in Mg-ATP salt solution, pH 6.7, 23°C, 2.15 x 10^{-5} M Ca^{++}.
Stretch activation was induced by a quick stretch of 3% L_i. The T_1 curves were obtained by quick length changes during the quick phase (A, schematically shown in the inset) and the delayed tension development (B, schematically shown in the inset).
The T_1 curves are corrected for passive parallel elasticity (cf. Methods) Note that stiffness (which is proportional to the slope of the T_1 curve) remains constant during the conditioning stretch (□→x in A) and then, during the tension drop in the quick phase, increases while the abscissa intercept is diminished.
In B it is shown that during the delayed tension development (▲ → ▼ in B) stiffness remains constant while the abscissa intercept is increased (parallel left shift of the T_1 curve).

vation of Schädler (1967) that during Ca^{++}- activation tension and actomyosin ATPase activity increase in proportion. On the other hand, the proportionality between tension and ATPase is in contrast to the fact that cross-bridge kinetics are influenced by Ca^{++} ions (Abbott, 1973, Herzig and Herzig, 1974, Rüegg *et al.*, 1975).

In terms of a two state model of cross-bridge action the proportionality between tension and stiffness means that all the attached cross-bridges are necessarily in a force generating state (cf. Podolsky and Nolan, 1973). Huxley and Simmons (1971, 1973) proposed a three or more state model (Fig. 7): During its cyclic motion the cross-bridge attaches to actin, perhaps rectangularly, (cf. Beinbrech *et al.*, 1976; Goody *et al.*, 1975) in a non-force generating state

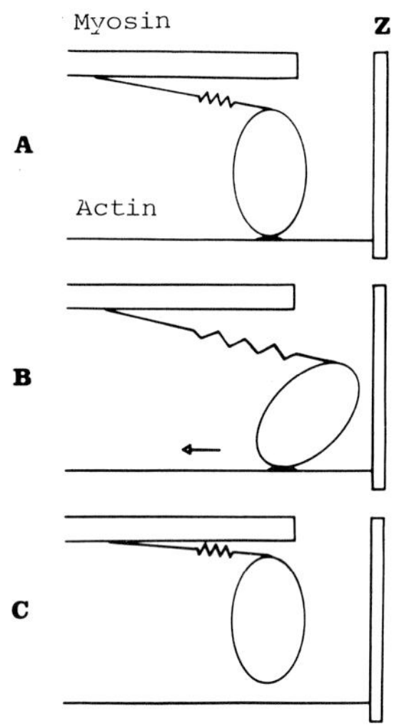

Fig. 7 Three state model of cross-bridge action
A: The cross-bridge attaches rectangularly to the actin filament. The series elastic component (schematically shown in the "neck" of the cross-bridge) is not yet distorted.
B: The attached cross-bridge performs a "rowing" motion thereby distorting its elastic component. The mechanical potential energy stored in the elastic component appears as force development in the direction of the arrow.
It is assumed that under certain conditions the cross-bridge can rotate "backwards" from position B to A thereby discharging the elastic component (cf. text).
C: The cross-bridge is detached. The elastic component is discharged.
cf. also Huxley and Simmons (1971, 1973); Julian *et al.* (1973).

(Fig. 7 A). The cross-bridge then performs a "power stroke" by bending to an acute-angled position such as is typical for the rigor state (Reedy *et al.*, 1965; Miller and Tregear, 1972; Goody *et al.*, 1975; Beinbrech *et al.*, 1976) thereby stretching its elastic component (Fig. 7 B). The mechanical potential energy stored in the distorted elastic element then induces sliding of filaments if the fibre is allowed to shorten, while under isometric conditions it gives rise to isometric force. Subsequent detachment of the cross-bridge starts a new cycle. According to this three state model (Huxley and Simmons, 1971; Julian *et al.*, 1973) the total number of attached bridges has to be assumed to increase during Ca^{++}- activated contraction but at the same time a definite constant steady state ratio must be maintained between force generating and non-force generating attached cross-bridges (cf. Fig. 7).

3. During the quick phase following quick stretch of a Ca^{++}- activated *Lethocerus* DLM fibre tension falls but stiffness increases slightly (Fig. 6 A) or

remains constant (Fig. 5). This finding is in contrast to the observation that in glycerol extracted heart muscle stiffness increases during stretch and decreases during the tension drop in the quick phase (Herzig, 1975). On the other hand, rabbit psoas (unpublished experiments) behaves like insect flight muscle. The constant or increasing stiffness during the quick phase may be explained in terms of a three state model (cf. Fig. 7). In the isometric case a steady state population of cross-bridges consisting of force generating and non-force generating but attached bridges is attached to actin at any one moment. Quick stretch increases the probability of attachment in the non-force generating state, (Fig. 7 A) i.e. force generating cross-bridges detach or "rotate back" into a position where they are discharged and/or new cross-bridges attach without generating force, leading to a tension fall during the quick phase associated with constant or rising stiffness. This is also suggested by the observation that the intercept at which the T_1 curve reaches the abscissa (Fig. 6 A) decreases during the quick phase, indicating that the mechanical potential energy stored in the contractile system is diminished during the quick phase.

4. The time course of stiffness during the delayed rise of tension following the quick phase depends on the amplitude of the stretch inducing stretch activation. In response to quick stretch of less than 1% L_i a delayed tension development is observed which is always associated with a parallel increase in stiffness. If stretch activation is induced by a greater stretch stiffness always reaches a constant plateau immediately after the onset of the delayed tension development. This amplitude dependence may explain the difference in the results obtained by Schädler *et al.* (1971) and White and Thorson (1973). Thus, in the case of stretch activation in response to a small stretch immediately after the quick phase a new equilibrium population of force generating and non-force generating attached cross-bridges may be established while the total amount of bridges attached to actin at any one moment increases in proportion to the delayed tension development. On the other hand, in the case of stretch activation in response to a large stretch immediately after the quick phase, the probability of attachment in the force producing state seems to increase. Thus, the cross-bridges which attached to actin in the non-force producing state (Fig. 7 A) during the quick phase now perform their power stroke, thereby producing the delayed tension development by addition of their forces (Fig. 7 B). This is also demonstrated by the fact that during the delayed tension rise the T_1 curve is shifted in parallel to the left, indicating that it is not the number of attached bridges but the mechanical potential energy stored in the contractile system which is increased.

In rabbit psoas muscle very similar results were obtained (unpublished experiments) whereas in heart muscle this "synchronization" of cross-bridges seems to be absent (Herzig, 1975).

5. The tension recovery following quick release is always associated with a corresponding change in stiffness. This may mean that in response to quick release, detachment and attachment processes seem to be significant rather than conformational changes of attached cross-bridges.

The results presented in this paper are entirely consistent with the three-state model proposed by Huxley and Simmons (1971). Moreover, there is new evidence that at least two mechanically distinguishable states of attached cross-bridges are present during contraction, both attached to actin at different mechanical potential energy (cf. also Beinbrech *et al.*, 1976; Julian *et al.*, 1973).

REFERENCES

Abbott R.H. (1973) *Cold Spring Harb. Symp. quant. Biol.*, 37, 647.

Beinbrech G., Kuhn H.J., Herzig J.W. and Rüegg J.C. (1976) *Cytobiologie*, 12, 385.

Dickhaus H., Herzig J.W., Kuhn H.J. and Rüegg J.C. (1974) *Experientia*, 30, 678.

Goody R.S., Holmes K.C., Mannherz H.G., Barrington-Leigh J. and Rosenbaum G. *Biophys. J.*, 15, 687.

Herzig J.W. (1973) *Pflügers Arch.* 343, R 23.

Herzig J.W. and Herzig U.B. (1974) *Symp. Biol. Hung.* Ed: E.N.A. Biró and N. Garamvölgyi, Akadémiai Kiadó, Budapest, 17, 85.

Herzig J.W. (1975) *Pflügers Arch.*, 355, R 13.

Herzig J.W. (1976) *Pflügers Arch. Europ. J. Physiol.*, 362, R 25.

Herzig J.W., Güth K., Kuhn H.J. and Rüegg J.C. (1976) *Verh. Dtsch. Zool. Ges.* 240.

Huxley A.F. and Simmons R.M. (1971) *Nature*, 233, 533.

Huxley A.F. and Simmons R.M. (1973) *Cold Spring Harb. Symp. quant. Biol.* 37, 669.

Jewell B. R. and Rüegg J.C. (1966) *Proc. Roy. Soc.*, B164, 428.

Julian F.J., Sollins K.R. and Sollins M.R. (1973) *Cold Spring Harb. Symp. quant. Biol.*, 37, 685.

Miller A. and Tregear R.T. (1972) *J. Mol. Biol.*, 70, 85.

Podolsky R.J. and Nolan A.C. (1973) *Cold Spring Harb. Symp. quant. Biol.*, 37, 661.

Pringle J.W.S. (1967) *Prog. Biophys. molec. Biol.*, 17, 1.

Reedy M.K. Holmes K.C. and Tregear R.T. (1965) *Nature (Lond.)*, 207, 1276.

Rüegg J.C., Kuhn H.J., Herzig J.W. and Dickhaus H. (1975) *In: Calcium transport in contraction and secretion*, p.547, E. Carafoli (Ed. *et al*). North-Holland Publishing Company.

Schädler M., Steiger G.J. and Rüegg J.C. (1971) *Pflügers Arch.*, 330, 217.

White D.C.S. and Thorson J., (1973) *Prog. Biophys. molec. Biol.*, 27, 173.

STRETCH ACTIVATION AND TENSION TRANSIENTS IN CARDIAC, SKELETAL AND INSECT FLIGHT MUSCLE

Gerhard J. Steiger

The Department of Physiology and the American Heart Association Cardiovascular Research Laboratory, University of California at Los Angeles, Center for the Health Sciences, Los Angeles, California 90024.

INTRODUCTION

At present there is considerable evidence that stretch activation is a common property of muscle tissue and is not confined to insect fibrillar muscle. Auto-oscillations in extracted rabbit psoas fibres on an auxotonic lever system with a frequency of 2.8 Hz at room temperature have been observed by Goodall (1956) and by Lorand and Moos (1956). Oscillatory responses and delayed tension changes have also been described in living frog skeletal muscle fibres by Armstrong, Huxley and Julian (1966).

Other muscles in which delayed tension changes have been observed include cicada tymbal muscle (Aidley and White, 1969), frog sartorius (Heinl, 1972), tortoise ileofibularis muscle (Heinl, Kuhn and Rüegg, 1974), frog anterior tibial muscle (Julian and Sollins, 1975), intact and skinned crayfish fibres (Kawai and Brandt, 1975; Kawai and Orentlicher, 1976) and many types of cardiac muscles of different vertebrates (Heinl, Herzig, Sawaya and Steiger, 1974). In rabbit psoas (Rüegg, Steiger and Schädler, 1970; Abbott and Steiger, 1977) and rabbit papillary muscles (Steiger, 1971; Steiger, Brady and Tan, 1977) the mechanism of stretch activation has been studied in detail. Some of these details are presented in this paper, their mechanisms are discussed and related to similar experiments in insect flight muscle.

METHODS

A. Living Cardiac Muscle

Preparation of Muscle

Papillary muscles, trabeculae or strips of ventricular wall were taken from the right ventricle of rabbits (male, New Zealand white rabbit, 1-2 kg.), cats or rats (Sprague-Dawley). The animals were heparinized and anesthetized with sodium pentobarbital. The heart was rapidly excised through a sternotomy and placed in oxygenated perfusate at 30-37°C. The right ventricle was carefully opened and a suitable papillary or trabecular muscle was selected. For some experiments, frog trabeculae (*Rana catesbiana*) or strips of the ventricular wall (*Rana pipiens)* were used. The frogs were decapitated and spinalized and the heart rapidly excised .

With the aid of a dissecting microscope and opthalmological scissors the chosen preparation was carefully tied with monofilament silk thread, dissected, transferred into the muscle chamber on the mechanical apparatus and horizontally mounted between two stainless steel hooks. Physiological salt solution from an oxygenated reservoir entered the chamber at one end at the bottom and left the chamber with minimum agitation of the surface by overflowing through a wick into another chamber from which it was aspirated. The chamber perfusate contained 2.4-2.7 Vol% O_2. pH was 7.2-7.4 at 25°C.

Solutions

The solutions consisted of the following components in mM:

Standard Tyrode Solution: NaCl = 130; $CaCl_2 \cdot 2H_2O$ = 5; KCl = 4; $MgCl_2 \cdot 6H_2O$ = 1; $NaH_2PO_4 \cdot H_2O$ = 0.435; $NaHCO_3$ = 10; Dextrose = 5.56; the pH was adjusted to 7.4.

Standard Ringer Solution: NaCl = 106; $CaCl_2 \cdot H_2O$ = 5; KCl = 2.7; $MgCl_2 \cdot H_2O$ = 0.49; $NaH_2PO_4 \cdot H_2O$ = 0.32; $NaHCO_3$= 10; Dextrose = 5.56; the pH was adjusted to 7.2.

The standard solutions were modified in the following ways:

(i) Ca-free solution: The Ca-free solutions contained the same components as the standard solutions except that $CaCl_2$ was omitted.

(ii) Ba-contracture solution: $CaCl_2$ was replaced by 0-1.0 mM $BaCl_2$.

(iii) High K - low Na contracture solution: Na^+ was partially replaced by K^+ and $choline^+$, and 5-10 mM caffeine was added. If not otherwise specified the high K^+ standard solution (HPS-solution) contained 70 mM KCl (instead of 4); 65 mM NaCl (instead of 130); 5 mM caffeine.

(iv) Rigor solution: Standard solutions with 0.4-1.0 mM Iodoacetic acid (IAA) added.

Mechanical Apparatus

The experimental apparatus consisted of a servo-controlled vibrator permitting instantaneous or sinusoidal changes in muscle length and a tension transducer to record the muscle force. The apparatus has been described previously (Steiger *et al.*, 1977).

(i) <u>The tension transducer</u>: The transducer used in these experiments had a frequency of 2 kHz, a linear range up to 10 g and a compliance of 3μm/g. The sensitivity was such that 10 mg of tension could easily be resolved.

(ii) <u>Application and measurement of length changes</u>: A Ling-dynamics 101 vibrator (Ling Dynamic Systems) mounted on a micrometer sledge was used to impose step as well as sinusoidal length changes on the muscle fibres. The electronics were capable of producing steady as well as transient length change up to 2 mm. Step and sinusoidal length changes were controlled by a feedback circuit. These electronics enabled application of a sequence of length changes to the muscle with each length held for a selected time duration. The velocity of the length change between the levels could be selected as well. A maximum velocity length change without overshooting the transient length changes was completed in less than 2 msec. Sinewave amplitude was independent of frequency up to 200 Hz. The motion was monitored by another capacitance transducer (Schilling, 1960).

(iii) <u>Recording</u>: For both the step and the sinusoidal experiments, the length and tension outputs were displayed on a Tektronix storage oscilloscope (D 11) and photographed directly with a polaroid camera. A continuous chart record was kept of each experiment (Brush Chart Recorder). The data could also be sampled, averaged and stored on-line on a Linc 8 computer for subsequent analysis.

(iv) <u>Muscle chamber and temperature control</u>: The muscle chamber was built from a special heat conducting epoxy material and mounted on a Peltier unit. The temperature in the chamber (usually between 18-25°C) was sensed by a Constantan-Fe thermocouple. The temperature control of the incubation solution was better than $\pm$ 0.2^{o}C. Stimulation of the muscle was accomplished by passing current through parallel wire electrodes placed on either side of the muscle. The chamber was 10 mm in length, 3 mm in width and 2 mm in depth.

Experimental Procedure

Mammalian tissues were first allowed to equilibrate in warm solution (about 28°C) for about 10 min., and then allowed to equilibrate at room temperature. This procedure shortened the total equilibrating period. Frog heart preparations remained at room temperature. After 1 hour of perfusion, during which time the twitch tension became stabilized and resting tension declined, the length-tension characteristic of each muscle was obtained. For visual examination a stereo microscope was used, but no sarcomere length determinations were made.

Only thin and uniform muscle preparations were selected for experimentation. The length ranged from 3 to 8 mm and no muscle exceeded 0.7 mm in thickness. Muscle length, L_o and thickness Ø, were measured at that muscle length at which the experiment was performed (i.e. 0.8-0.9 L_{max}, where L_{max} = muscle length at which activity developed force was maximal). The muscle length was measured between the two tied ends with a micrometer and the muscle diameter was measured at mid length.

Steady-state contractures were induced by one of the following methods:

(i) Various combinations of KCl, NaCl, choline chloride and caffeine.

(ii) Replacement of Ca^{++} by Ba^{++}.

By means of a 3-way stopcock the three test solutions could be rapidly applied to the muscle fibres. In order to produce a sustained contracture in Ba^{++} the muscles were first soaked in Ca^{++}-free solution to avoid spontaneous beating which resulted in irregular contracture tension. In the test solution, the muscle was held isometrically until a steady-state tension had developed and thereafter sinusoidal or step changes in muscle length were applied.

For rigor conditions the muscles were held isometrically in IAA solution (∿ 0.4-1.0 mM iodoacetic acid added to a standard Tyrode or Ringer solution) for 90-120 min. and stimulated (about 30 beats/min) in this solution until there was no response to the stimuli and the rigor tension was maximal.

Analysis of the tension transients: The experimental curves following step perturbations were fitted by an eigenfunction expansion method. This method has been developed by Provencher (1976) for the analysis of data composed of random noise, an unknown constant baseline and a sum of exponential functions. The computer program provided by Dr. Provencher runs completely automatically without potentially biassed initial estimates.

B. Extracted Muscles *

Dorsal longitudinal muscles (DLM) of the giant water bug, *Lethocerus maximus*, rabbit psoas muscles and cardiac preparations were either directly glycerinated or briefly soaked in an ether-water mixture, washed several times, then immersed in glycerol extraction solution, stirred overnight in the cold-room and finally stored in fresh glycerol solution in the deep freeze (Steiger and Rüegg, 1969; Steiger, 1971).

Solutions

The solutions consisted of the following compounds in mM:

Extracted solution: 50 Vol% glycerol p.a., 50 Vol% H_2O bidest., Imidazole buffer = 20; NaN_3 = 10; the solution was adjusted to pH 6.5.

---

* The experiments with extracted preparations were performed in the Dept. of Cell Physiology, Ruhr-University, Bochum, Germany.

<u>Relaxation solution</u>: ATP = 15-17.5; $MgCl_2$ = 15-17.5; Imidazole buffer = 20; NaN_3 = 10; EGTA = 4; pH = 6.0, 6.5, 7.0.

<u>Activation solution</u>: ATP = 15-17.5; $MgCl_2$ = 15-17.5; Imidazole buffer = 20; NaN_3 = 10; Ca-EGTA and EGTA total 4 mM in an appropriate ratio to give the desired level of free calcium.

<u>Ca^{++} and Mg^{++}- poor solution (X-solution)</u>: ATP =5-12; $MgCl_2$ = 0; Imidazole buffer = 20; NaN_3 = 10; EGTA = 4mM; $CaCl_2$ = 0.

A modified version of the Perrin and Sayce (1967) computer program was used to calculate the concentrations of the free ligands, metals and those of the various complexes. The modification was provided by R.H. Abbott and was written in Algol.

Mechanical apparatus

The mechanical apparatus was the same as that described by Steiger (1971). The fibres were glued between two glass rods, one attached to the vibrator (Ling Dynamic Systems) and the other one to the anode pin of an RCA 5734 transducer valve. The force transducer employed a stabilized constant current power supply for the heating device and high voltage. During the course of the experiment, the signal variation was normally $\pm$ 10 mV/h, and the noise less than 2 mV. The resonance frequency of the transducer system was 800 Hz or better. The compliance of the transducer was 20 μm/g. The rise time of the length signal upon application of square waves was about 3 msec. The length and tension signal were displayed on a double beam oscilloscope (Tektronix 502A) and photographed. Occasionally the tension was continuously registered with a pen-recorder (Schwarzer Varioscript 242).

Temperature regulation

The temperature (usually between 20-25°C) was controlled by means of a Peltier element in combination with a temperature sensitive semi-conductor (Siemens Sirigor-Kühlblock, PKE 36 BO3, Siemens Heissleiter K17K, Gerber Temperaturregelgerät). The temperature regulation of the solution was better then $\pm$ 0.2°C.

Experimental procedure

After the preparation was mounted and the glue dried, the muscle tissue was first washed in relaxation solution in order to remove the remaining glycerol-extraction solution. In fresh relaxation solution the preparation was stretched by means of the micromanipulator until noticeable resting tension was displayed. The fibre was allowed to equilibrate and to develop isometric tension in the test solution for about 15 min before length changes were applied.

RESULTS

Stretch Activation in Extracted Cardiac Preparations

Extracted cardiac preparations suspended in a Mg-ATP-salt solution develop force under isometric conditions if the concentration of free Ca-ions in the bathing solution is in the range 10^{-7} to 10^{-5} M Ca^{++} at pH 6.5. Plotting muscle force versus the free Ca-ion concentration on a semi-logarithmic scale results in an S-shaped curve with a half maximal activation at 1×10^{-6} M (Fig.1. T_o-curve).

Quick step perturbations of muscle length during an isometric contraction yield a set of force transients which are shown in Fig.2. At the instant of lengthening the muscle, the force rises from the isometric tension level T_o to a force maximum T_1. There follows a fast tension drop which ends in a temporary minimum called T_2. For small amplitudes of stretch the tension may well drop below the initial isometric value before it reverses direction. The force rises to a new maximum which is considerably delayed with respect to the completion of the length change. Under the experimental conditions given in Fig.2. this delay is about 0.75 sec. The same phenomenon has first been described in glycerinated insect flight muscles and was there called stretch activation. It remains to be shown to which degree both are homologues to each other. Often some damped oscillations of the isometric tension can be observed at the end of the delayed tension rise. The final equilibrium value is usually somewhat below the delayed tension maximum.

In case of a quick release of muscle length, after the initial drop in force simultaneously with the reduction of length, there is a fast recovery of muscle force, a delayed "de-activation", and a final approach to a new

---

Fig.1. Influence of the free Ca-ion concentration on the initial isometric force T_o (■), elastic increment in tension T_1 (●) and the maximum stretch activated delayed tension T_3 (▲). The relation between T_o and T_1 is linear (inset).
Experimental conditions: Ether-glycerol extracted rabbit papillary muscle. pH 6.5; T = 24°C; 15 mM ATP; 17 mM $MgCl_2$; Myokinase = 0.1 mg/ml; $\Delta L = 2\% L_o$; 10 mg $\simeq$ 15 g/cm^2.

Fig.2. Tension-time course following step changes in length. Top: Step change in length. The duration of the length step was approximately 3 msec. Bottom: Tension-time course accompanying the length perturbation. The isometric tension level is about 150 mg.
Experimental conditions: Ether-glycerol extracted rabbit papillary muscle. pH 6.0; $Ca^{++} = 7.5 \times 10^{-7}$ M; T = 21.5° C; 15 mM Mg ATP; 50 mg $\simeq$ 15 g/cm^2.

1

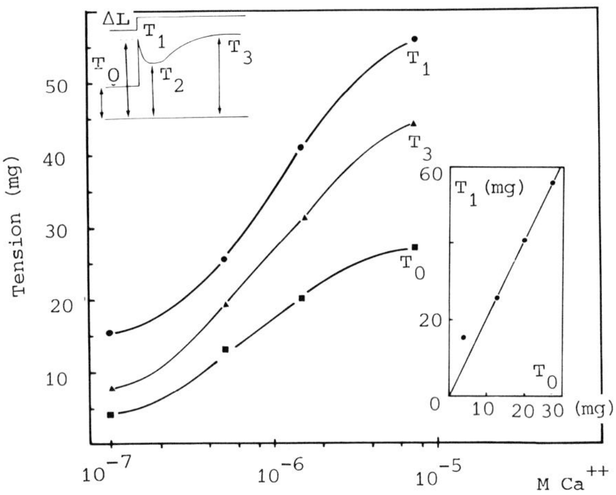
ΔL
T_1
T_3
T_0
T_2
Tension (mg)
50
40
30
20
10
T_1
T_3
T_0
60
T_1 (mg)
20
0
10 20 30 (mg)
10^{-7}
10^{-6}
10^{-5}
M Ca^{++}

2

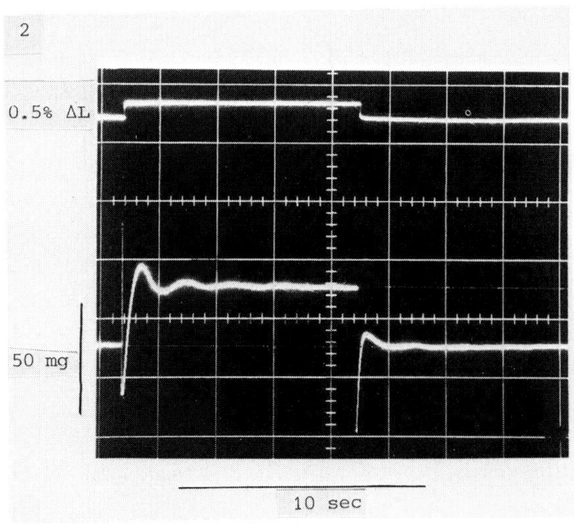
0.5% ΔL
50 mg
10 sec

equilibrium value. In general with decreasing length perturbations the force transients for stretch and release become more mirror-image like. As shown for intact cardiac preparations the early recovery phases, i.e. the fast tension drop after stretch and the fast tension rise after release are composed of at least 2 components and can be fitted with two rate constants. The delayed tension rise for small amplitude perturbations can also be approximated with an exponential function.

T_1, the initial tension maximum or, respectively, minimum reached at the end of the step change in length is the best experimental representation of the instantaneous elasticity of the contractile systems which in skeletal muscle seems to be localized mainly in the crossbridges (Huxley and Simmons, 1972). As Fig.1. demonstrates the isometric tension T_o and the tension T_1, which is a relative measure of muscle stiffness, both increase in cardiac muscle with increasing concentrations of free Ca-ions and both follow the same sigmoid relationship. Indeed, as the inset of Fig.1. shows, there exists a linear relationship between T_o and T_1. In relaxation solution containing 4 mM EGTA, little resting tension and only a small elastic phase can be observed. Delayed tension changes are completely absent. Similar to T_o and T_1, the delayed tension maximum T_3 is a sigmoid function of the log of the free Ca^{++} concentrations.

Fig.3. displays the effect of amplitude of length changes on the T_1 and T_2 tension in rabbit papillary muscle. T_1 is a linear function of the amplitude of stretch and release when the amount of length change is less than 1%. In skeletal muscle one assumes that the point of intersection of T_1 with the abscissa gives an estimate of the maximal motion of a crossbridge stroke and also the maximum release necessary to reduce crossbridge force to zero. Since the cardiac muscle is much more compliant than the skeletal muscle, elastic properties and stiffness changes are not necessarily attributable to crossbridges. Whereas in living preparations a compliance between 5-10% has been reported (Brady, 1971), in glycerinated preparations zero tension can be reached by quick releases of about 1.5% of muscle length which is within a factor of 2 of the same magnitude as in skeletal muscle.

The amplitude dependence of the T_2 tension is more complicated. It is apparent that the T_2 tension, if the muscle is stretched, first decreases below the isometric level and then at higher amplitudes of stretch increases with about the same slope at T_1. The distance between the T_1 and T_2 tension is then about 1.5% ΔL. In the case of release, T_2 decreases continuously.

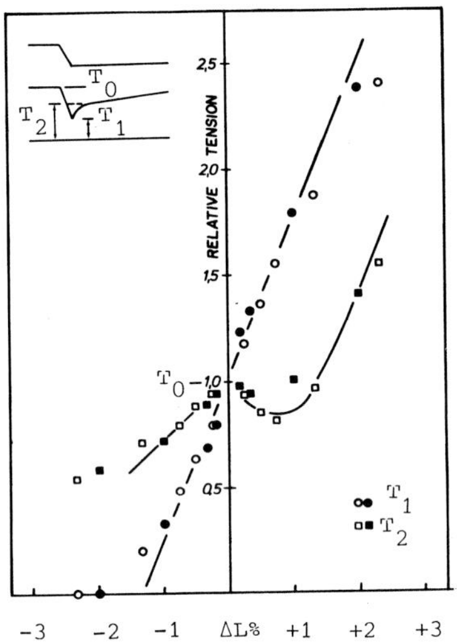

Fig.3. Amplitude dependence of the tension values T_1 and T_2. T_1 describes the extreme tension reached during the step in length; T_2 describes the tension approached at the end of the quick recovery phase after release and the point of tension minimum between the rapid tension drop and the delayed tension rise after stretch; T_o is the initial isometric tension value, as indicated by the inset. The relative tension is plotted on the ordinate, referenced to T_o. The amplitude of stretch and release in % L_o is on the abscissa.
Experimental conditions: Ether-glycerol extracted rabbit papillary muscle: pH 7.0; 2 x 10^{-6} Ca^{2+}; T = 22°C; 15 mM MgATP.

Due to the stretch activation mechanism, activated cardiac muscle, like insect flight muscle, is able to perform work in a certain frequency range when stretched and released sinusoidally. Fig.4. shows the frequency dependence during forced sinusoidal oscillations at pH 6.0 and pH 7.0. At pH 6.0 (Fig.4A) the length tension is written counterclockwise (i.e. the work producing mode) between 0.2 and 0.7 Hz and in clockwise direction (i.e. the work absorbing mode) at 0.8 Hz and higher frequencies. Fig.4B shows that the frequency range in which the muscle produces work is only slightly different at pH 7 (counterclockwise rotation between 0.2 and 0.6 Hz). There is, however, a significant enhancement in the contractile activity and tension with increasing pH. At a constant Ca-concentration the isometric tension and the dynamic stiffness (i.e. the amplitude of tension change during sinusoidal

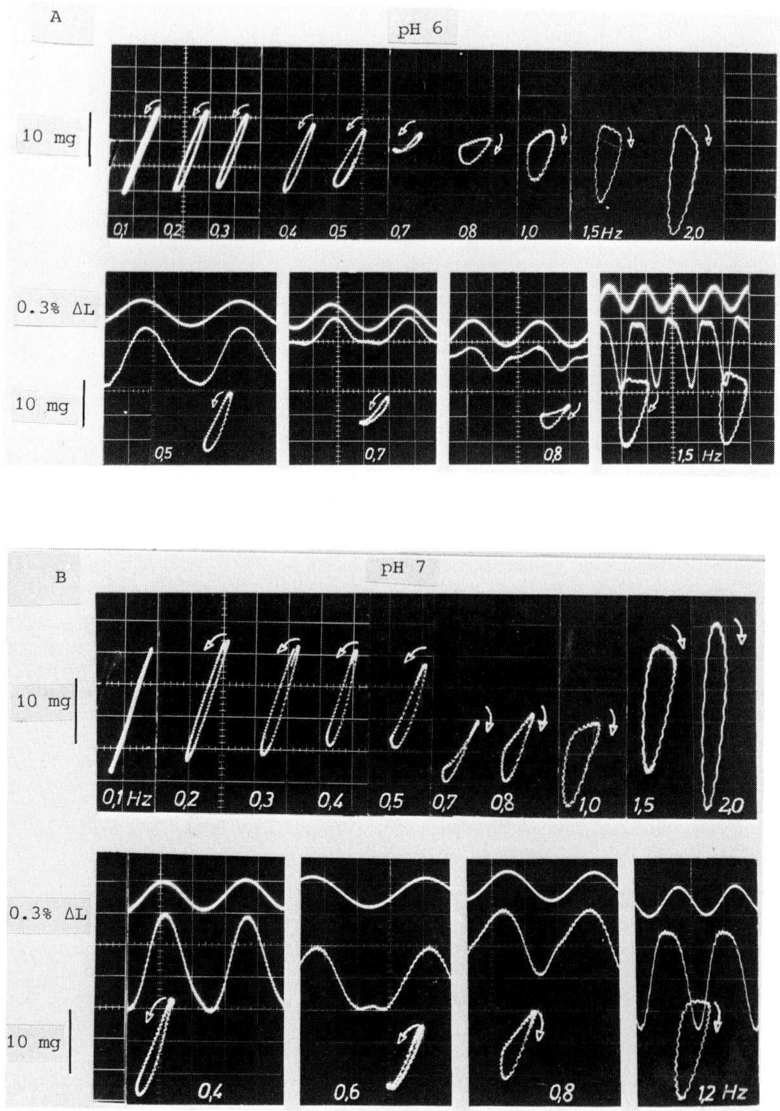

Fig.4. A. The time course of length and tension changes (lower set) and length tension phase plot (upper traces) during sinusoidal length perturbations in extracted rabbit papillary muscles at pH 6.0. The arrow describes the direction in which the length tension diagram is written. Experimental conditions: $T_2 = 22^{\circ}C$, 15 mM Mg-ATP, $\Delta L = 0.3\%$, $Ca^{++} = 2 \times 10^{-6}$ M, 100 mg $\approx$ 30 g/cm^2, initial isometric tension $\approx$ 150 mg.

B. Same as A except pH = 7.0, initial isometric tension $\approx$ 240 mg.

stretches and releases) rose considerably when the pH was changed from 6.0 to 7.0. Some attention should be paid to the non-linearity of the tension response in these preparations which becomes more and more prominent as the frequency of the sinusoidal length change increases. Reproducible non-linearities have been observed many times in other muscle as well (Steiger and Rüegg, 1969). These non-linearities are illustrated in Fig.4. Their phase shift increases by about 180° with each doubling of the frequency. For 0.8 Hz at pH 6.0 length and tension (and the length-tension-loop) show the following characteristics. Starting at the maximum length of a cycle, releasing the muscle causes a drop in contractile tension but the minimum of the tension is reached earlier than the minimum of length change. There is a rising phase in tension until the length reaches its minimum level. Stretching the muscle results in almost no tension change until about medium length, then tension increases again approximately in phase with the length change. The situation is similar at 1.5 Hz but the phase relation between the length and tension has changed. One might speculate that under the given experimental conditions 0.75 Hz is the corner frequency, i.e. the frequency at which the crossbridges are just able to follow the length tension changes, whereas at 1.5 Hz the crossbridges are cycling with the wrong phase lock to produce work.

Cardiac actomyosin ATPase has been found to be slightly but significantly increased when the cardiac preparation is stretched and released by about 1% of its length in contraction solution (Table I).

TABLE I

Extra-ATPase Activity in Cardiac Am-Systems

ATPase (Isometric) %	ATPase (during stretch & release) %
100	109 ± 5.2

(14 values in 5 exp.)

Experimental conditions: Ether-glycerol extracted rabbit papillary muscle; pH 6.8; $Ca^{++} = 3 \times 10^{-6}$M; T = 26°C; 15 mM ATP; 17.5 mM $MgCl_2$; myokinase = 0.1 mg/ml. 1% rectangular extensions at 0.2Hz.

The time course of the delayed tension rise and the time until the delayed tension maximum is reached varies for different species whereas the variations of different cardiac preparations of the same species are only minor. The delay time was the largest in the slowly beating pig muscle, about 3.5 sec (22°C) and becomes increasingly shorter as the heart beat rate increases, 0.5 sec (22°C) in the rat and about 150 msec (17.5°C) in the humming bird heart (Fig.5). The approximate rate constants are given in Table II. as the rate constant increases, so does the frequency range in which the cardiac muscles of the investigated vertebrates are able to perform work. At a glance it might seem that the frequency at which these effects appear is too slow to be of physiological importance. However, it should be recalled that the measurements were performed at room temperature and as already shown (Steiger, 1971) the time course of delayed rise in tension and also the frequency of optimal work output and working range are very temperature sensitive. Normalized to the functional body temperature, the frequency at optimal power output occurs coincidentally in the same range as that of the normal heart beat (Fig.6).

TABLE II

Rate Constant of the Stretch Induced Delayed Tension Rise for Different Vertebrates

Species	Rate Constant (sec^{-1})	Temperature (°C)
Humming bird	14.3	17.5
Rat	8.3	23
Guinea pig	4.8	23
Rabbit	3.3	22
Pig	0.7	23
Frog	5.0	20

Experimental conditions: Glycerol or ether-glycerol extracted cardiac preparations, pH 6.5-7.0; 15-17.5 mM ATP; 17.5 $MgCl_2$; $Ca^{++} = 1 \times 10^{-6} - 1 \times 10^{-5}$M. Heinl, Herzig, Sawaya and Steiger, 1973). 3.5% amplitude stretches except for the rabbit, which was 0.4%.

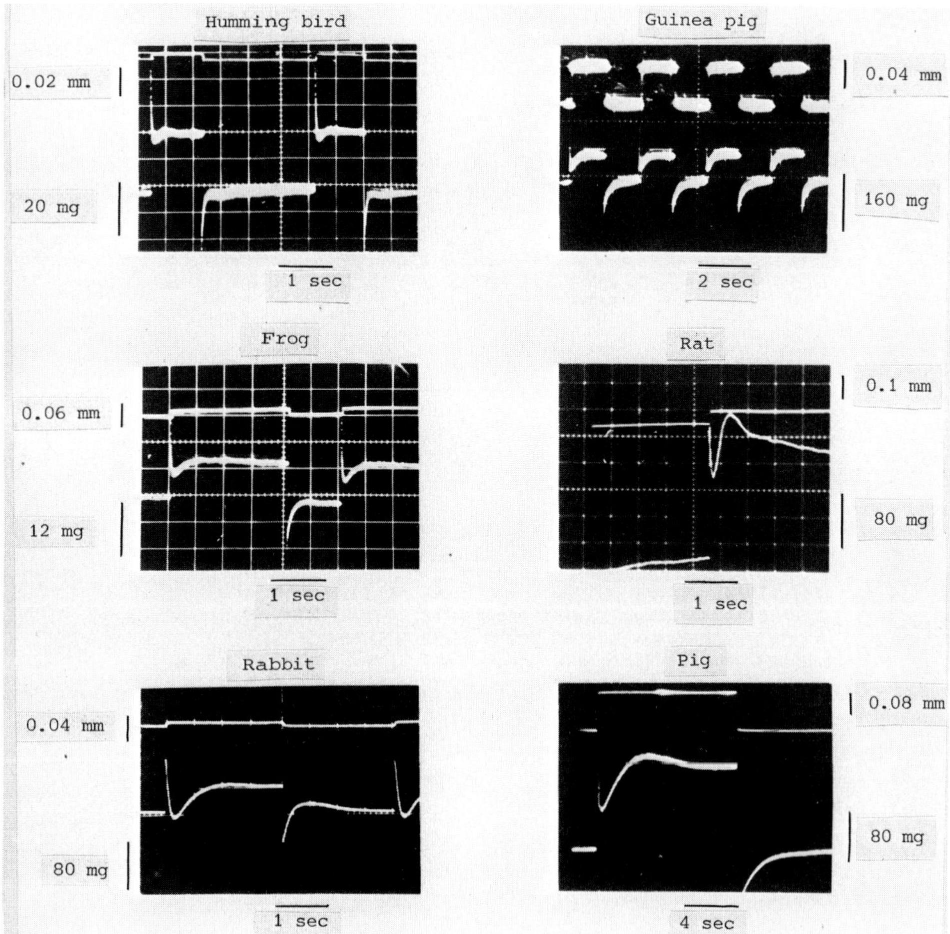

Fig.5. Tension transient of cardiac preparations of different vertebrates. For details see Table 2.

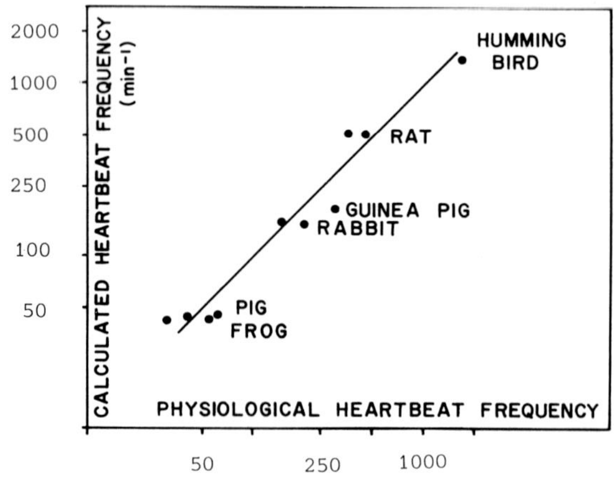

Fig. 6. Calculated *versus* natural heartbeat frequency in different vertebrates. From the rate constant of the delayed tension rise or from the frequency at which work output was optimal we calculated a theoretical heart beat frequency for several vertebrates. A Q_{10} of 3.5 was assumed since most of the experiments were performed at room temperature. Experimental conditions: see TABLE II.

Stretch Activation in Living Cardiac Preparations

Steady state contractures in living cardiac preparations have been induced by Ba^{++}-Tyrode solutions or by isotonic solutions with elevated potassium concentrations. If not otherwise stated, the high-potassium-standard solution (HPS-solution) was used, containing 70 mM KCl, 65 mM NaCl 5 mM Ca and 5 mM Caffeine. Fig.7 (left side) shows the response of the muscle incubated in standard tyrode solution and stimulated at a rate of 20 beats per minute. The stimulus was then turned off and the muscle was subjected to step perturbations in length of about 0.5% of its length. The amplitude of the tension response of the relaxed inactive muscle incubated in the standard tyrode solution is rather small and no delayed tension transients are visible. After switching the solution from standard tyrode to high K-low sodium solution, the basic tension rises to a new equilibrium value.

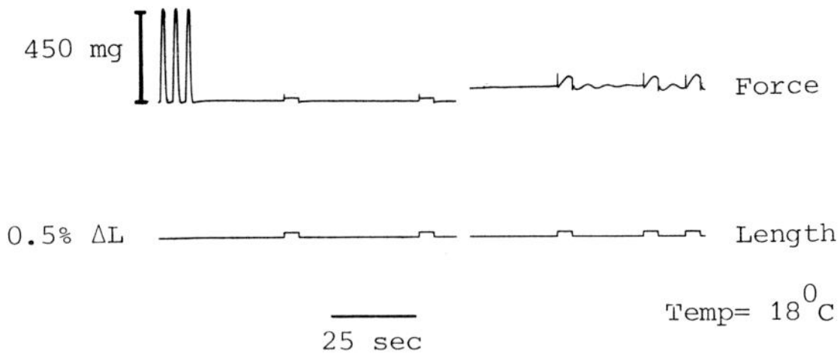

Fig.7. Left. Control twitch tension in standard tyrode solution with 5 mM Ca. Step changes in length do not show any delayed tension, the amplitude of the overall tension response is comparitively small.
Right. High Potassium standard solution (70 mM K, 65 mM Na, 5 mM Ca, 5 mM caffeine). After the rise in isometric tension was completed a sudden length change was applied to the cardiac muscle. Note that the amplitude of the length change was only 0.5% L_0. (Rabbit papillary muscle, 100mg ≃ 50 g/cm^2).

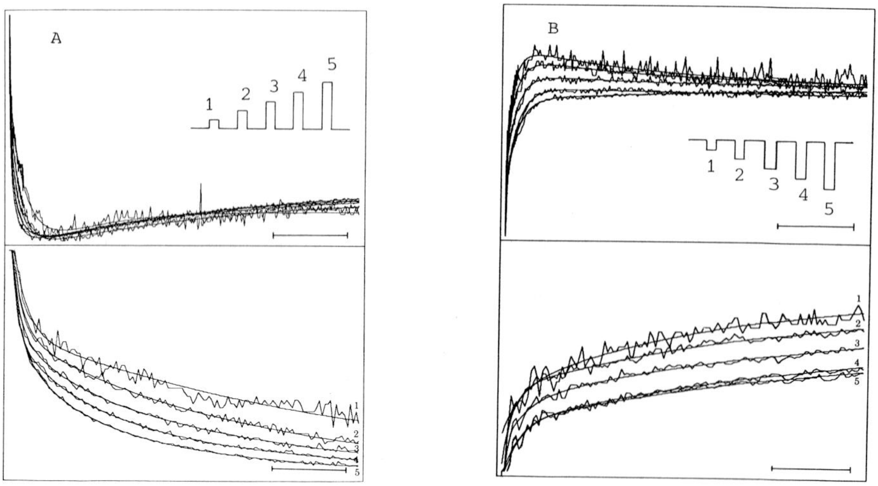

Fig.8. Tension transients for stretch and release.
A. Tension response following stretch at $9^{o}C$.
B. Tension response following release at $16^{o}C$.
Top traces show the whole response, bottom traces show the first 200 msec of the response only.
(1) 0.1% ΔL (2) 0.3% ΔL (3) 0.5% ΔL (4) 0.7% ΔL (5) 0.9% ΔL.
The fitted curve is shown as a solid line. Force is given in rel. units.
Time bar: top, 1 sec.
bottom, 40 msec
Experimental conditions: Rabbit papillary muscle, HPS-solution.

During this period the muscle was kept isometric. A step change in length after the completed change in isometric tension (Fig.7, right) now results in a polyphasic tension response, sometimes accompanied by isometrical fluctuations in tension especially in muscles in which stretch activation is particularly well pronounced. In general, and in particular for small amplitudes of length change, the tension transients are rather linear and can be analysed by a set of exponential functions. Fig. 8 shows the tension transients in HPS-solution for stretch at 9°C (Fig.8A) and for release at 16°C (Fig.8B). The force amplitude is plotted in relative units. This makes it easy to visualise changes of the rate constants or of their relative amplitudes with the amplitude of stretch or release. The data have been analysed by an eigenfunction-expansion-method for the analysis of exponential decay curves (Provencher, 1976).

Depending on the temperature, up to four exponential components have been identified. As for rabbit psoas and insect flight muscle (Abbott and Steiger, 1977), it was found that the instantaneous tension recovery after the completion of the length change is a two phase process. Each component has a different characteristic. The fastest component is called phase 1, and is described by rate constant K_1. Phase 1 seems to be responsible for roughly 50% of the tension drop after stretch, and has the lowest temperature sensitivity (Fig.9). In case of a release, the amplitude of K_1 seems to decrease with an increasing amplitude of release.

Phase 2, a further fall of force after stretch, and a further rise after release, is represented by the rate constant K_2. The amplitude dependence of K_2 is best seen at 9°C for stretch and at 16°C for release (Fig.8). At 9°C, K_2 increases from a value of 1.8 sec^{-1} at 0.9% release to 14.1 sec^{-1} at 0.9% of stretch or from 6.8 sec^{-1} to 20.5 sec^{-1} at 16°C (Table III). K_2 is also highly temperature sensitive as illustrated in Fig.9.

Following Phase 1 and 2, the tension course reverses and a delayed rise of tension is observed. The rate constant of the development of the delayed tension is called K_3. K_3 has the highest temperature coefficient but does not seem to change with the amplitudes of the length perturbations. The amplitudes of all three experimental components are a linear function of the amplitude of the stretch within 0.1 and 0.9% of the muscle length but not of the release. The delayed tension is usually not fully maintained. It decays slowly to a final equilibrium with a rate constant K_4, which at 24°C is about 1 sec^{-1}.

TABLE III

AMPLITUDE AND TEMPERATURE DEPENDENCE OF THE RATE CONSTANTS K_1, K_2 and K_3

EXPERIMENTAL CONDITIONS AS IN FIG.8.

AMPLITUDE	%	K_1 9°	K_1 16°	K_1 24°	K_1 36°	K_2 9°	K_2 16°	K_2 24°	K_2 36°	K_3 9°	K_3 16°	K_3 24°	K_3 36°
STRETCH	0.9	158	201	191	324	14.1	20.5	60	162	0.21	0.8	3.8	29.2
	0.7	189	253	295		11.7	18.2	43.7	153	0.21	0.9	3.4	29.9
	0.5	180	221	247		9.15	17.0	37.4	155	0.27	0.8	4.2	29.5
	0.3	159	334	310		6.75	15.6	32.0	154	0.38	0.8	3.3	29.3
	0.1	139	279			4.5	11.8	30.2	184	0.79	0.9	2.9	23.
RELEASE	0.1	121	142	184		2.9	10.8	31			0.5 (+)	3.4	
	0.3	104	236	359		2.2	8.3	25			0.6 (+)	4.9	
	0.5	70	158			2.1	7.4				0.8 (+)		
	0.7	86.5	156			1.8	6.7				0.42 (+)		
	0.9	176	164			1.8	6.8				0.75 (-)		

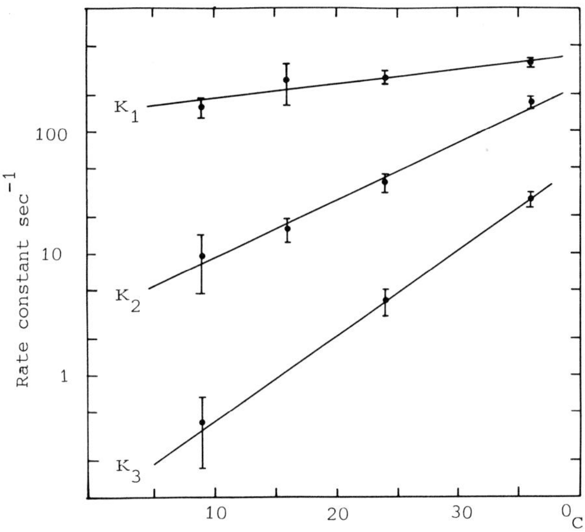

Fig.9. The effect of temperature on the rate constants for rabbit papillary muscle; means $\pm$ SE; the Q_{10} values are $K_1 = 1.3$; $K_2 = 3.0$; $K_3 = 5.0$.

As Fig.10 shows, the rate constant for the delayed rise of tension is affected by the level of activation. The shift in the kinetics of K_3 was observed irrespective of whether the activation was induced in a high K - low sodium contracture, by various amounts of Ca^{++} or adding caffeine, or by different concentrations of Ba^{++} ions. The time constant of the delayed tension rise decreases with increasing contractile activity. This was also found by Herzig and Herzig (1974) on chemically treated cardiac preparations. The changes in the contractile kinetics are larger if the level of activation is less than 50%. Above 50% there are only minor changes. The influence of the contractile activity on K_1 and K_2 has not yet been analysed. The amplitude of both the instantaneous and the delayed tension increases with

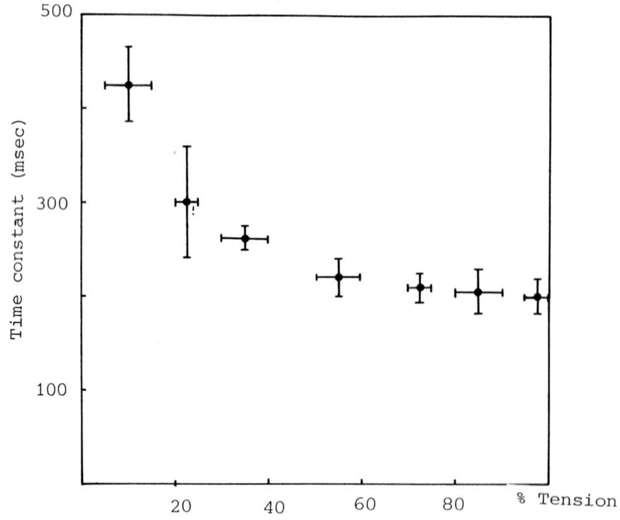

Fig.10 Approximate time constant of the delayed tension rise in relation to the level of activation in rabbit papillary cardiac muscle; means and SD from 6 experiments. Temperature ≃ 20-22°C.

increasing muscle length (Fig.11). The amplitude of the stretch activation reaches its maximum at the length at which the twitch tension reaches its maximum. This demonstrates that delayed tension is not an effect limited to the ascending limb of the Starling length tension curve. The rate constant for the rise of the delayed tension and the time to peak tension do not appear to be significantly affected by muscle length.

Fig.11 Effect of muscle length on the amplitude of delayed tension changes in rabbit papillary muscle.

A. Twitch tension at varied muscle lengths.

B. Tension transients at different muscle lengths with activation by high frequency stimulation (50Hz). Temperature = 20°C; Ca^{++} = 2.5 mM. The step change in length was the same at each muscle length (ΔL = 0.04 mm). 100 mg ≃ 50 g/cm^2.

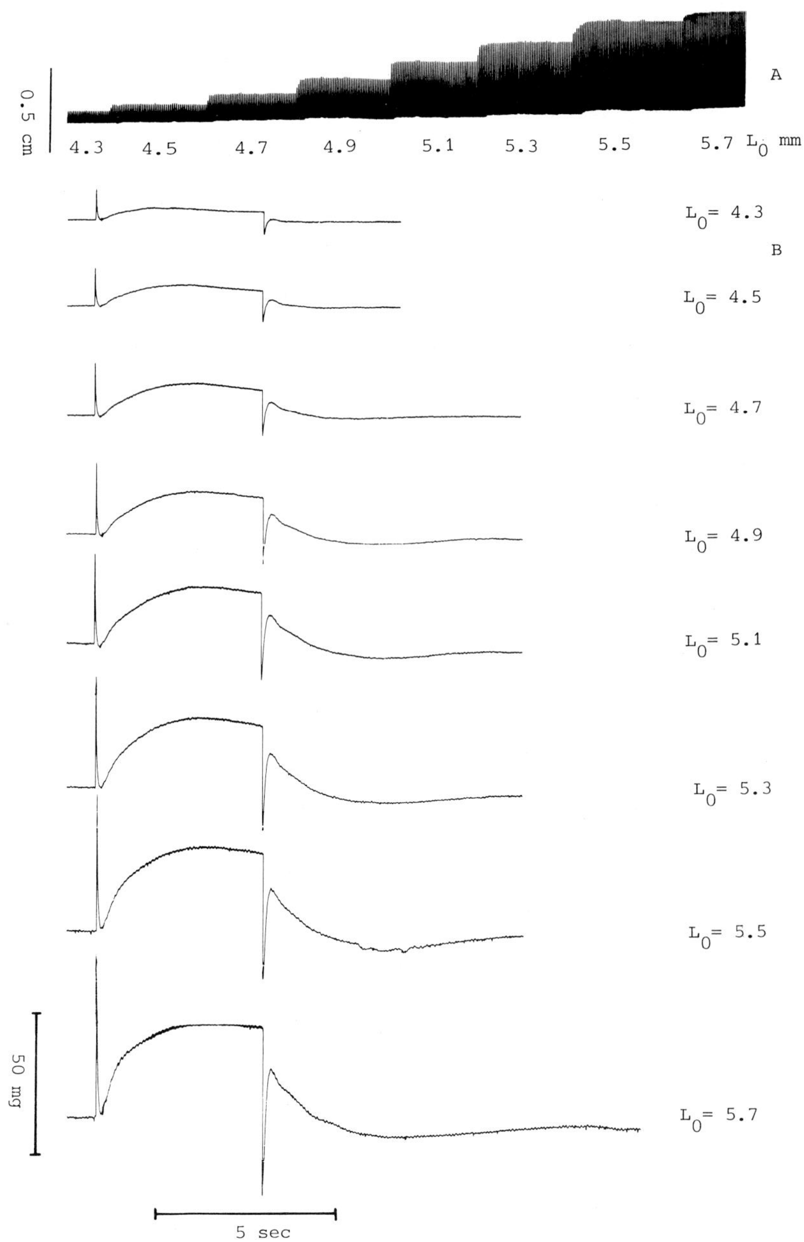
A
0.5 cm
4.3
4.5
4.7
4.9
5.1
5.3
5.5
5.7
L_0 mm
B
L_0= 4.3
L_0= 4.5
L_0= 4.7
L_0= 4.9
L_0= 5.1
L_0= 5.3
L_0= 5.5
L_0= 5.7
50 mg
5 sec

Tension Responses to Sinusoidal Perturbations in Living Cardiac Muscles

The tension response to sinusoidal perturbations in muscle length was studied between 0.05 and 30 Hz. Length and tension changes at each frequency were recorded and displayed in the Nyquist plot (Fig.12) in relaxed muscle, in muscle during sustained contractures, and in muscle in rigor. When sinusoidal length changes were applied during sustained contracture, the cardiac myofibrillar system was able to perform work between 0.05 and 1.2 Hz.

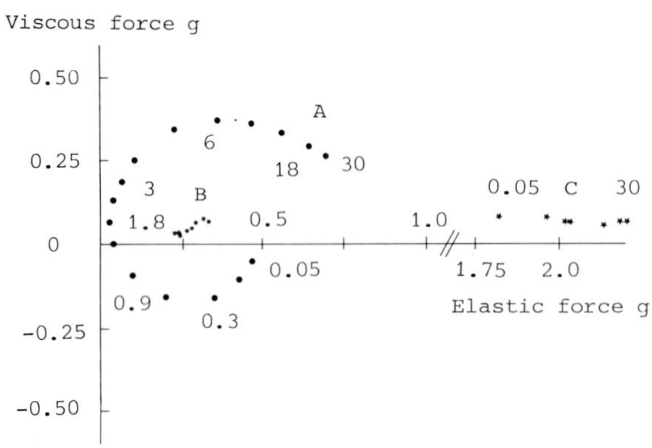

Fig.12 Viscous and elastic properties of living Cat papillary cardiac muscle at various frequencies (Nyquist plot). When the viscous modulus is negative, the heart fibre does work on the apparatus; T = 22°C.
A. Sustained contracture: 0.5 mM Ba^{++}
B. Relaxed muscle: Normal tyrode solution, unstimulated
C. Rigor muscle: 1 mM Ba^{++} + IAA

Fig.13A Work per cycle of cat papillary cardiac muscle as function of the amplitude of length change. The work output per cycle increases with the square of the amplitude of length change. The curve was fitted with the relation to $Z_1 + Z_2 \cdot x + Z_3 \cdot x^2$ with $Z_1 = -0.023$; $Z_2 = 108.757$; % of error = 0.0039 (Experimental conditions:0.5 mM Ba^{++}; T ≃ 22° C; 100 mg ≃ 10 g/cm^2).

Fig.13B Frequency of optimal work output of rabbit papillary cardiac muscle at 22° and 35°C. Left: Temperature 22°C. Right: Temperature 35°C. Top: Length and tension changes versus time. Note the phase lag between length and tension. Bottom: Length-tension diagram. Note that the frequency of optimal work output is 0.4Hz at 22°C (left) and 2.7 Hz at 35°C (right). (Experimental conditions: High K^+ (70 mM) Low Na^+ (65 mM); 5 mM Ca^{++}, 5 mM caffeine solution; 100 mg ≃ 16g/cm^2).

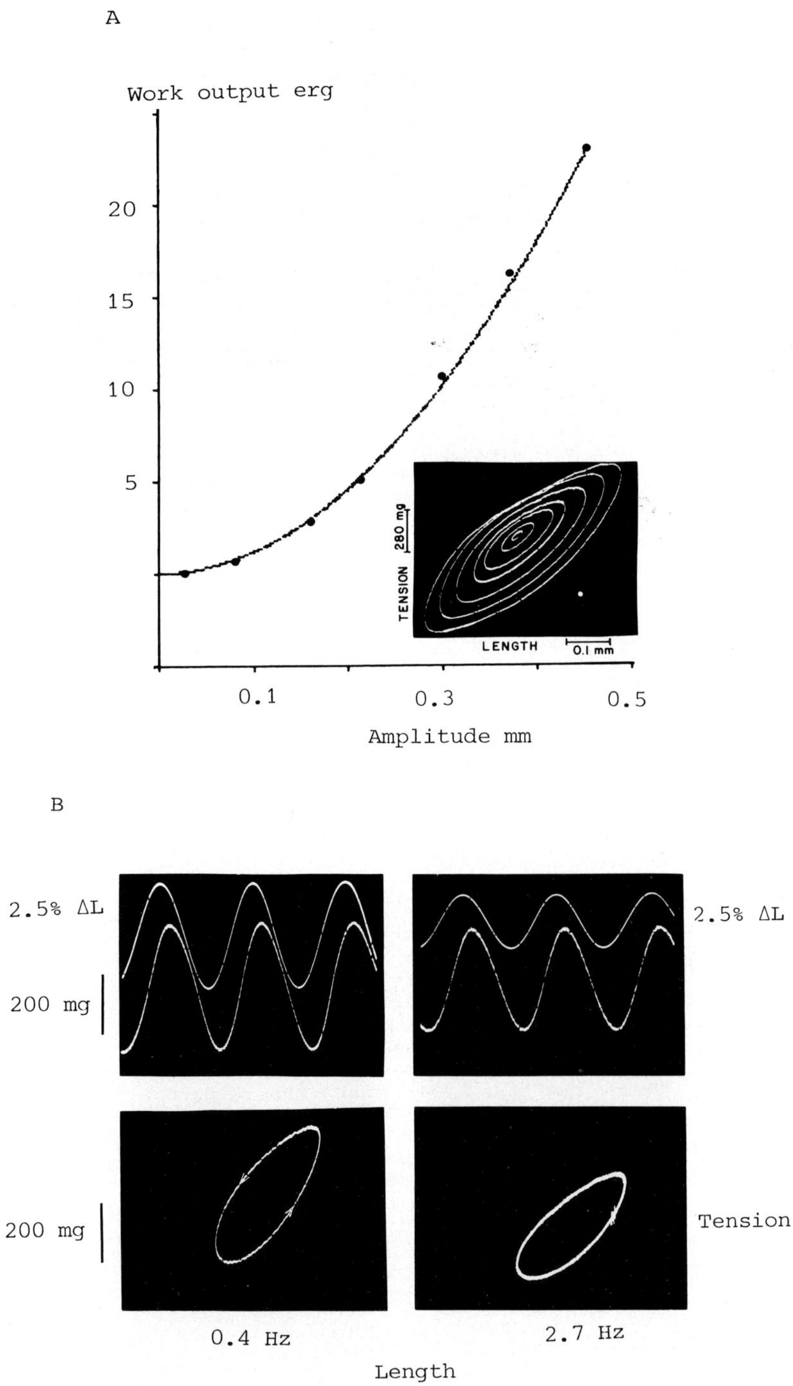
A
Work output erg
20
15
10
5
TENSION
280 mg
LENGTH
0.1 mm
0.1
0.3
0.5
Amplitude mm
B
2.5% ΔL
200 mg
2.5% ΔL
200 mg
Tension
0.4 Hz
2.7 Hz
Length

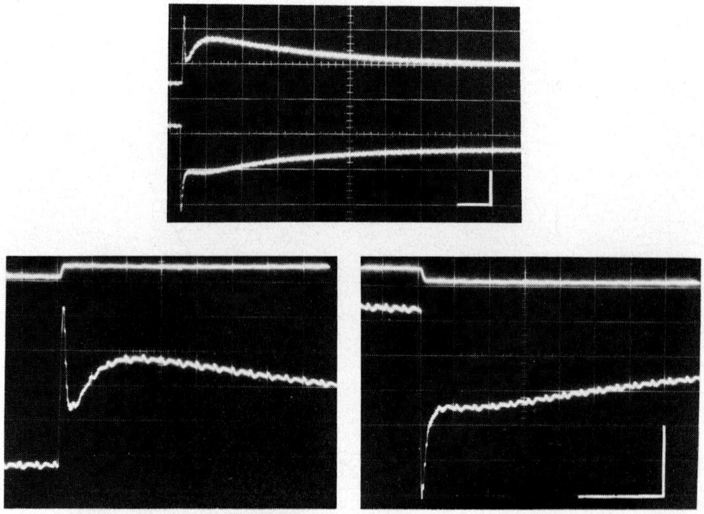

Fig.14 Stretch activation in rabbit psoas muscle. Top and bottom show the same experiment. Middle: length signal. Tension bar: 12 mg. Time bar: 100 msec. Amplitude of length change: 0.6% of muscle length. Isometric tension: ∿90 mg. Experimental conditions: Glycerol extracted rabbit psoas muscle; T = 18°C; 17.5 mM MgATP; $Ca^{++} = 5 \times 10^{-7}$ M; pH 6.5; Myokinase = 0.1 mg/ml.

Fig.15 Tension transients in rabbit psoas muscles under different ionic conditions.

A. left: Rabbit psoas muscle incubated in regular contraction solution with 5×10^{-5} M Ca^{++}, 17.5 mM MgATP; isometric tension: 370 mg. Time bar: 5 sec; tension bar: 60 mg.
right: Rabbit psoas muscle incubated in Ca- and Mg-poor ATP-solution; p Ca ≃8-9; p Mg ≃ 7-8; 12 mM ATP; isometric tension ≃ 300 mg; time bar: 5 sec; tension bar: 60 mg.

B. Same as Fig.15 A-left. Time bar: 100 msec; tension bar: 30 mg.

C. Same as Fig.15 A-right. Time bar: 100 msec; tension bar:30 mg.

D. Rabbit psoas in rigor solution. Time bar: 100 msec; tension bar: 115 mg. isometric tension: 370 mg.

Experimental conditions: Amplitude of length change 0.7% of muscle length; T = 18° C; pH 6.5; Myokinase 0.1 mg/ml.

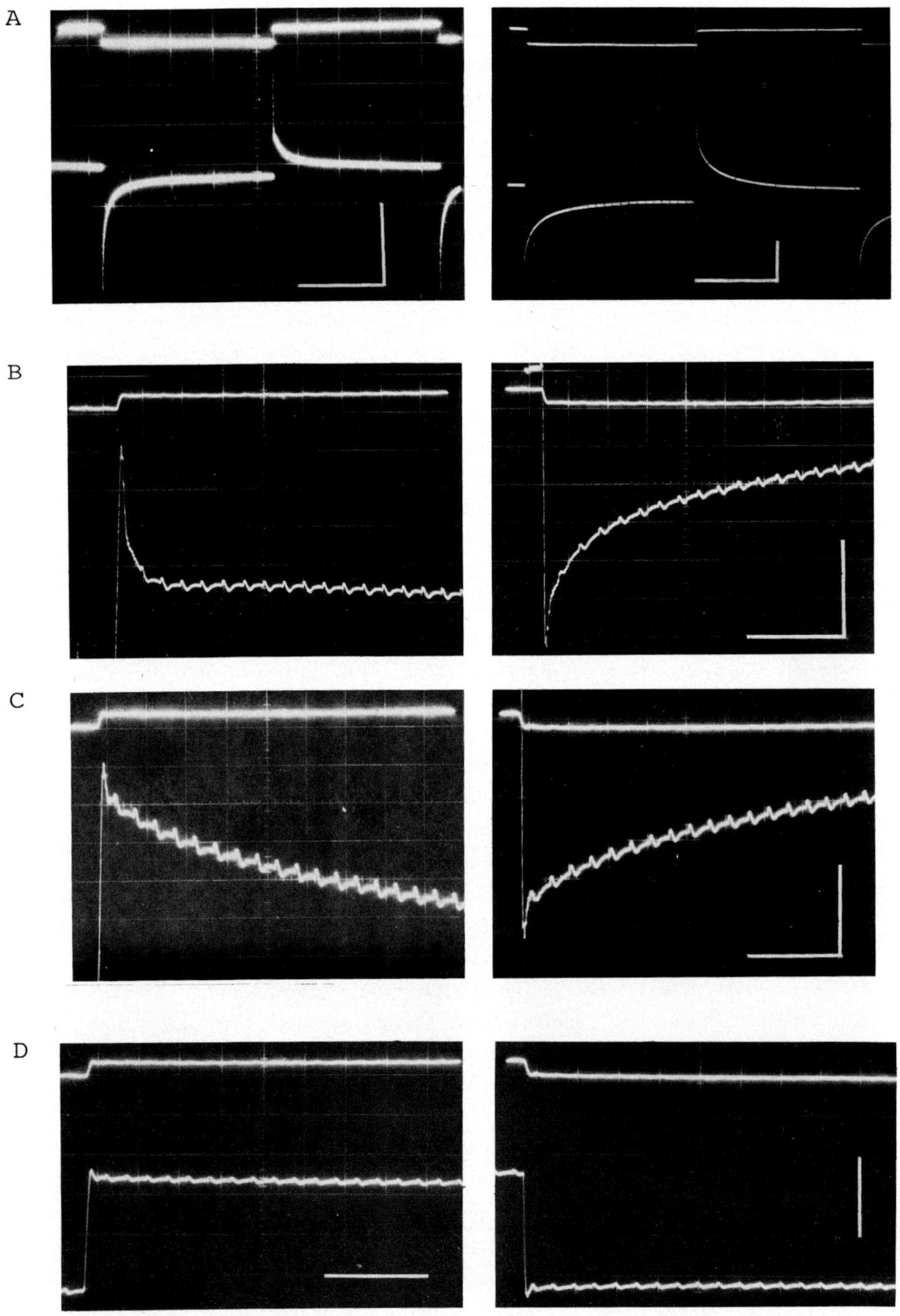

The frequency for the maximal work output was 0.6 Hz for cat papillary muscle and 0.4 Hz for rabbit papillary muscle, corresponding to a rate constant of 3.7 sec^{-1} or 2.5 sec^{-1}. This rate constant is equivalent to K_3 in the case of rectangular length changes (Table III). At 1.2 Hz the loop closes and then opens again at higher frequencies with rotation in the clockwise direction, i.e. displaying a positive viscosity.

The elastic modulus decreases when the frequency increases from 0.05 to 1.2 Hz. At 1.2 Hz the elastic tension amplitude is less than 50 mg, compared with ∿ 500 at 0.05 Hz. At frequencies above 1.2Hz, stiffness rises again but the median tension decreases. The Nyquist plot has a maximum at about 6 or 8 Hz corresponding to a second rate constant of 40-50 sec^{-1}. This rate constant is equivalent to K_2 found in quick skeletal experiments (Table III). Similar experiments performed on relaxed and muscles in rigor where rate constants can be extracted show a straight line with only a positive viscosity, instead of half circles. This implies that there is no single rate constant, but rather many. The phase angle in passive muscle is approximately constant over the entire frequency range. The elastic modulus increases slightly.

Work and power output per cycle (Fig. 13A) depend on the product of length amplitude and delayed tension. Thus the work output per cycle is proportional to the square of the amplitude of length change, since at constant frequency the amount of stretch activation and the width of the length-tension loop increases linearly with increasing amplitude. The amplitude of 0.45 mm is equivalent to 7.5% length change. The curve was fitted to the relation $Z_1 + Z_2 \cdot x + Z_3x^2$ with $Z_1 = -0.023$; $Z_2 = 1.0$ and $Z_3 = 108$ which shows that the square term is the major component. In rabbit cardiac muscle the frequency for optimal mechanical work output increases from 0.4 Hz at 22° to 2.7 Hz at 35° which is incidentally very close to the natural heart beat frequency (Fig.13B).

Stretch Activation in Rabbit Skeletal Muscle

Fig.14 shows a "standard response" to step perturbations in rabbit psoas muscle at 18°C in a medium with 17.5 mM MgATP plus myokinase and 5 x 10^{-7}M free Ca^{++} ions. Stretch activation in glycerinated rabbit psoas has already been reported in previous papers (Rüegg, Steiger and Schädler, 1970) and a detailed analysis of the tension transients has been done by Abbott and Steiger (1977). They showed that the tension time course in rabbit psoas muscle following small length perturbations may be composed of three different phases, the same number of phases as found in insect fibrillar and cardiac muscle. The two phases during the initial recovery period following stretch are not readily distinguishable by eye in Fig.14 but become clearly visible at lower temperatures (Fig.1 in Rüegg, *et al.*, 1970, or Fig.3 in Abbott and Steiger, 1977) or when the psoas muscle is in a high tension state (Fig.15 left).

Sometimes during the course of an experiment the isometric tension rises very steeply to values of 30-35 mg/fibre, corresponding to about $3Kg/cm^2$,and delayed tension changes completely disappear. The transients under these circumstances are shown in Fig.15A, left and 15B. The isometric tension rose from about 90 to 300 mg. The initial elastic phase is now about 85 mg compared with 25 mg before (Fig.14). The two components of the early recovery are now clearly separated. The kinetics of the fastest component seem to be unaffected and its amplitude seems to have proportionally increased. The second phase is now very slow. It decays with a time constant of about 1 sec. The final tension value is still very close to the isometric value before the length change. This state is called the high tension state. It is not a rigor state because the tension recovers almost completely whereas in rigor (Fig.15D) the changes in tension are dominantly elastic and only little tension recovery takes place.

In principle, the high tension effect is reversible. After a resting period in relaxation solution the preparation usually shows delayed tension again when placed in contraction solution. Many fibres, however, break because of the high force generated. It is thought that the high tension state is due to diffusional limitation of the ATP supply (Steiger and Rüegg, 1969) or due to an excess of ADP relative to ATP in the region of the myofibrils (Pringle, 1967; Abbott and Mannherz, 1970). A similar response should therefore be obtained if the substrate MgATP available to the muscle fibre is reduced by lowering the concentration of Mg-ions in the incubation medium. Fig.15A (right) and Fig.15C show the tension response when no $MgCl_2$ is added to the contraction solution. The contractile system is activated under those conditions even if the solution contains 4 mM EGTA.

The isometric tension is about as high as during the high tension state. In contrast, however, the tension recovery following a stretch or a release seems to be composed of only one component (Fig.15A, right). The time constant of this component has about the same value as the slower portion of the recovery phase of the high tension state described above (Fig.15A, left).

ATPase Measurements in Extracted Rabbit Psoas Muscles

The ATPase activity of extracted rabbit psoas muscles has been studied during periods of isometric contraction and periods during which the preparation was periodically stretched and released either sinusoidally or in a rectangular fashion. The frequency of the length perturbations was 3 Hz and the amplitude between 1.2-2.0%. Only muscles which showed good stretch activation and/or did oscillatory work were used. Fig.16 and Fig.17 show an example and the sequence of the two types of experiments performed. The results are summarized in Table IV. In Fig.16 after about 30-45 minutes of

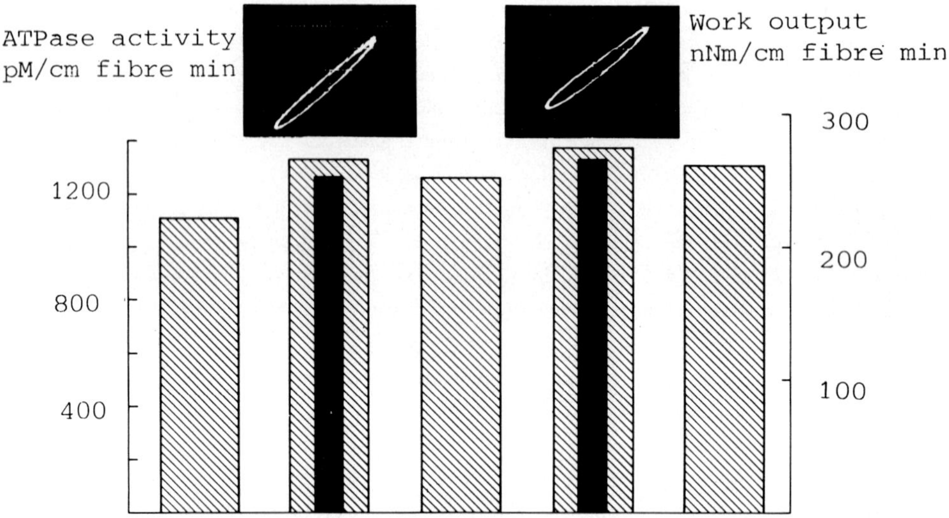

Fig.16 Oscillation induced extra-ATPase and power output in glycerinated rabbit psoas muscle.
The preparation was incubated for successive periods of 20 minutes duration. The P_i accumulated during this period was determined. The shaded column represents the ATPase activity, the dark column the power output during oscillation. The photograph shows the length-tension loop during stretch and release periods.
Experimental conditions: Glycerinated rabbit psoas muscle. T = 21°C, 20 mM ATP, 15 mM $MgCl_2$, $Ca^{++} = 7 \times 10^{-7}$ M; pH 6.5; 1.7% ΔL; 0.1 mg/ml Myokinase; isometric tension ∿ 80 mg; oscillation frequency 3 Hz.

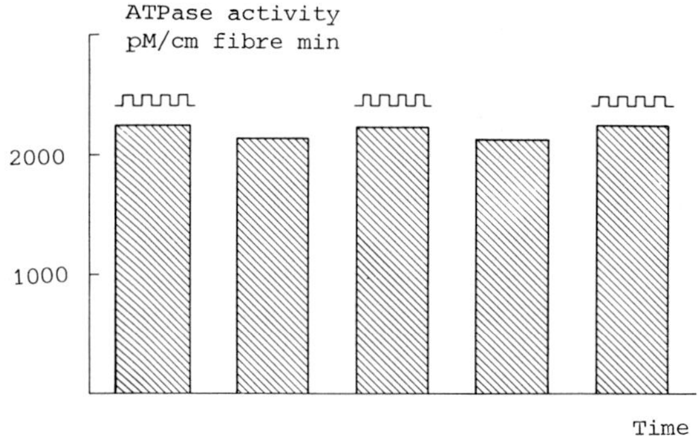

Fig.17 Stretch induced extra ATPase activity in glycerol-extracted rabbit psoas muscle. The preparation was incubated for successive periods of 20 minutes duration. The P_i accumulated during this period of time was determined.
<u>Experimental conditions</u>: T = 20.5° C; 17.5 mM MgATP; 2 x 10^{-6} Ca^{++}; isometric tension ≃ 150 mg.

equilibration the preparations were incubated for successive periods of 20 minutes duration in contraction solution. In the first 20 minutes the fibres were kept isometrically. After 20 minutes the incubation bath was replaced by a fresh one and the amount of P_i accumulated during this time was determined. In the sequence, the incubation periods in which the fibres were kept isometric were alternated with periods in which the fibres were stretched and released sinusoidally at 3 Hz. During this period the muscle oscillated and produced work output, i.e. the loops are running in the counter-clockwise direction. The ATPase activity is enhanced during periods in which work was produced by the psoas preparation by an average of more than 12% (Table IV). The work output with max. 870 nJ/cm fibre min was about 15-20% of what insect flight muscle is producing under similar conditions.

TABLE IV

Extra ATPase in Rabbit Psoas Muscles Oscillated at 3Hz

Mode of Oscillation	Work $\frac{nJ}{cm \cdot fibre \cdot min}$	Extra-ATPase $\frac{pmole}{cm \cdot fibre \cdot min}$
Sinusoidal	325 ± 67	175 ± 34 (24 values; 8 exp.)
Rectangular	-	102 ± 30 (5 values; 2 exp.)

Experimental conditions: Glycerol extracted rabbit psoas muscle: T = ∿20°C; 15-20 mM ATP; 15-17.5 mM $MgCl_2$; pCa 5.7-6.1; ΔL = 1.2-2.0%; pH 6.5; myokinase 0.1 mg/ml; average isometric ATPase value = 1366 pmole/cm fibre; minimal work performed ≃ 100 nJ/cm fibre·min; maximal work performed ≃ 900 nJ/cm fibre·min; median ± SE.

Some extra ATPase activity was also found, when the preparation was stretched and released by rectangular length changes (Fig.17). To exclude diffusional and/or mechanical effects the bath was rocked continuously. The solution contained up to 20 mM ATP and 17.5 mM $MgCl_2$ as well as myokinase, in order to guarantee a sufficient MgATP supply.

Tension Transients in Fibrillar Insect Flight Muscles

The phenomenon of stretch activation in insect flight muscle and the tension transients following length perturbations have been studied extensively by several authors (Rüegg, 1968; Pringle and Tregear, 1969; Steiger and Rüegg 1969; Abbott, 1972; White, 1972; Breull, Steiger and Rüegg, 1973). Fig.18A represents a typical response of glycerinated DLM-fibres of the water bug *Lethocerus maximus* in a standard activation solution at 20°C. In solutions containing 4 mM EGTA with no Ca^{++}ions added, the muscle preparation is usually relaxed, the amplitude of the tension response to length perturbations is greatly diminished and delayed tension changes have completely disappeared. In contrast, when the actual amount of Ca^{++} and Mg^{++}-ions is limited to traces due to impurities of the substances used, in particular ATP, activation is evident (Fig.20). For the calculations of the concentrations of the free ligands, metals and those of the complexes it has been assumed that the traces of Ca^{++} and Mg^{++} might be between 1 and 10 μM resulting in an estimated pMg of between 7 and 8, a pCa between 8 and 9, and a MgATP concentration of between 0.94 and 9.4 μM. Under those conditions the contractile system was activated and distinctive tension transients were observed (Fig.18B) in reaction to length perturbations.

Four differences to a "normal" response are easily visualized:

1. The amplitude of the initial tension drop after completion of the stretch is very small but nevertheless it seems to be composed of two components.
2. There is a large delayed tension rise but the delayed tension maximum, and hence stretch activation, is transient and not maintained.
3. Releases trigger a very pronounced recovery phase, having an amplitude far exceeding the isometric tension value before the release. This recovery phase is also transient.
4. The time to reach the delayed tension maximum is >500 msec compared with about 70 msec in regular solution (Fig.18A). The maximum of the tension recovery phase is reached after ∿900 msec in comparison to ∿10-2- msec in a solution with normal concentration of calcium and magnesium.

18

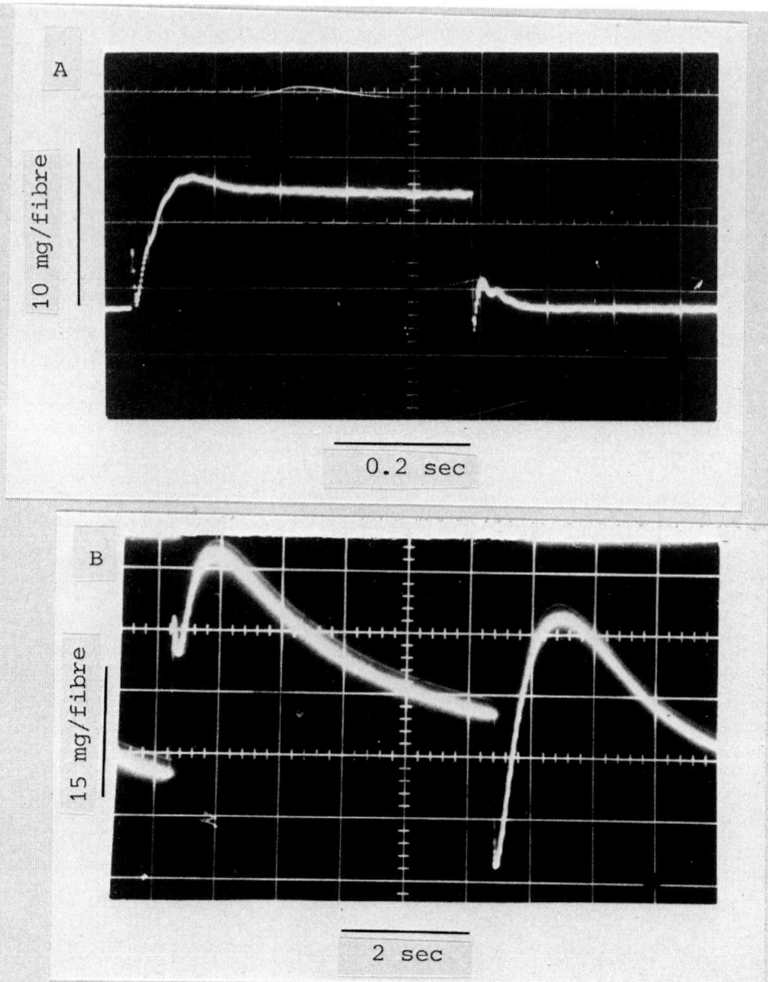
A
10 mg/fibre
0.2 sec
B
15 mg/fibre
2 sec

19

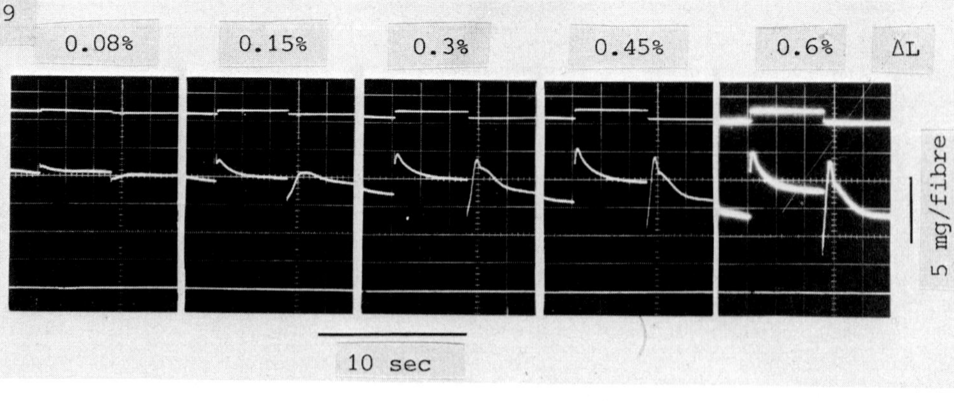
0.08%
0.15%
0.3%
0.45%
0.6%
ΔL
5 mg/fibre
10 sec

Figs.19 and 20 show that the tension recovery following a release seems to be at least a biphasic process. Interestingly, the second phase increases in amplitude very strongly with increasing amplitudes of release thus masking the faster first component almost completely (Fig.19). It also shows a clearly non-linear characteristic during the period in which the tension declines again after reaching the maximum of the tension recovery phase. The final decay of the recovery phase has a time constant of about 2 sec. Fig.20 demonstrates also that the immediate tension fall following stretch contains a faster and a slower phase but both of relatively small amplitude. The kinetics of the fastest phase seem to be of the same order of magnitude as in regular contraction solution. The kinetics of the second phase are evidently considerably slower. The decay of the stretch activation also has a time constant of about 2 sec.

In Ca^{++} and Mg^{++} poor solutions zero tension was reached when the muscle preparation was released at between 0.7 and 1.0% of its actual length (Fig.21). At 0.8% release a fast recovery phase was still visible. At an amplitude of release of 1.4% the muscle remains slack for about 80 msec but tension recovers to a value still larger than the isometric value before the length change. At 1.75% the recovery maximum is of approximately the same size as the isometric tension value. The tension then decays to a final equilibrium.

Fig.18 Tension transients in insect flight muscle in regular and Ca^{++} and Mg^{++} poor contraction solution.

A. Tension transients following quick stretch and release in regular contraction solution. Experimental conditions: Glycerol extracted DLM, *Leth.max.*, T ≈ 20°C, 10^{-5} M Ca^{++}, pH 6.5, L ≈ 1.5% L_o.

B. Tension transients in Ca^{++} and Mg^{++} poor contraction solution. Experimental conditions: Glycerol extracted DLM, *Leth. max.*, T ≈ 18°C; 5 mM ATP; pCa 8-9; pMg 7-8; pH 6.5; L ≈0.8%.

Note: different time scale.

Fig.19 The amplitude dependence of the tension transients in *Lethocerus maximus* flight muscle in Ca^{++} and Mg^{++} poor solution.
Top: stretch and release signal.
Middle: Tension transients.
Bottom: Zero tension.
Experimental conditions: T ∿ 23°C; 15 mM ATP; pMg 7-8; pCa 8-9; pH 6.5.

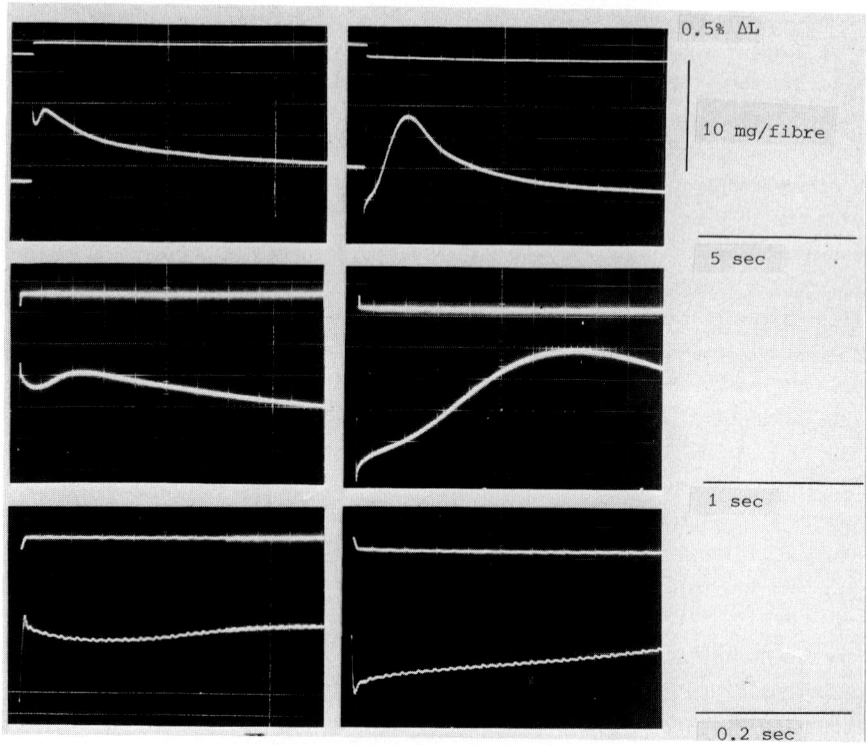

Fig.20. Time course of the tension transients in *Lethocerus maximus* flight muscle in Mg^{++} and Ca^{++} poor solutions. The same set of transients is displayed at three different recording speeds. The left column shows the transients following stretch. The right column shows the transients following release. Top: Length trace. Middle: Tension trace. Experimental conditions: T $\approx$ 10°C; 15 mM ATP; pMg 7-8; pCa 8-9; L $\sim$ 0.5%; pH 6.5; 0.1 mg/ml Myokinase.

If the muscle preparation is stretched again at a time when the tension approaches a new steady state it responds with a sizeable delayed tension rise which is transitory, as already mentioned (Fig.22A). If however, the stretch occurs at a time where the tension recovery of a preceding quick release reaches its maximum, the delayed tension amplitude is rather poor. A similar observation can be made in the case of stretch (Fig.22C). If a second stretch is applied at a time when the delayed tension of a preceding stretch reaches its maximum only a very small further delayed tension rise

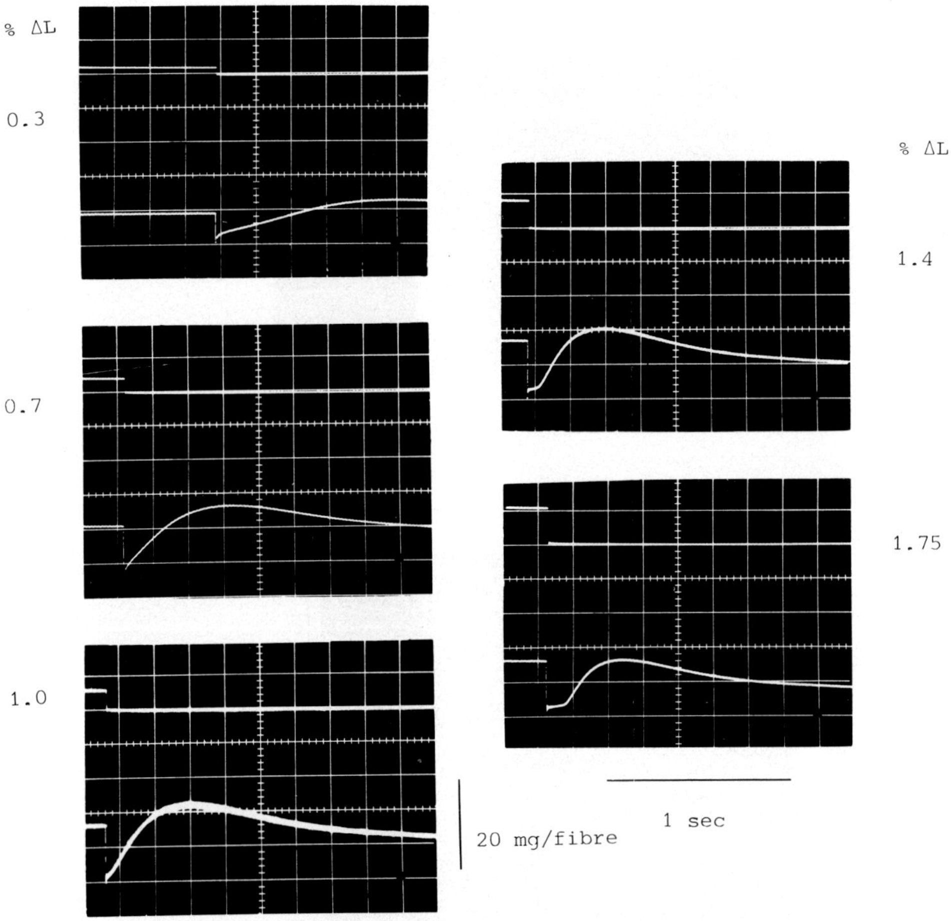

Fig.21 Release to zero tension in *Lethocerus maximus* flight muscle in Ca^{++} and Mg^{++} poor solution. Zero tension level is reached between 0.7% and 1% release. Experimental conditions: T ∿ 20.5°C; 5 mM ATP; pMg 7-8; pCa 8-9; pH 6.5.

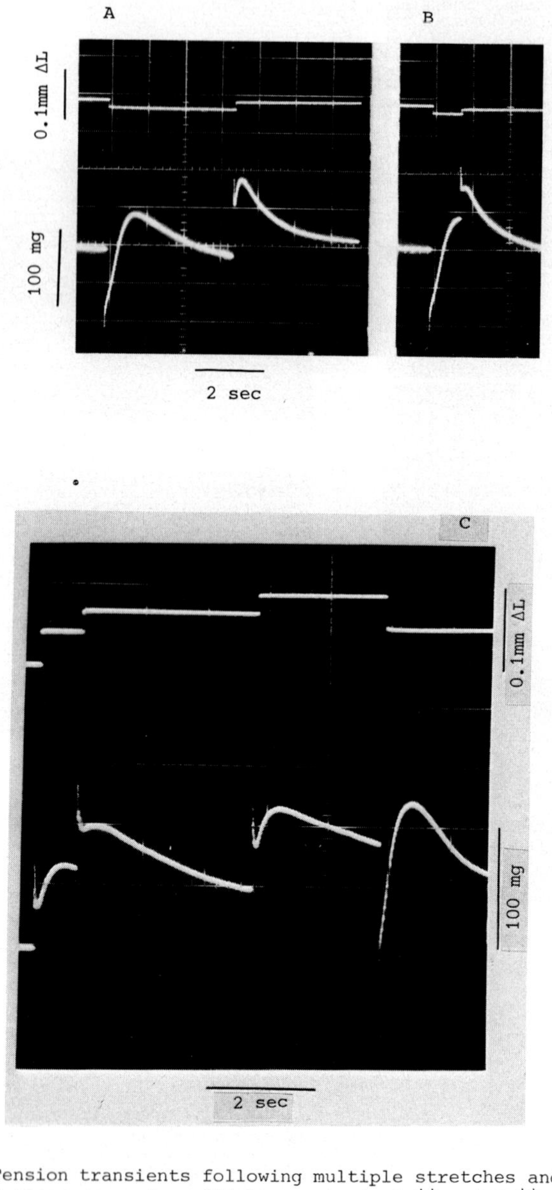

Fig.22 Tension transients following multiple stretches and releases to *Lethocerus maximus* flight muscle in Ca^{++} and Mg^{++} poor solution. Experimental conditions: 5 mM ATP; pMg 7-8; pCa 8-9; pH 6.5.

A and B: T = 20.5° C;

C: T = 15.5° C.

can be observed, whereas a third stretch applied after the decay of the stretch activation results in substantial amplitude of delayed tension rise

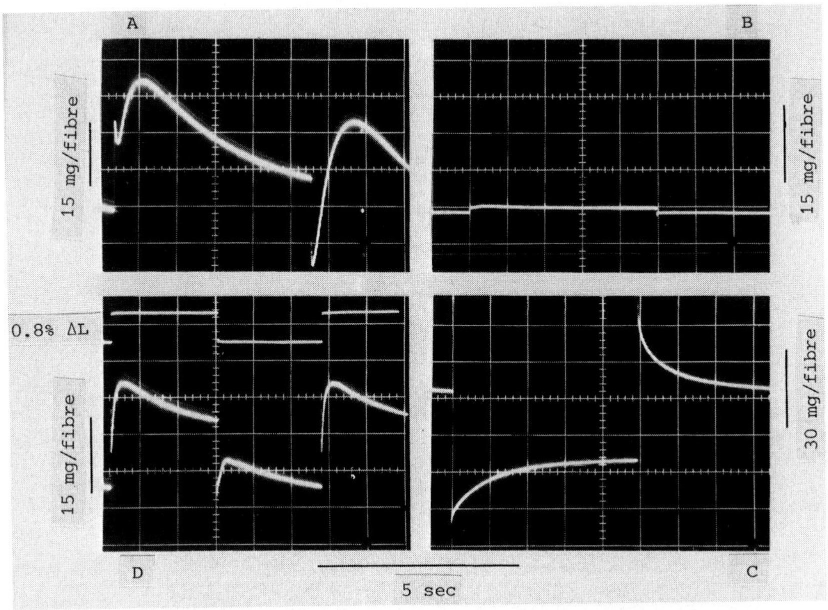

Fig.23 Ca^{++} and Mg^{++} sensitivity of the contractile system of *Lethocerus maximus* flight muscle.

A. Tension transients in Ca^{++} and Mg^{++} poor contraction solution, pMg 7-8, pCa 8-9.

B. Tension transients in the same solution after adding $MgCl_2$ up to a total concentration of 0.2 mM corresponding to a pMg of ∿ 5.

C. Tension transients in the same solution after adding $CaCl_2$ up to a total concentration of 0.2mM corresponding to a pCa of ∿ 6.8.

D. Tension transients in the same solution after adding both $CaCl_2$ and $MgCl_2$ up to a total concentration of 0.2 mM corresponding to a pMg of ∿ 5 and a pCa of ∿ 6.8. Top: amplitude of length change. <u>Experimental conditions</u>: T = 18.5°C; 5 mM ATP; pH 6.5.

It seems as if Mg^{++} ions greatly vary the Ca^{++} sensitivity of the contractile system and also the kinetics of portions of the contractile cycle (Fig.23). Adding $MgCl_2$ to the Ca^{++} and Mg^{++} poor contraction solution to a total concentration of 0.2 mM (Fig.23B), corresponding to a pMg of about five, results in a complete relaxation of the muscle preparation. Relaxation in Mg^{++}

and Ca^{++} poor solution is possible also by adding more EGTA (e.g. 10mM) instead of small amounts of $MgCl_2$. Small amounts of Ca^{++} (total 0.2 mM) corresponding to a pCa of ∿ 6.8, lead to a steep increase in the isometric tension as well as in muscle stiffness and a high tension state, as already described for the rabbit psoas muscle. Delayed transients are absent. The slow recovery phase seems to be dominant as in the case of the rabbit psoas. Because of the slow but expressed tension recovery this state is not at all comparable with a rigor state (Fig.23C). Adding both Ca- and $MgCl_2$ in small concentrations (0.2 mM) to the solution results again in delayed tension responses following small length perturbations but the form and kinetics of the tension transients have been altered; the kinetics and shape are a mixture of the transients observed in regular contraction solution and in Ca^{++} and Mg^{++} poor solution.

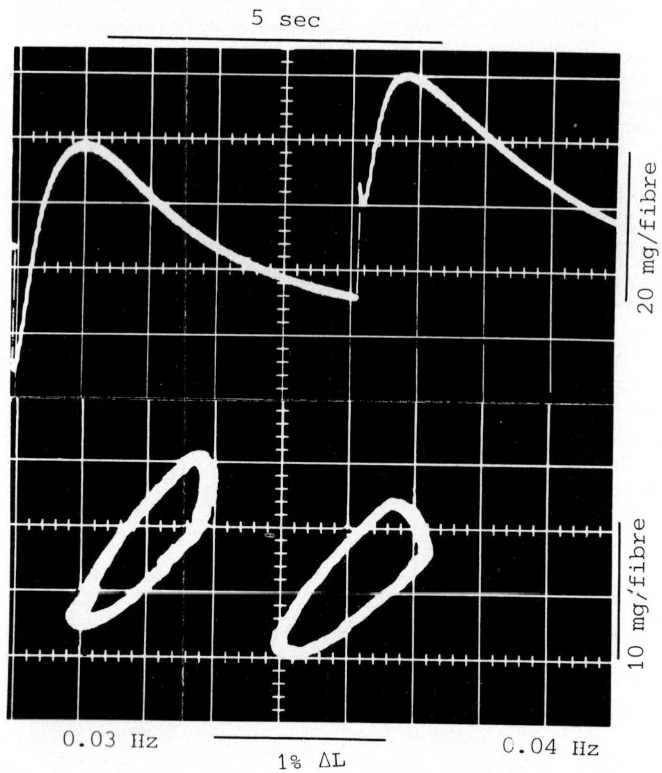

Fig.24 Length-tension-loops from *Lethocerus maximus* flight muscle in Ca^{++} and Mg^{++} poor solutions. Top: Tension transients after step changes in length. Bottom: Length-tension-loop during sinusoidal length changes. The loop is written counter-clockwise but notice the very low frequency. Experimental conditions: T = 18°C; 5 mM ATP; pMg 7-8; pCa 8-9; ΔL ∿ 1%; pH 6.5.

Fig.24 demonstrates that the preparation is able to perform work in Ca++ and Mg++ poor solution but with a frequency which is 100 times slower than in regular contraction solution. The ATPase activity is also strongly reduced under those conditions (on the average about 90%). The isometric ATPase activity in X-solution had about twice the value of the ATPase activity in relaxation solution or about 10% of the value in the normal contraction solution (Fig.25).

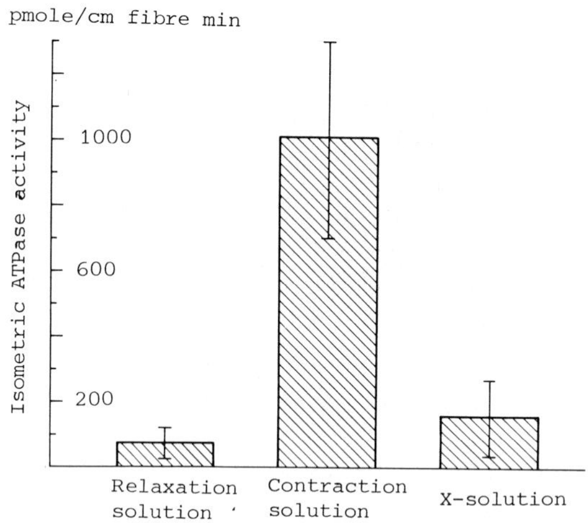

Fig.25 Isometric ATPase activity of *Lethocerus maximus* flight muscle in regular and Ca^{++} and Mg^{++} poor contraction solution (x-solution).
Experimental conditions: T ∿ 18° C; median ± SE.

DISCUSSION

Stretch Activation in Cardiac Muscle

The phenomenon of stretch activation in cardiac muscle was first described in preparations in which the membrane system was destroyed by glycerol or ether-glycerol treatments (Steiger, 1971). But recent experiments with living cardiac tissue have revealed that delayed tension changes are even more pronounced in living muscle when the time dependence of the processes of the excitation contraction coupling are eliminated. Kinetics-wise no significant differences have been detected between a rabbit papillary muscle preparation which has been extracted (Fig.26A) and a living preparation during a steady state contracture (Fig. 26B). Fig.26C shows the tension response in insect flight muscle. With the exception of the different time scale the transient responses following a quick stretch and release look remarkably similar in both types of muscles even though they vary considerably in structure and function. The data presented here and previously show that the similarity is not limited to the optical impression. Many of the features of the stretch activation mechanism first described and discovered in insect flight muscle have been rediscovered in cardiac tissue.

Despite the fact that detailed and carefully conducted enzymatic studies are still missing for cardiac muscles, preliminary results (Table II) and changes in muscle stiffness during the delayed rise in tension (Fig.27) seem to indicate that delayed tension rise is due to a recruitment of more cross-bridges caused by the length change. Most of the delayed tension response seems to be maintained over a considerable period of time in contrast to stretch activation responses in skeletal muscle (Armstrong, Huxley and Julian, 1966; Heinl, Kuhn and Rüegg, 1974; Abbott and Steiger, 1977), where the delayed tension decays almost completely after reaching its maximum. Therefore, we might assume that as in insect muscle, the isometric ATPase activity should be length dependent. ATPase measurements, which are in progress, will hopefully give an answer to those open questions. With respect to stretch activation it is striking that cardiac and insect flight muscle have structurally two features in common, which are not shared by skeletal muscles: (a) the high resting stiffness and (b) the fact that neither muscle can be extended much beyond its optimal length.

Fig.26 Stretch activation in cardiac papillary and fibrillar insect flight muscle.

A. Ether-glycerol extracted rabbit papillary muscle, T ∿ 20°C, 17.5 mM MgATP, 10^{-6} M Ca^{++}, pH 6.0.

B. Living rabbit papillary muscle activated in HPS-solution, T ∿ 22°C.

C. Glycerol extracted fibrillar insect flight muscle, T ∿ 20°C, 15 mM MgATP, 10^{-5} M Ca^{++}, pH 6.5.

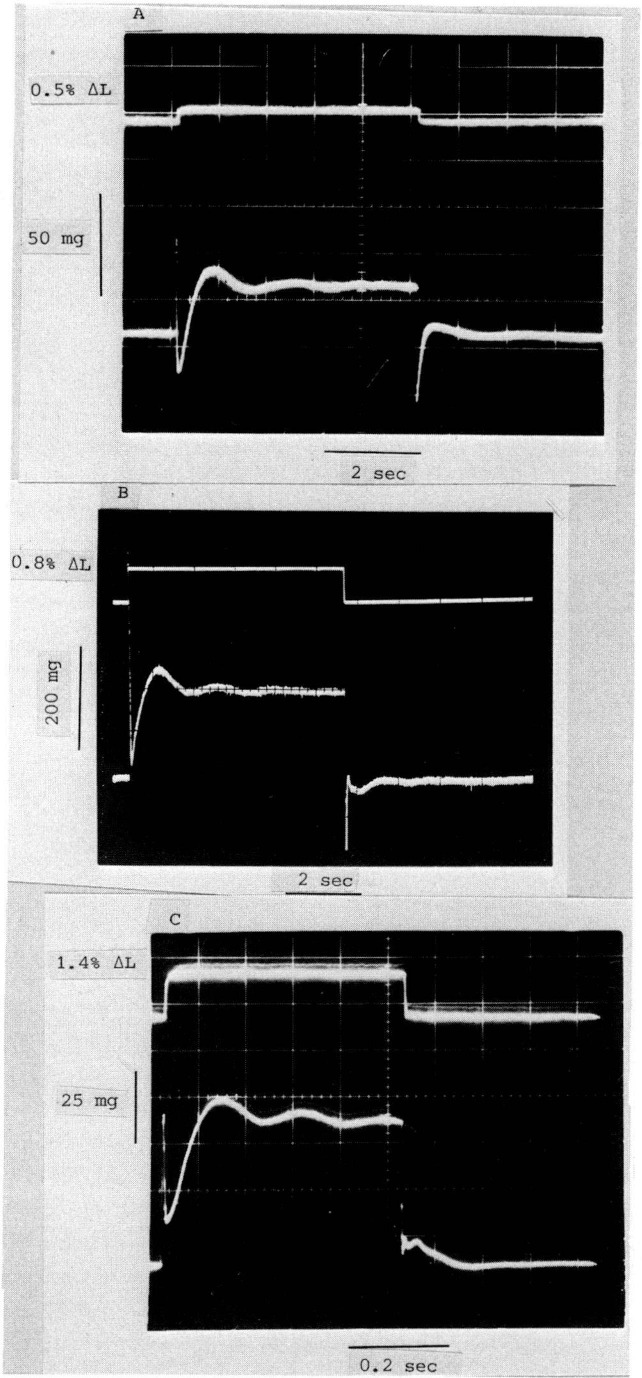
A
0.5% ΔL
50 mg
2 sec
B
0.8% ΔL
200 mg
2 sec
C
1.4% ΔL
25 mg
0.2 sec

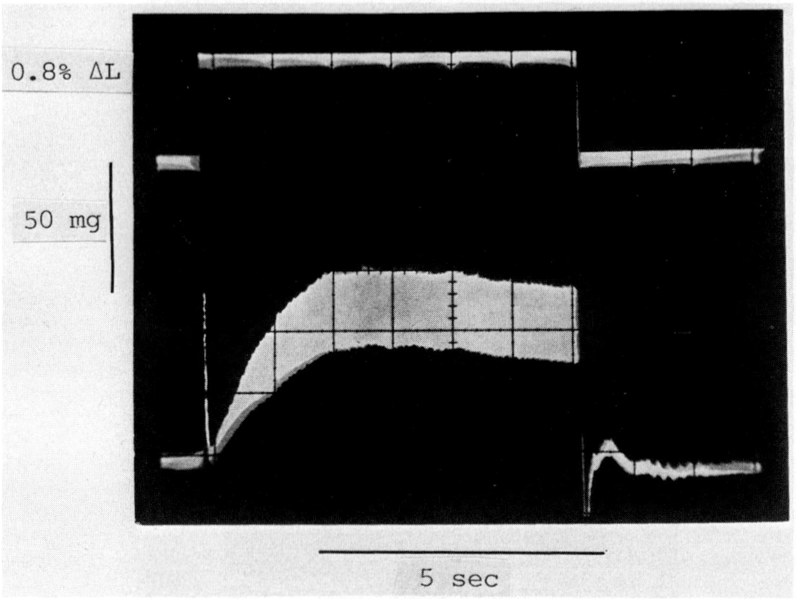

Fig. 27 Change in muscle stiffness of rabbit papillary cardiac muscle concomitant with a delayed rise in tension. Small high frequency (100 Hz) sinusoidal length changes (about 0.1% ΔL) were superimposed on the step change in length. Note that there is no significant change in stiffness during the fast transient periods following a stretch but later a remarkable increase in oscillatory amplitude occurs with the delayed rise in tension. (Experimental conditions: high K (70 mM), low Na (65 mM), 10 mM caffeine, 10 mM Ca solution, T = 20°C). (100 mg ≈ 40 g/cm^2).

Effect of Length and Ca-ions on the Stretch Activation

As shown in Fig.11, stretch activation is most pronounced in the optimal overlap region where further muscle lengthening no longer augments active twitch tension. If stretch activation were only due to changes in the actomyosin overlap one would expect delayed tension changes only on the ascending limb of the length-tension curve, and, in particular, the delayed tension amplitude should be larger at short muscle length and should be diminished or even disappear at lengths in the optimal overlap region.

In both cardiac and insect flight muscle changes of muscle length or of perturbation amplitude affect the amplitude of the stretch activation but not the rate constant of the delayed tension, whereas Ca^{++} acts on the amplitude as well as on the rate constant (Herzig and Herzig, 1974; Abbott, 1972). In living cardiac muscle it was not directly possible to relate the rate constant of the delayed rise in tension directly to the free concentration of Ca. The time-course, however, was affected when the steady contractile activation was modified by increasing concentrations of Ca-ions, by variation of (Ca^{++})in a high K^{+}-low Na^{++} solution or by addition of caffeine. This seems to indicate that activation induced by depolarization or by ionic intervention, e.g. Ca^{++}-ions, is due to a different mechanism than the activation induced by length changes.

In order to conclude that Ca^{++} influences crossbridge kinetics in proposed 2-state models (Huxley, 1957; Thorson and White, 1969; Abbott and Steiger, 1977), it is necessary to assume that Ca^{++} changes the probability of crossbridge attachment to actin. On the other hand the effects of muscle length and amplitude could be explained by an increase of the total number of crossbridges which are able to cycle.

Physiological Significance of the Stretch Activation in Cardiac Muscle

Both the ability to perform work and the changes in dynamic muscle stiffness during steady-state activation which fall in a relatively narrow frequency range are readily measurable and their physiological implications are of interest. The work output increases as the square of the amplitude of the perturbation and can exceed 20 μJ per cycle. Normalized to muscle weight the work output comes close to 1/3 of the amount which cardiac muscle performs during a normal beat. The changes in the dynamic stiffness are not minor. In the same frequency range that the muscle produces work the dynamic stiffness may drop by as much as a factor of 3. The frequency at optimal power output and the frequency at which maximal stiffness changes occur at 37°C are, coincidentally, in the same range of those of the normal heart beat. Delayed tension changes and these oscillatory responses were also found in living frog, bull frog, cat and rat cardiac preparations (Steiger, *et al.*, 1976) as well as in extracted preparations of different vertebrates as reported previously (Heinl, *et al.*, 1973). The time constant of the delayed tension was specific for each animal and seemed to be correlated with the natural heart beat frequency (Fig.6). It is therefore quite likely that despite the fact that the heart beat is triggered electrically there is an optimal tuning between the biochemical and mechanical cycle of the crossbridges consistent with the optimal frequency range of performance for each species.

The physiological significance of this myofibrillar feedback system is well established in fibrillar insect flight muscle as a critical factor in the oscillatory motion of the insect wing (Pringle, 1967). As documented here, a similarly powerful feedback mechanism exists in the living cardiac muscle but tuned to a much lower frequency, i.e. near that of the natural heart beat.

It seems, however, that this feedback mechanism might not play a major role during a normal twitch, since during the major portion of a twitch length changes deactivate rather than activate contractile tension. A strong probability exists, however, that this mechanism may serve as a regulatory function in heart muscle to modulate the resting tonus of cardiac muscle, as postulated by Wezler (1973) and substantiated more recently by McClellan, *et al.*, (1976).

Stretch Activation in Rabbit Psoas Muscles

The delayed tension response of glycerinated rabbit psoas muscle fibres with respect to fibrillar insect flight and cardiac muscle shows two major differences: (1) the amplitude of the delayed tension rise is much smaller and generally not maintained, (2) as already mentioned, rabbit psoas does not have a high resting stiffness as insect and cardiac muscle do and the sarcomeres can be stretched until actin and myosin do not further overlap. There do not seem to be any significant connections between Z-lines and the myosin filaments as postulated by White and Thorson (1973) for insect flight muscle. If so, overall muscle length can only exert stress on the filaments through the crossbridges. Since the crossbridges are being continually broken and remade, such an effect cannot be permanent. The transient nature of the delayed tension seems to be a general feature of skeletal muscle fibres, as it has been observed in frog semitendinosus (Armstrong, *et al.*, 1966), frog anterior tibial (Julian and Sollins, 1975), frog sartorius (Heinl, 1972) and tortoise ileofibularis muscle (Heinl, 1974).

Crossbridge Turnover

In simplified two state models the rate of the delayed tension rise reflects the average cycling rate of the crossbridges. The minimum of the Nyquist plot reveals the "resonance" frequency of the contractile system. At this frequency the phase lage between length and tension is maximal and an optimal number of crossbridges are able to follow the rhythmical length changes to perform maximal work. In rabbit papillary cardiac muscle this frequency at 20°C is about 0.4 Hz (Fig.13). This coincides with the frequency of the isometric fluctuations in tension (Fig.26) and it corresponds also to the molecular ATP splitting rate expressed as molecules ATP per molecule of myosin per second, measured in rabbit cardiac actomyosin gels (Bárány, Gaetjens, Bárány and Karp, 1964).

In insect flight muscle the optimal frequency of power output is at least 10 times faster than in rabbit papillary muscles (Fig.28) as is the frequency of the isometric oscillations (Fig.26C). The frequency of the

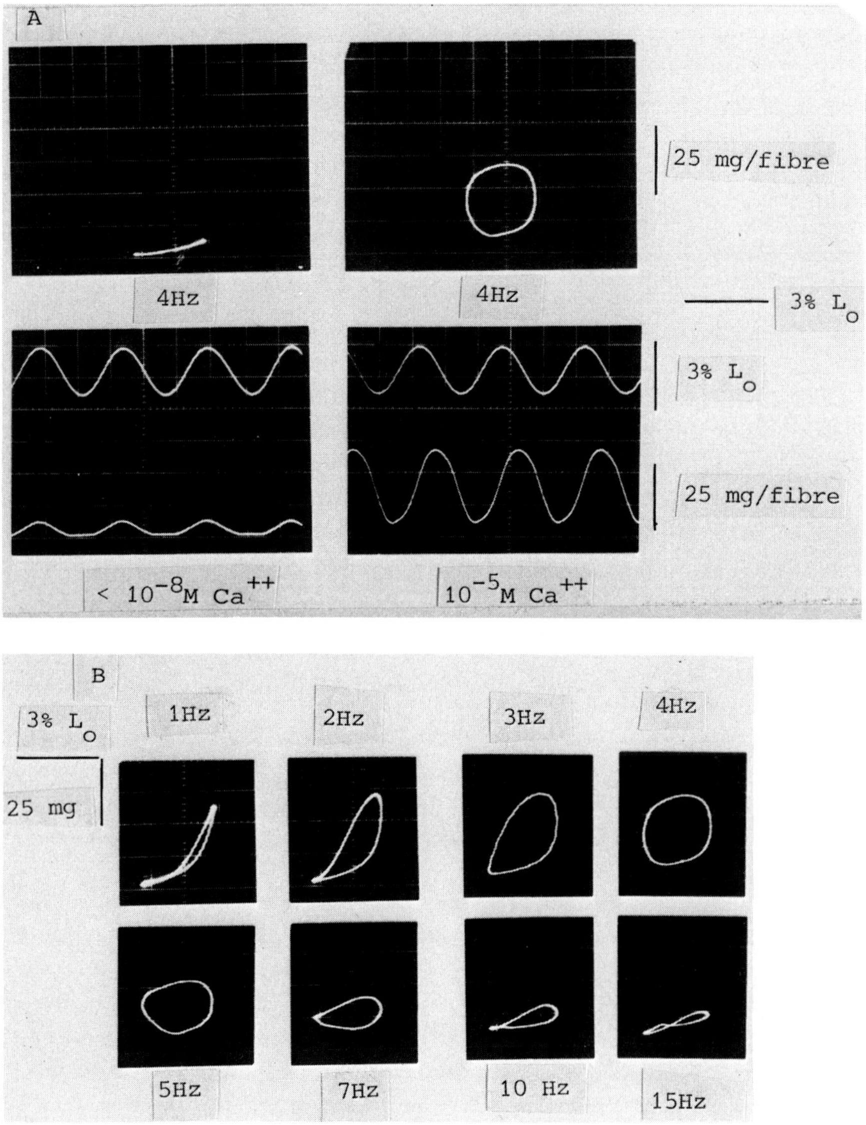

Fig.28 Frequency dependence of sinusoidal oscillations in fibrillar insect flight muscle.

A. Left: Ca^{++} 10^{-8} M
Right: Ca^{++} 10^{-5} M
Top: Length-tension curve
Bottom: Sinusoidal length (above) and tension changes (below).

B. Frequency dependence at 10^{-5} M Ca^{++}

Experimental conditions: 21° C, 15 mM ATP, 17.5 mM Mg^{++}, pH 6.8.

isometric oscillations has been found to correspond to the molecular ATP turnover rate per myosin per second observed at maximum tension (Breull, *et al.*, 1973). From this triple coincidence in both types of muscles it is concluded that the observed tension responses most probably reflect the same type of mechanism and hence that the molecular turnover rate of the myosin crossbridges in rabbit cardiac fibres is about one order of magnitude slower than in DLM fibres of *Lethocerus maximus* under comparable experimental conditions.

Fast and Slow Tension Transients in Cardiac, Skeletal and Insect Flight Muscle

Following the fast step perturbations in length the resulting tension response could always be described adequately by a maximum of four exponential terms. The sinusoidal analysis in cardiac muscle with frequencies between 0.05 and 30 Hz revealed two rate constants. These rate constants are identical with K_2 and K_3 obtained by the fitting procedure of the step responses. This is an excellent confirmation of the usefulness of the exponential analysis. Unfortunately the sinusoidal analysis of the step responses was only carried out up to 30 Hz so that it was not possible to detect K_1, but it is obvious from the results that if K_2 and K_3 have been measured correctly there must be another faster component. K_4, the slow decay following the delayed tension maximum, was usually outside the range which could be measured accurately, and even at 24° or 36° was of small amplitude.

Comparing the values of K_1, K_2 and K_3 found in cardiac muscle with the values for psoas and insect flight muscle (*Lethocerus cordofanus*) given by Abbott and Steiger (1977), the values for K_1 are about 4 to 6 times slower in cardiac muscle, the values for K_2 roughly 15 times and the values for K_3 are approximately 25 to 50 times slower in the heart than in the psoas or insect flight muscle.

It was mentioned earlier that the kinetics of the delayed tension rise in *Lethocerus maximus* fibres are approximately 10 times faster than of rabbit papillary muscle. The above data, however, indicate that K_3, the rate constant of the delayed tension rise, is 25 to 50 times faster. This apparent discrepancy is due to the fact, that in those experiments DLM fibres of *Lethocerus cordofanus* have been used which are about twice as fast as *Lethocerus maximus* fibres and further, 20 mM phosphate buffer was used instead of histidine or imidazole. Phosphate is known to increase the oscillation frequency, the frequency of the isometric oscillation and the kinetics of the delayed tension rise (Rüegg, *et al.*, 1971). It is not known whether P_i also affects K_1 or K_2. As shown in Figs. 15 and 18, the form and time course of the tension transients vary widely with the ionic conditions.

The rate constants and their amplitudes are altered in a specific way. A careful analysis of the tension transients and their components under these various conditions should reveal precious information about the link between the biochemical and mechanical cycles in contraction.

ACKNOWLEDGEMENTS

I am greatly indebted to Mrs. Elizabeth Hanke, Mrs. Iris Lindemann, Mrs. Liesel Mintert and Mrs. Bärbel Poeschmann for their excellent technical assistance, to Professor Allan J. Brady for reading the draft and for his very helpful criticisms and suggestions and to Bradford A. Lubell for some programming. Supported by Grant Ru 154/3-7 of the Deutsche Forschungsgemeinschaft and Grant HL 11351-10 from the Public Health Service of the United States.

REFERENCES

Abbott R.H. and Mannherz H.G. (1970) *Pflügers Arch. ges. Physiol.*, 321, 233.

Abbott R.H. (1972) *Cold Spring Harbor Symp. Quant. Biol.*, 37, 647.

Abbott R.H. and Steiger G.J. (1977) *J. Physiol.*, 266, 13.

Aidley D.J. and White D.C.S. (1969) *J. Physiol.*, 205, 179.

Armstrong C.F., Huxley A.F. and Julian F.J. (1966) *J. Physiol.*, 186, 26P.

Bárány M., Gaetjens E., Bárány K., and Karp E. (1964) *Arch. Biochem. Biophys.*, 106, 280.

Brady A.J. (1971) *Cardiovasc. Res.*, 1, 11.

Breull W., Steiger G.J. and Rüegg J.C. (1973) *J. Mechanochem. Cell Motility*, 2, 91.

Goodall M.C. (1956) *Nature*, 177, 1238.

Heinl P., Herzig J., Sawaya M.C.B. and Steiger G.J. (1973) *Pflügers Arch.*, 339, R.12.

Heinl P., Kuhn H.J. and Rüegg J.C. (1974) *J. Physiol.*, 237, 243.

Herzig J.W. and Herzig U.B. (1974) *Symp. Biol. Hung.*, 17, 85.

Huxley A.F. (1957) *Prog. Biophys. Biophys/chem.*, 7, 255.

Huxley A.F. and Simmons R.M. (1972) *Cold Spring Harbor Symp. Quant. Biol.*, 37, 669.

Julian F.J. and Sollins M.R. (1975) *J. Gen. Physiol.*, 66, 287.

Kawai M. and Brandt P.W. (1975) *Biophys J.*, 15, 154a.

Kawai M. and Orentlicher M. (1976) *Biophys J.*, 16, 152a.

Lorand L. and Moos C. (1956) *Nature*, 177, 1239.

McClellan G., Lai N.P. and Winegrad S. (1976) *Biophys J.*, 16, 72a.

Pringle J.W.S. (1967) *Prog. Biophys.*, 17, 1.

Pringle J.W.S. and Tregear R.T. (1969) *Proc. Roy. Soc. Lond. B.*, 174, 33.

Provencher S.W. (1976) *J. Chem. Phys.*, 64, 2772.

Rüegg J.C. (1968) *Experientia*, 24, 529.

Rüegg J.C., Steiger G.J. and Schädler M. (1970) *Pflügers Arch.*, 319, 139.

Rüegg J.C., Schädler M., Steiger G.J. and Müller G. (1971) *Pflügers Arch.*, 325, 359.

Schilling M.O. (1960) *Rev. Sci. Instr.*, 31, 1215.

Steiger G.J. and Rüegg J.C. (1969) *Pflügers Arch.*, 307, 1.

Steiger G.J. (1971) *Pflügers Arch.*, 330, 347.

Steiger G.J., Brady A.J. and Tan S.T. (1976) *Biophys J.*, 16, 206.

Steiger G.J., Brady A.J. and Tan S.T. (1977) *Circ. Res.* (in press).

Thorson J. and White D.C.S. (1969) *Biophys J.*, 9, 360.

Wezler K. (1973) *Nova Acta Leopoldina*, 38, 10.

White D.C.S. (1972) *Cold Spring Harbor Symp. Quant. Biol.*, 37, 201.

White D.C.S. and Thorson J. (1973) *Prog. Biophys. Mol. Biol.*, 27, 173.

References added in proof:

Heinl, P. (1972) *Pflügers Arch. gen. Physiol.*, 333, 213.

Perrin, D.D. and Sayce, I.G. (1967) *talanta.*, 14, 833.

THE RELATIONSHIP BETWEEN BIOCHEMICAL KINETICS AND MECHANICAL PROPERTIES

Roger H. Abbott

A.R.C. Unit, Department of Zoology, Oxford, U.K.

For many years an explanation of the mechanical behaviour of muscle has been sought in terms of states of the cross-bridges and the rate constants of the transitions between them. Most of this work is based upon the ideas put forward by Huxley (1957). It has long been recognised that these postulated states must correspond to the states, or groups of states, of the muscle proteins as identified by kinetic biochemical studies. (Lymn and Taylor, 1971; White, 1972). A fundamental problem in correlating the mechanical and biochemical studies is that proteins in solution neither shorten against a load nor can the rate constants be altered by mechanical conditions. Both these things are important and certainly take place in working muscle.

There are several ways of by-passing this difficulty. The mechanical experiments can be confined to essentially isometric conditions by using only small changes of length. It is then at least possible that the rate constants in the muscle fibres will be the same as those measured in solutions. Length changes large or fast enough to cause an asymmetric response to stretch and release are almost certainly altering the rate constants. Another possibility is to make parallel studies on muscle fibres and isolated proteins, changing conditions such as chemical concentrations in each, and looking for effects which could have a common molecular origin. Probably the most potentially powerful method, because it can be used with large length changes, is to adapt some of the biochemical techniques such as fluorescent labelling to intact fibres. So far the practical difficulty that the signal to noise ratio is very low has prevented any useful quantitative results being obtained in this way. Whichever approach is chosen a common model must be constructed which mimics both the mechanical and chemical behaviour of the intact muscle and the kinetic properties of the isolated proteins. The first of these approaches is used in this paper.

The experiments were carried out on single fibres or small bundles of fibres of the dorsal longitudinal muscle of the giant water bug *Lethocerus cordofanus*. The dominant feature of the tension response of this fibrillar flight muscle to length changes is an exponentially delayed change of active tension; this is the stretch activation phenomenon which accounts for the myogenic oscillation of this type of muscle. A sudden length change is a disturbance of the system and the resulting tension change can be considered as a reflection of the approach of the system to a new equilibrium state. The rate constant of this tension change can be measured by applying sinusoidal length changes as well as step changes; both methods have been used in obtaining the results which will be discussed here. The rate constant will be called the mechanical rate constant to distinguish it from the kinetic rate constants obtained in studies on the isolated proteins. The delayed tension response to length changes is also observed in vertebrate skeletal muscle, but the stretch activation is not maintained and its amplitude relative to other components of the transient response is not so great. However, the tension transient rate constants obtained with *Lethocerus* flight muscle and rabbit psoas muscle are very similar; the vertebrate muscle is marginally slower (Abbott and Steiger, 1977).

The measurements we shall be concerned with are of the average tension, the ATPase activity and the mechanical rate constant. All were measured at room temperature, pH 7 and a KCl concentration near 70 mM. The conditions varied were mean muscle length (also called stretch) and calcium ion concentration. The small test length changes which were superimposed on the mean length to measure the mechanical rate constant had an aplitude of 0.2% peak to peak.

The following results must be accounted for:

(i) Static stretch increases the ATPase activity and tension. The ratio ATPase activity/tension, known as tension cost, does not vary with stretch (Pybus and Tregear, 1975). The mechanical rate constant is also not affected (Abbott, 1973).

(ii) When the calcium ion concentration is increased the ATPase activity and tension rise. The mechanical rate constant is also increased (Abbott, 1973), but the tension cost remains constant or is slightly reduced (Abbott and Loxdale, paper in preparation).

(iii) The turnover rate of the ATPase ranges up to about 10 s^{-1}. The rate constant of the delayed tension varies from about 15 - 200 s^{-1} as the calcium ion concentration is varied.

All these findings except one can be accounted for by a two-state model, first discussed in relation to the delayed tension response by Thorson and White (1969). The cross-bridges are either attached and producing tension or are not attached. The attachment rate is denoted by f and the detachment rate by g. In such a cycle the proportion of attached bridges is given by $f/(f + g)$, the turnover rate is $fg/(f + g)$ and the rate constant of approach to equilibrium when the system is disturbed (for example by length changes) is $f + g$. It can be seen that the ratio of turnover rate to the fraction of attached bridges is simply g. Under isometric conditions the tension cost will therefore be proportional to g, under the reasonable assumption that isometric tension is proportional to the number of attached bridges.

The only two parameters which can be varied are f and g. Since stretching the fibres does not change the tension cost, it cannot change g. But stretching does increase both ATPase turnover rate and tension so f must be increased. Since the static length does not alter the mechanical rate constant the sum f + g is not altered by length changes. Therefore f must be much smaller than g. Since an increase in calcium ion concentration increases the mechanical rate constant, it should also increase the tension cost. It has been found that tension and ATPase activity increase in proportion when the calcium ion concentration is raised implying a constant tension cost (Schädler, 1967), but in these experiments the mechanical rate constant was not measured. The result has been checked using a new technique which allows simultaneous measurement of the ATPase activity, the tension and the mechanical rate constant (Loxdale, 1976). It has been confirmed (Abbott and Loxdale, in preparation) that calcium ion concentration increases the mechanical rate constant simultaneously with a constant or even a slightly reducing tension cost. The two-state model is, therefore, inadequate for explaining the isometric behaviour of glycerinated insect fibrillar muscle.

There are many possible elaborations of the simple two-state cycle. Recent work by Sleep and Taylor (1976) and White and Taylor (1976) indicates that there are probably four significantly populated states in the actomyosin ATPase at room temperature when there is a plentiful supply of ATP.

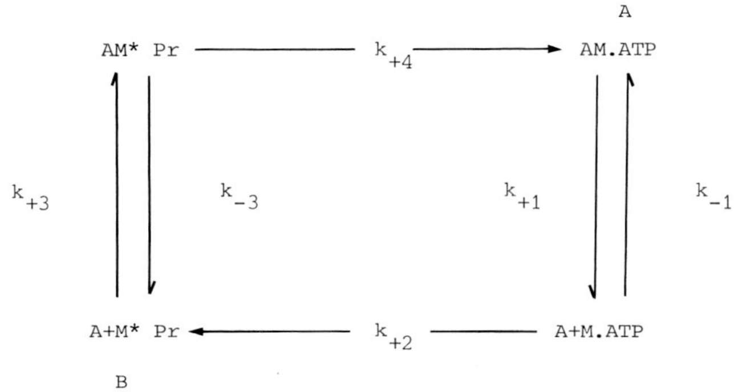

Fig.1. Proposed four-state cycle.

This is basically the cycle proposed by Lymn and Taylor (1971). The state labelled A is a mixture of various states including AM.ADP and AM, which will be present at much lower concentration because product release is rapid relative to k_{+4} and the ATP concentration is 2 - 3 orders of magnitude higher than the apparent K_m for ATP. State B is a rapid equilibrium between M*Pr and M*ATP, the equilibrium constant being around 10. This kinetic scheme can be used to model the properties of muscle fibres if an assumption is made about the contribution of each state to tension generation. For simplicity I have assumed that both attached states (AM*Pr and AM.ATP) generate the same tension and the others none.

The previously mentioned experimental results provide a surprisingly tight constraint on the possible values of the rate constants and the parameters which may be altered by calcium ion concentration and muscle length. If any of the rate constants are altered sufficiently to change the tension or ATPase activity (or both) then the mechanical rate constant is altered. It is possible to mimic the effect of stretch by altering the effective actin concentration. This is not to say that stretch must necessarily affect the actin filaments; there must be some change in geometry which makes more actin available for interaction with myosin, and this could be an effect of muscle length on either set of filaments or both. If either k_{+2} or k_{+4} is changed, the ATPase activity and tension move in opposite directions, contrary to the effects of length and calcium ion concentration. So they are almost certainly unaltered and calcium ion concentration must alter one or both equilibrium steps. The only combination which mimics the effect of calcium ions is to make step 1 a rapid equilibrium, $k_{+2} \geqslant 100\ s^{-1}$, K_3 approx. $2 \times 10^3\ M^{-1}$ and $k_{+4} = 10 - 50\ s^{-1}$. The effect of calcium ion activation can be simulated by changing both k_{+3} and k_{-3} without changing the equilibrium constant. This is interesting in view of the fact that calcium does not change the dissociation constant of acto-subfragment-1 (Marston and Weber, 1975). k_{+1} and k_{-1} may also change but they do not have to. Calcium could also change the effective actin concentration but again this is not a necessary assumption.

These rate constants are entirely consistent with the ones suggested by Taylor and his co-workers. k_{+1}, the dissociation rate of AM.ATP is about $10^3\ s^{-1}$ and the association rate is probably about $10^7\ M^{-1}\ s^{-1}$, giving an equilibrium constant of $10^4 M^{-1}$. This value is not critical to the fit of the mechanical and fibre ATPase activity data. k_{+2} may be the rate of the fluorescence change after dissociation which is 150 s^{-1} at 20°C. White and Taylor (1976) found that the product $K_3 k_{+4}$ (which they called $k_4{}^a$) at 20°C is about $10^4 M^{-1} s^{-1}$ at 100 mM KCl and 10^5 at 40 mM KCl. The mechanical experiments are usually done at around 70 mM KCl and the product $K_3 k_{+4}$ of

$2 \times 10^4\ M^{-1}s^{-1}$ deduced from mechanical experiments, which is fairly critical for obtaining a fit to the results, agrees well. k_{+4} has a major influence in determining the ratio of turnover rate to fraction of attached bridges and must be in the range 10–50 s^{-1}.

This agreement between rate constants of a four state model as deduced from both mechanical and biochemical studies offers the exciting possibility that the same states have indeed been identified in both types of experiments. The idea is readily testable because KCl concentration has a marked effect on the equilibrium constant identified with K_3, which should produce easily detected changes in the relationship between ATPase activity, tension and rate constant in the mechanical experiments. Temperature has big effects as well, but the interpretation may be complicated because at temperatures near $0^{\circ}C$ there will probably be more significantly populated states. In particular, the equilibrium between M*ATP and M*Pr is not rapid and the release of ADP from M.Pr is much slower than at room temperature. The cycle also offers an explanation of the mechanical effects of ADP, which are consistent with a prolonging of the attachment of the crossbridges (Pringle, 1967; Abbott and Steiger, 1977). This can be done by reducing k_{+1}. An effect of ADP in slowing the observed rate of dissociation of actomyosin by MgATP has been reported by White (1976), who ascribed it to simple competition between MgADP and MgATP for the nucleotide binding site.

REFERENCES

Abbott R.H. and Steiger G.J. (1977) *J. Physiol.*, 266, 13.

Abbott R.H. (1973) *J. Physiol.*, 231, 195.

Huxley A.F. (1957) *Prog. Biophys*, 7, 255.

Loxdale H.D. (1976) *J. Physiol.*, 260,4P.

Lymn R.W. and Taylor E.W. (1971) *Biochemistry*, 10, 4617.

Marston S. and Weber A. (1975) *Biochemistry*, 14, 3868.

Pringle J.W.S. (1967) *Prog. Biophys*, 17, 1.

Pybus J. and Tregear R.T. (1975) *J. Physiol.*, 247, 71.

Schädler M. (1967) *Pflüg. Arch. gesamte Physiol.*, 296, 70.

Sleep J.A. and Taylor E.W. (1976) *Biochemistry*, 15, 5813.

Thorson J. and White D.C.S. (1969) *Biophys J.*, 9, 360.

White D.C.S. (1972) *Cold Spring Harb. Symp. quant. Biol.*, 31, 201.

White H.D. (1976) *Abstracts of the International Symposium on contractile systems in non-muscle tissues, Bressanone, Italy, September 1976.*

White H.D. and Taylor E.W. (1976) *Biochemistry*, 15, 5818.

BIOCHEMICAL KINETICS

THE DUAL SYSTEM OF CALCIUM-ACTIVATION IN ARTHROPOD MUSCLES

DR. WILLIAM LEHMAN

Department of Physiology, Boston University School of Medicine, 80, East Concord Street, Boston, Massachusetts 02118.

My first encounter with a member of Professor Pringle's ARC Unit occurred after I presented a paper on thin filaments at the 1972 meeting of the Biophysical Society in Toronto. Richard Tregear pointed out that I had not only mispronounced *Lethocerus* but had misspelled it on one of my slides. We thereupon agreed that my joining the ARC Unit might improve my appreciation of the English language and culture; furthermore, my participation in the characterization of calcium-dependent regulatory systems in insect muscles might even prove to be useful. As has been often the case, an understanding of the properties of insect muscle has had more general relevance, and our results could lead to a re-evaluation of our concepts on the evolution of calcium-regulation in the animal kingdom.

The role of calcium in activating the actomyosin ATPase and its function in tension development of isolated muscle fibres is well established. (Perry and Grey, 1956; Weber and Winicur; 1961; Portzehl *et al.*, 1964; Leenders, 1966; Rüegg and Tregear, 1966; Vom Brocke, 1966; Chaplain, 1967; Schädler, 1967; Maruyama *et al.*, 1968a; Maruyama *et al.*, 1968b; Portzehl *et al.*, 1969; Brandt *et al* , 1972; Kendrick-Jones *et al.*, 1970). Ebashi (1963) first demonstrated that the calcium-dependency of the interaction of vertebrate actin and myosin is due to the complexing of troponin and tropomyosin to actin (Ebashi and Endo, 1968). Szent-Györgyi, Kendrick-Jones and I showed the presence of a second regulatory mechanism, in molluscs and other invertebrates, in which the calcium-dependency resides in the myosin molecules (Kendrick-Jones *et al.*, 1970; Lehman and Szent-Györgyi, 1975).

This myosin-linked regulatory system can be shown in two ways (Lehman and Szent-Györgyi, 1975). The isolated myosin fraction is mixed with tropomyosin-troponin free actin and is then analysed directly for a calcium-dependent ATPase reaction. The results are then correlated with calcium-binding studies. Prior to my arrival in Oxford, myosin-regulation had been shown, using this approach, in muscles from molluscs and brachiopods. An indirect and more controversial assay, "the competitive actin binding assay" was used to test for myosin-linked regulation in other animals and indicated its presence in several other phyla including annelids and insects. In this

TABLE I

ATPase Activities Conferred by Locust Thin Filaments and Locust Myosin*

	Mg-ATPase		Calcium-sensitivity ‡
	EGTA	Calcium	
	µmol ATP split/min/mg myosin		%
Rabbit actin & locust myosin	0.023	0.083	72
Rabbit actin & rabbit tropomyosin & locust myosin	0.068	0.27	75
Locust thin filaments & rabbit myosin	0.1	0.5	80
Locust thin filaments & locust myosin	<0.008	0.31	>97

* Data from Lehman et al. (1974)

‡ $\left(\frac{ATPase_{Ca} - ATPase_{EGTA}}{ATPase_{Ca}}\right) \times 100$

assay, pure rabbit actin is mixed with myofibrils or actomyosin. If the pure actin, in the presence of EGTA, activates the ATPase rate to the level found in the presence of calcium, the preparation is solely actin-regulated; if on the other hand the ATPase is low in EGTA, the preparation is myosin-regulated. The competitive actin binding assay is based on myosin being free to interact with the added pure actin, in the absence of calcium, in actin regulated systems, but inhibited from interacting in myosin-regulated systems. This competitive assay requires only milligram amounts of protein and avoids the effects of losses and protein denaturation, which can occur during more extensive purification methods.

At the time these experiments were performed, it was already known that insects and annelids have a thin filament-linked calcium-regulatory system (Maruyama *et al.*, 1968b; Lehman *et al.*, 1972). The competitive actin binding experiments suggested the presence of an additional myosin-linked regulatory system; thus a dual-regulatory system was proposed in these species (Lehman *et al.*, 1972). Because of the possible errors in interpretation of this indirect assay, it was felt that direct demonstration of myosin-regulation in a dual-regulated muscle was imperative. Belinda Bullard and I (Lehman *et al.*, 1974) found locust flight muscles particularly suitable for preparing purified myosin which when mixed with pure actin forms a calcium-dependent ATPase (Table I). This direct demonstration of a calcium-regulated myosin in an insect confirmed the competitive actin binding assay. As would be expected from other studies (Maruyama et al., 1968b; Lehman *et al.*, 1972), reciprocal experiments using locust thin filaments indicated that they are also calcium-regulated (Table I). Taken together these results confirm a dual system of calcium-regulation.

The presumed advantage of dual-regulation is enhancement of calcium-sensitivity which indeed was demonstrated in locusts and later confirmed in *Limulus* (Lehman *et al.*, 1974, Lehman and Szent-Györgyi, 1975). The individual calcium-sensitivities of either their myosin or thin filaments, when assayed as actomyosin with the appropriate calcium-insensitive pure rabbit protein, are substantially lower than mixtures of both calcium-dependent thin filaments and calcium-dependent myosin (Table I).

Later extensive comparative studies (Lehman and Szent-Györgyi, 1975) were carried out to categorise the different muscles of the animal kingdom according to their calcium-regulatory systems. These studies showed that most invertebrate muscles are actually dual-regulated with the exception of molluscan and brachiopod muscles which show no evidence of thin filament-regulation. On the other hand, solely thin filament-regulation was found in all vertebrate striated muscles tested, but only rarely in invertebrates, e.g. decapod-crustacean muscles with fast ATPases and in another minor crustacean order.

The mechanism by which calcium regulates insect myosin or, for that matter, myosin from any other arthropod is not known. In molluscs, which have been more extensively studied, a specific class of myosin light chain subunits is found to be necessary for myosin-regulation (Szent-Györgyi *et al.* 1973), but the details of their function have not been elucidated. The thin filament-linked troponin-tropomyosin systems found in insects and other arthropods appear to have the same general characteristics as those in vertebrates, and both troponin and tropomyosin are required for conferring the calcium-sensitivity (Lehman *et al.*, 1972; Bullard *et al.*, 1973; Regenstein and Szent-Györgyi, 1975). Like vertebrate troponin, lobster troponin consists of three components, TN-I, TN-C and a third is presumed to be analagous to TN-T (Regenstein and Szent-Györgyi, 1975). The stoichiometry of these proteins on *Limulus* thin filaments is analagous to that in vertebrates, there being 7 actin molecules for one tropomyosin, TN-T and TN-C. However, two TN-I's may occur in the complex (Lehman *et al.*, 1976).

The general similarity in behaviour of arthropod and vertebrate troponin does not necessarily distinguish between two possible phylogenetic pathways. In one there could be convergent evolution from independent origins; alternatively there could be a common ancestor. If the latter were the case, it would be likely that the troponin components of each phyla would be similar in structure and interaction properties. If so, troponin subunits from one species should interact with the complementary ones from a species of the other phyla. We demonstrated such interaction by making functional hybrids from troponin components of arthropod and vertebrate muscle (Lehman 1975). Troponin-C, from *Limulus* can substitute functionally for vertebrate troponin-C, since when mixed with rabbit troponin I and T and tropomyosin it confers a calcium-dependency on the vertebrate actomyosin ATPase. In a complementary experiment lobster troponin I and chicken troponin C and T were mixed and again a functional hybrid conferred calcium-sensitivity (Figures 1 and 2).

Richard Tregear, in his inimitable way, found dual-regulation, to say the very least, exciting, and in light of the similarities of the thin filament-linked regulatory systems throughout the animal kingdom thought that with improved techniques and understanding we might find all muscles to be dual-regulated. Belatedly I have begun to share his views. The following set of experiments indicate that dual-regulation may have been overlooked in decapod-crustacea and in vertebrate striated muscle because of the assay conditions employed. As already mentioned, the only large group of invertebrate muscles tested showing solely thin filament regulation are the fast muscles of certain decapods. In lobster, in which both fast and slow muscles were tested, the slow muscles in fact showed dual-regulation and

TABLE II

ATPase Activities of Various Lobster Preparations
in the Presence of Rabbit Actin*

	mM NaCl	Mg-ATPase EGTA	Mg-ATPase Calcium	Calcium-sensitivity
		µmol ATP split/min/mg lobster proteins		%
Lobster Myofibrils	27	1.47	1.9	23
& Rabbit Actin	87	0.036	0.7	95
Lobster Actomyosin	26	0.9	0.9	0
& Rabbit Actin	86	0.24	0.74	68
Lobster Myosin	20	1.13	1.13	0
& Rabbit Actin	84	0.23	0.57	60

* Data extracted from Lehman (1977); in these selected results a calcium-dependency is observed at raised ionic concentration but is partially or completely obliterated at very low NaCl concentration. The ratio of rabbit actin: lobster proteins was 1:3(mg/mg) in each case.

TABLE III

ATPase Activities of Rabbit Preparations
in the Presence of Rabbit Actin*

	mM NaCl	Mg-ATPase EGTA	Mg-ATPase Calcium	Calcium-sensitivity
		µmol ATP split/min/mg		%
Rabbit Myofibrils	26	0.33	0.38	13
& Rabbit Actin	88	0.073	0.30	76
Rabbit Myosin	23	0.47	0.47	0
& Rabbit Actin	83	0.37	0.37	0

* Data extracted from Lehman (1977); in these selected results a calcium-dependency is observed only in mixtures of rabbit myofibrils and rabbit actin at raised ionic concentration but not in mixtures of purified rabbit myosin and actin.

Rabbit actin was added in a 1:3 ratio (mg/mg) in each case.

the fast muscles only thin filament-regulation (Lehman and Szent-Györgyi, 1975). The physiological significance of those results was unclear. Compounding this confusion were results of the competitive actin binding assay showing crayfish muscles, homologous to lobster fast muscles, to be dual-regulated (Lehman 1977).

One feature distinguishing most marine decapod fast muscles from other arthropod muscles is their high actomyosin ATPase activity, being four to five times the rate found for other arthropod actomyosin ATPases. This maximum ATPase activity is obtained in a solution of low ionic concentration (20-40mM NaCl). The question therefore arises whether myosin-linked regulation can be observed at a higher ionic concentration (80-100mM NaCl), which, although closer to the physiological level (Hinke, 1959), leads to a decrease in the ATPase activity. Under this condition myosin-linked regulation was indeed observed (Lehman 1977; Table II). My interpretation of these results is that the association of the fast decapod myosin with actin is so increased at low ionic strength (20-40mM) that binding to pure actin can occur in the absence of interaction of myosin and calcium, whereas at more physiological ionic strengths the binding of actin and myosin is reduced and modulated by calcium.

In my view, these studies show that probably all invertebrate muscle myosins are calcium-regulated. Once again the question is raised whether myosin-regulation might also occur in vertebrate striated muscle, the only apparent exception, since it is now recognized that several vertebrate smooth muscles, with relatively low ATPase rates, are also myosin-regulated (Bremel, 1974; Mwra and Rüegg, 1975; Sobieszek and Small, 1976). By analogy to the analysis of decapod fast muscles, evidence of myosin-regulation in vertebrate striated muscle might be obtained at more physiological ionic concentrations. My results using different rabbit muscle preparations are, however, difficult to interpret. The competitive actin binding assay on rabbit myofibrils shows myosin regulation at ionic concentrations of 80mM or greater, but purified rabbit myosin when recombined with pure actin is calcium-independent at all ionic strengths tested (Table III) (Lehman 1977). One explanation for these disparate results is that the myosin in vertebrate striated muscle is labile and its calcium-regulating component is lost or denatured using standard preparative techniques. We are currently pursuing this problem further.

These comparative studies of invertebrate muscles, including arthropods have not only revealed specific information but have contributed broadly to an improved understanding of calcium-regulation of muscle contraction. The arthropod phylum still offers a wealth of unexplored avenues, which should lead to both a better understanding of the evolution of the regulatory systems and help determine the molecular mechanism of the regulation of contraction.

REFERENCES

Brandt P.W., Reuben J.P., and Grundfest H (1972) *J. Gen. Physiol.,* 59, 305.

Bremel R.D. (1974) *Nature (Lond.),* 252, 405.

Bullard B., Dabrowska R. and Winkelman L. (1973) *Biochem. J.* 135,277.

Chaplain R.A. (1967) *Biochim. Biphys. Acta* 131, 385.

Ebashi S. (1963) *Nature (Lond.),* 200, 1010.

Ebashi S. and Endo M. (1968) *Progr. Biophys. Mol. Biol.,* 18, 123.

Hinke J.A.M. (1959) *Nature* 184, 1257.

Kendrick-Jones J., Lehman W. and Szent-Györgyi A.G.(1970) *J. Mol. Biol.,* 54, 313.

Leenders H.D. (1966) *Naturwissenschaften* 53, 617.

Lehman W. (1975) *Nature,* 255, 424.

Lehman W. (1977) *Biochem. J.* 163, in press.

Lehman W., Bullard B., and Hammond K. (1974) *J. Gen. Physiol,* 63, 553.

Lehman W., Kendrick-Jones J. and Szent-Györgyi A.G. (1972) *Cold Spring Harbor Symp. Quant. Biol.* 37, 319.

Lehman W. and Szent-Györgyi A.G. (1975) *J. Gen. Physiol,* 66, 1.

Lehman W., Regenstein J.M. and Ransom A.L. (1976) *Biochim. Biophys. Acta.* 434, 215.

Maruyama K., Nagashima S. and Drabikowski W. (1968a) *Comp. Biochem. Physiol.* 25, 1107.

Maruyama K., Pringle J.W.S. and Tregear R.T. (1968b) *Proc. Roy. Soc., Ser. B.,* 169, 229.

Mwra V. and Rüegg J.C. (1975) *F.E.B.S. Letters,* 60, 81.

Perry S.V. and Grey T.C. (1956) *Biochem. J.* 64, 5P.

Portzehl H., Caldwell P.C. and Rüegg J.C. (1964) *Biochim. Biophys. Acta,* 79, 581.

Portzehl H., Zaoralek P., and Gaudin H. (1969) *Biochim. Biophys. Acta,* 189, 440.

Regenstein J.M. and Szent-Györgyi A.G. (1975) *Biochem.* 14, 917.

Rüegg J.C. and Tregear R.T. (1966) *Proc. Roy. Soc., Ser. B.* 165, 497.

Schädler M. (1967) *Pfluegers Arch. Gesamte Physiol. Menschen Tiere,* 296, 70.

Sobieszek A. and Small J.V. (1976) *J. Mol. Biol.* 102, 72.

Szent-Györgyi A.G., Szent Kiralyi E.M. and Kendrick-Jones J. (1973) *J. Mol. Biol.* 74, 179.

Vom Brocke H.H. (1966) *Pfluegers Arch. Gesamte Physiol. Menschen Tiere* 290, 70.

Weber A. and Winicur S. (1961) *J. Biol. Chem.* 236, 3198.

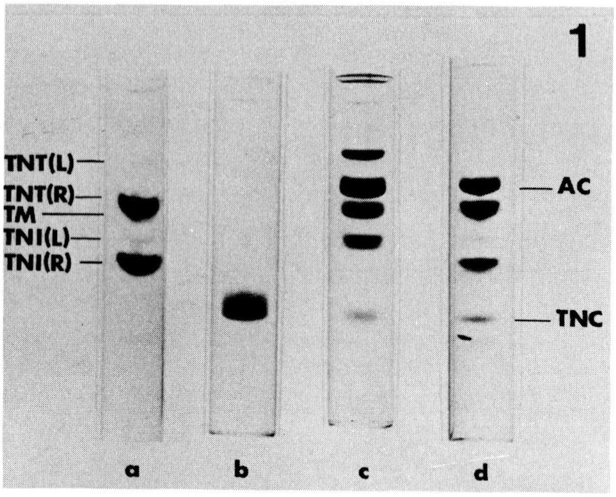

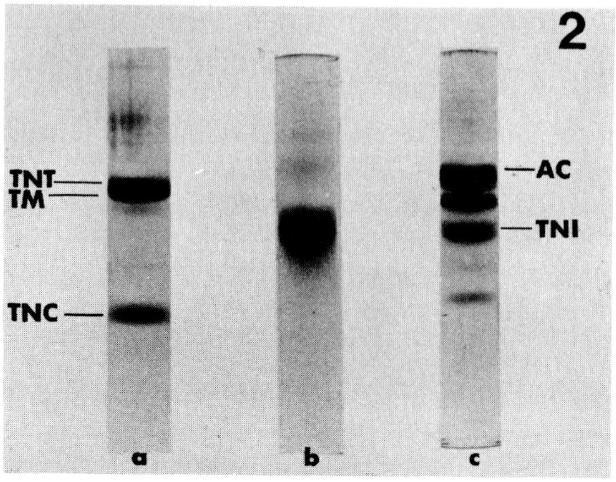

Fig. 1 10% SDS polyacrylamide gels using Coomassie Brilliant Blue stain. a, Rabbit TN-I-T. b, *Limulus* TN-C. c, Native *Limulus* thin filaments. d, Reconstituted thin filament complex of rabbit actin, chicken tropomyosin, rabbit TN-I-T, and *Limulus* TN-C. All proteins resolve into separate bands except tropomyosin and rabbit TN-T on gel d. The hybrid thin filament complex shown on gel d. conferred a 52% calcium-dependency on the Mg-ATPase of rabbit myosin. Data extracted from Lehman (1975).

Fig. 2 10% SDS polyacrylamide gels using Coomassie Brilliant Blue stain. a, Chicken TN-C-T. b, Lobster TN-I. c, Reconstituted thin filament complex of rabbit actin, chicken tropomyosin, chicken TN-C-T, and lobster TN-I. The hybrid thin filament complex shown on gel c. conferred a 58% calcium-dependency on the Mg-ATPase of rabbit myosin.

PHOSPHORYLATION OF A LIGHT SUBUNIT OF LOCUST MYOSIN

Lowell Winkelman and Belinda Bullard.

A.R.C. Unit, Department of Zoology, South Parks Road, Oxford.

INTRODUCTION

The actomyosin ATPase of insect flight muscle has dual control: the thin filaments contain a troponin complex and the myosin is itself capable of imparting calcium sensitivity to the system (Lehman *et al.*, 1974). The calcium sensitivity of the myosin is, however, very labile and is lost if the insect myosin is precipitated.

Recently several investigators have been studying the phosphorylation of myosin light chains (Frearson and Perry, 1975; Perry *et al.*, 1975; Adelstein and Conti, 1976), and it seemed possible that the loss of calcium sensitivity of the insect myosin might be related to a change in the state of phosphorylation. We have begun an investigation of the role of phosphorylation in the activity of flight muscle myosin from the locust *Schistocerca gregaria*.

METHODS

Protein preparations

Flight muscles were dissected from 5-15 locusts for each myosin preparation. The myofibrils were washed and the myosin extracted as described by Lehman *et al.*, (1974). Myosin extracts obtained in this manner were normally in the concentration range 5-8 mg/ml. Some of the myosin was precipitated by dialysing against 40 mM KCl, 15 mM tris pH 7, 8 mM $MgCl_2$, 5 mM DTT, 10^{-4}M phenylmethylsulfonylfluoride, and either 0.1 mM $CaCl_2$ or 1 mM EGTA. The whole light chain fraction was prepared from unprecipitated myosin by alcohol fractionation in the presence of 5M guanidine hydrochloride (Perrie *et al.*, 1973).

Phosphorylation of locust myosin

Prior to phosphorylation the myosin was dialysed overnight against 0.1 M KCl, 25 mM tris pH 7.5, 5 mM $MgCl_2$, 2 mM DTT, 0.1 mM $CaCl_2$, 10 mM NaF. Samples (0.2 ml) of this myosin suspension were incubated at 27°C for 30 min with 0.25 mM γ-^{32}P ATP (approx. 5 μCi/assay) in the presence and absence of an added mixture of protein kinase and phosphorylase kinase (0.1 mg/ml each; both kinases were from rabbit muscle and were obtained from Sigma Chemical Co., London). The reaction was stopped by the addition of solid urea to 8 M and the samples were dialysed against 10 M urea, 0.1 M β-mercapotethanol (βMCE) prior to electrophoresis.

ATPase activities

The Mg-activated ATpase activity of locust myosin was measured in the presence of added rabbit actin and tropomyosin using the pH-Stat method previously described (Szent-Györgyi *et al.*, 1971). The ATPase activity of precipitated myosin was measured using a suspension of myosin not separated by centrifugation.

Electrophoresis

Electrophoresis was carried out on 10% polyacrylamide gels in 8 M urea at pH 8.6 as described by Perrie and Perry (1970). Prior to loading the samples were dialysed against 10 M urea, 0.1 MβMCE and then mixed with the marker dye Bromophenol blue.

SDS gel electrophoresis on 10% polyacrylamide gels in 0.1% SDS was performed according to Weber and Osborn (1969). Gels were stained with Coomassie brilliant blue and scanned at 555 nm in a Gilford densitometer attached to a Unicam SP500 recording spectrophotometer.

To determine the location of radioactively labelled polypeptides on these gels, the stained gels were frozen in powdered dry ice and cut into 1 mm slices on a Mickle gel slicer (Mickle Laboratory Engineering Co., Gomshall, nr. Guildford, Surrey). The slices were added to 2 ml of H_2O_2 and incubated overnight at 50°C in tightly sealed scintillation vials to dissolve the gel. 5 ml water was then added to each sample and ^{32}P determined by the Cerenkov method.

RESULTS AND DISCUSSION

ATPase activities

Our measurments of Mg-activated ATPase of locust myosin when combined with rabbit actin and tropomyosin confirm those of Lehman *et al.*, (1974) showing that the myosin is calcium sensitive when fresh but loses this control upon precipitation. In one preliminary experiment, addition of 0.1 mM $CaCl_2$ to the dialysis solution for precipitation protected the locust myosin from loss of calcium sensitivity (Table 1).

TABLE 1. ATPase ACTIVITIES OF LOCUST MYOSIN (μmoles/mg/min)*

	Ca^{++}	EGTA	Ca^{++}/EGTA
fresh	0.022	-	∞
1 day old	0.022	0.017	1.3
precipitated (Ca^{++})	0.07	0.033	2.1
precipitated (EGTA)	0.067	0.067	1.0

*Locust myosin was combined with rabbit actin and tropomyosin in the ratio 8:2:1 in 0.5 M KCl.

Electrophoresis of light chains

On SDS polyacrylamide gels locust myosin has two light subunits of molecular weights 30,000 and 20,000 daltons (Fig. 1). On 8 M urea gels also only two light chains are evident in freshly prepared, unprecipitated locust myosin. Neither light chain is removed by treatment of the myosin with 5,5'-dithiobis (2-nitrobenzoic acid) (Winkelman, unpublished results).

After precipitation, SDS gels of locust myosin show no change in light chain composition compared to unprecipitated myosin. However, on 8 M urea gels at pH 8.6, precipitated myosin has an additional band of slightly lower mobility than the slower band of unprecipitated myosin (Fig. 2). In these conditions a phosphorylated polypeptide migrates slightly more rapidly than its dephosphorylated counterpart (Perrie *et al.*, 1973) and the change that is seen in the precipitated locust myosin on 8 M urea gels would be expected if the lower mobility light subunit were being partially dephosphorylated by precipitation. The fact that no change is seen on the corresponding SDS gels makes this plausible.

Phosphorylation of a light subunit

When locust myosin was incubated with γ-^{32}P ATP under conditions favoring phosphorus incorporation in other systems (Adelstein and Conti, 1975), ^{32}P was bound to the myosin in the presence and absence of added phosphorylase kinase and protein kinase. This indicated that there was an endogenous kinase associated with the locust myosin that was capable of phosphorylating the myosin. When SDS gels of the labelled myosin were scanned and counted, the bound ^{32}P was found to be only on the 30,000 dalton light chain. Only the lower mobility light chain on urea gels was labelled with ^{32}P, (Fig. 3) therefore this corresponds to the 30,000 dalton light chain.

CONCLUSION

There is in locust muscle an endogenous kinase capable of phosphorylating the 30,000 dalton light chain of locust myosin. Upon precipitation, locust myosin loses its calcium sensitivity and its 30,000 dalton light undergoes a change, seen on 8 M urea gels, that is consistent with partial dephosphorylation. It remains to be proved that the loss of calcium sensitivity is caused by dephosphorylation, but circumstantial evidence is building up that makes it seem a possible explanation.

REFERENCES

Adelstein, R.S. and Conti, M.A. (1975) *Nature* 256, 597.

Adelstein, R.S. and Conti, M.A. (1976) *Cell Motility: Cold Spring Harb. Conf. on Cell Proliferation (Cold Spring Harbor Laboratory, Cold Spring Harbor, N.Y.) vol. 3, p. 725.*

Frearson, N. and Perry, S.V. (1975) *Biochem. J.*, 151, 99.

Lehman, W., Bullard, B. and Hammond, K.A. (1974) *J. Gen. Physiol.*, 63, 553.

Perrie, W.T. and Perry, S.V. (1970) *Biochem. J.*, 119, 31.

Perrie, W.T., Smillie, L.B. and Perry, S.V. (1973) *Biochem. J.*, 135, 151.

Perry, S.V., Cole, H.A., Morgan, M., Moir, A.J.G. and Pires, E. (1975) *Fed. Eur. Biochem. Soc. Meet. (Proc.)* 31, 163.

Szent-Györgyi, A.G., Szentkiralyi, E.M. and Kendrick-Jones, J. (1971) *J. Mol. Biol.*, 74, 179.

Weber, K. and Osborn, M. (1969) *J. Biol. Chem.*, 244, 4406.

Fig. 1. Electrophoresis of locust myosin light chains on 10% polyacrylamide gels in 8 M urea and in 0.1% SDS.

Fig. 2. Electrophoresis of locust myosin on 10% polyacrylamide gels in 8 M urea at pH 8.6 (a) unprecipitated myosin (b) precipitated myosin.

Fig. 3. Densitometer trace of 10% polyacrylamide gel in 8 M urea at pH 8.6. The sample is phosphorylated locust myosin. The gel was sliced and the radioactive label found only in the place indicated.

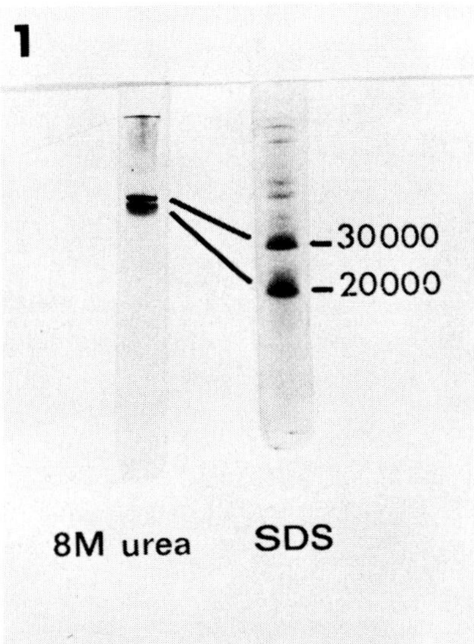
1
30000
20000
8M urea
SDS

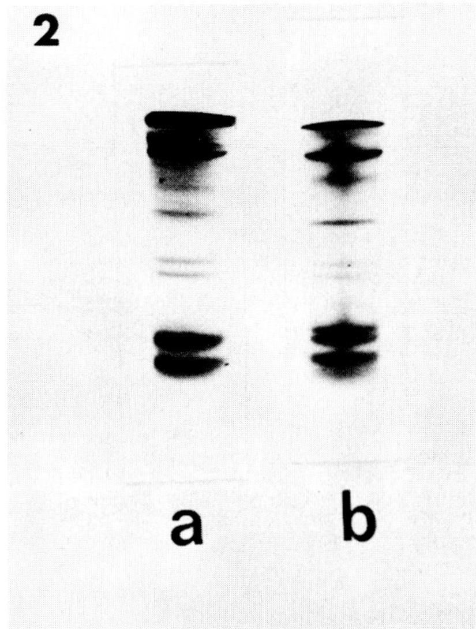
2
a
b

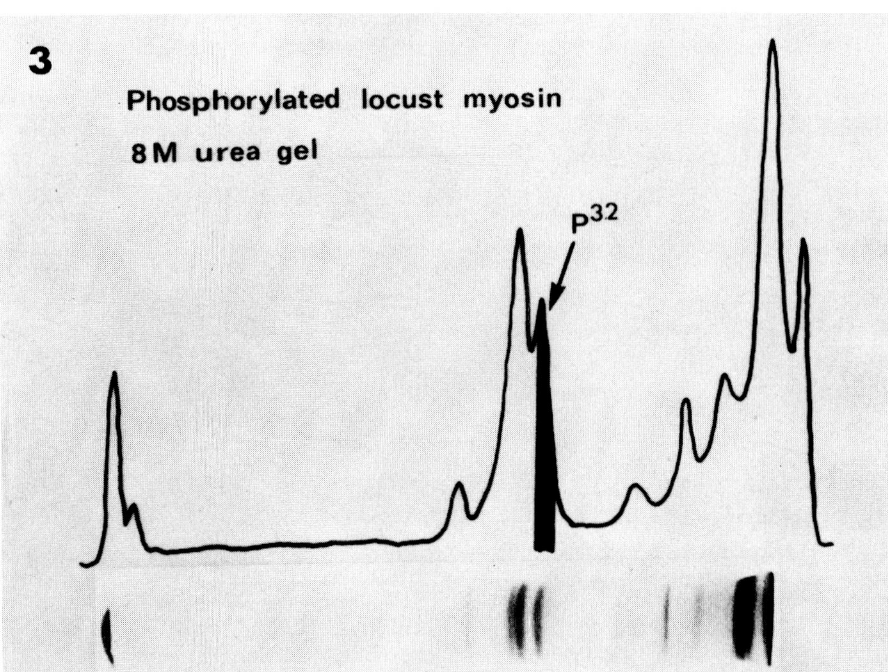
3
Phosphorylated locust myosin
8M urea gel
P32

PHOSPHORYLATION AND ITS INFLUENCE ON THE ATPase OF THE MYOSIN OF INSECT FLIGHT MUSCLE

G. Beinbrech

Universität Münster, Zoologisches Institut, Hindenburgplatz 55, D-4400 Münster, BRD, West Germany.

Dipteran flight muscles with shortening frequencies of several hundred Hz belong to the muscles with the highest energy consumption. Purified myosin of these muscles however shows little ATPase activity, suggesting that something has happened during preparation (Lehman, *et al.*, 1974). In order to find out whether this loss of activity could be restored by phosphorylation our group (M. Breuer, R. Achazi and G. Beinbrech) isolated myosin after the method of Lehman, *et al.*, (1974) from purified myofibrils of *Phormia terrae-novae*. In the presence of a protein kinase (4 μg/ml) and theophyllin 8 mM in an ATP solution phosphate was incorporated into the 29000 - 30000 Dalton light chain and to a minor amount into the heavy chain of myosin. This incorporation was stimulated by the presence of 0.2 mM cyclic AMP. The phosphorylation was correlated with an increase of the myosin ATPase activity (in the presence of 5 mM Mg^{++}, 2 mM ATP, 0.2 mM Ca^{2+}) by about 80% (0.0042 μM PO_4/mg. min. to 0.0076 μM PO_4/mg.min.).

REFERENCES

Lehman W., Bullard B. and Hammond K. (1974) *J.gen.Physiol.*, 63, 553.

ATPase CYCLE IN INSECT FLIGHT MUSCLE

Steven B. Marston

A.R.C. Unit, Department of Zoology, Oxford.

INTRODUCTION

Insect flight muscle, like any other muscle, is a molecular machine in which ATP hydrolysis induces structural changes in the actin-myosin complexes, which are manifested macroscopically as contraction. Most of this symposium is devoted to these structural changes, but it is necessary to set down at the beginning what we know about the mechanism of ATP hydrolysis which powers the structural changes in insect flight muscle. When compared with other muscles (Canfield, 1971; Marston and Taylor, 1977) the ATPase activity of IFM fibres (*Lethocerus* dorsal longitudinal muscle) falls between that of vertebrate fast skeletal muscle and slow skeletal or cardiac muscle (Rüegg and Tregear, 1966); Marston and Tregear, 1974). Work on *Lethocerus* actomyosin has confirmed that the ATPase activity is a little less than vertebrate fast muscle actomyosin (Bullard, Dabrowska and Winkelman, 1973). Our current knowledge of the mechanism of ATP hydrolysis is mostly based on transient kinetic studies of rabbit fast skeletal muscle but recent experimentation on a wide range of vertebrate muscle types has shown that the same mechanism prevails throughout (Marston and Taylor, 1977). Although no work has actually been done on *Lethocerus* ATPase kinetics, it is reasonable to propose as a working hypothesis that the mechanism is the same as in vertebrates.

According to the current model (Bagshaw, Eccleston, Eckstein, Goody, Gutfreund and Trentham, 1974), myosin hydrolyses ATP thus

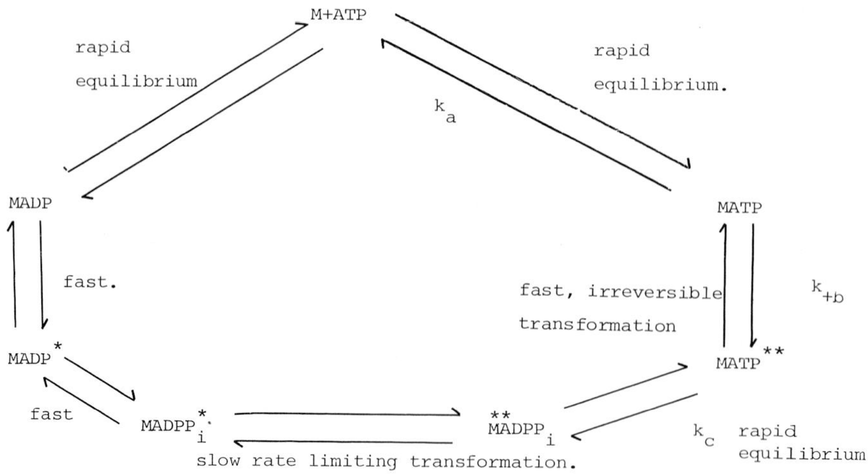

The most recent formulation of the actomyosin ATPase cycle has been made by White and Taylor,(1976) and Sleep and Taylor, (1976), based on the original cycle proposed by Lymn and Taylor, (1971).

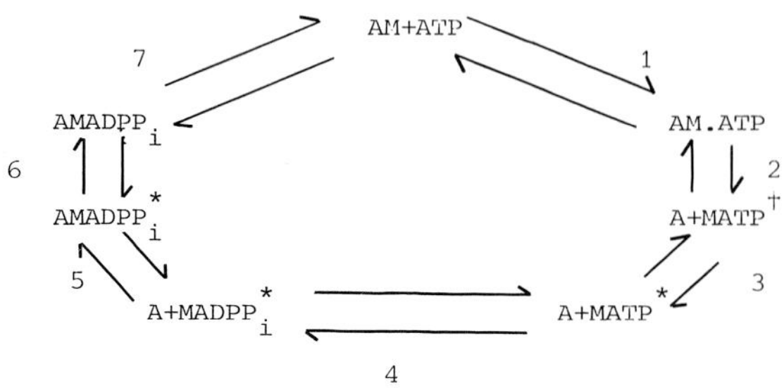

In vivo, ATP concentrations are saturating (Marston and Taylor, (1977) so the cycle may be simplified to a minimum of four steps

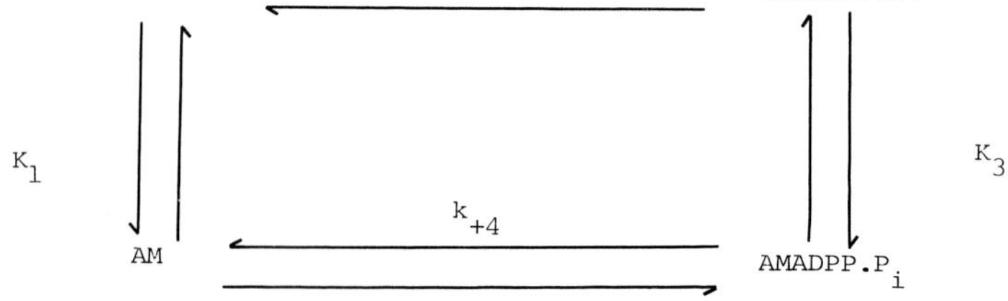

It is probable that steps 1 and 3 are rapid equilibria, whilst step 4 is the slowest step in the cycle (White and Taylor, 1976; Marston and Taylor, 1977). This scheme immediately suggests an interpretation of biochemical steps in terms of the crossbridge cycle model (Huxley, 1969; Lymn and Taylor, 1971) and, consequently, correlation with mechanical experiments. For instance, step 4 could correspond to the tension-producing step; indeed White and Taylor (1976) have shown that at least half the total free energy change of ATP hydrolysis by acto - S1 occurs in this step.

Without some supporting data it is quite unjustifiable to apply the ATPase cycle, determined with actin and S-1 in solution, to the intact myofibrillar matrix of a contracting muscle. Actin and myosin are present in very high concentrations in muscle and they are integrated into filaments so that the range of intermolecular contacts is quite specific. Does a muscle fibre behave, biochemically, like a tube filled with actomyosin, or is the situation more complex? The series of experiments described here were initiated in order to answer this question.

PART I: THE STEADY STATE INTERMEDIATE OF ATP HYDROLYSIS IN INTACT MUSCLE FIBRES

Techniques are not yet available to measure the faster reactions of the ATPase cycle in a fibre, but it is possible to measure the nature and concentration of the steady-state intermediates. According to the actomyosin ATPase mechanism in solution the muscle is in the AM state in the absence of ATP and mostly in the M.ADP.Pi and AM.ADP.Pi states in the presence of ATP. In relaxed muscle the actin-myosin interaction is inhibited (e.g. by altering K_3 in the solution scheme) so that M.ADP.Pi is the predominant intermediate, whilst in contracting muscle, AM.ADP.Pi also becomes a significant intermediate. Therefore, if a fibre is incubated in ATP and free ADP and Pi are rigorously excluded one expects to find an excess concentration of Pi and ADP equivalent to the concentration of myosin active sites in the fibre and the excess Pi and ADP should have the same properties as the postulated steady-state intermediate. The method described below has successfully demonstrated these features in glycerol-extracted muscle from several phyla including the arthropod, *Lethocerus* (Marston, 1973; Marston and Lehman, 1974).

METHOD

Single glycerol-extracted muscle fibres were dissected out and incubated in a radioactive ATP solution (1 to 100 μM) together with a highly active ATP regenerating system (10 mg/ml creatine kinase and 3 mM creatine phosphate) which maintained the free ADP concentration in the fibre below 2 μM during steady ATP hydrolysis. The solution also contained ^{14}C mannitol as an inert volume marker. After equilibration the fibres were transferred to 10 μl of 10% trichloroacetic acid which stopped the ATPase and released the bound intermediates. Alternatively, the fibres were transferred to a bath containing 10 μl of 5 mM non-radioactive ATP and 5 mM non-radioactive ADP which chased the labelled intermediates from their binding sites. By varying the time in each chase bath, the time course of release of the radioactive species from the fibre could be measured. The eluted radioactive ATP, ADP and mannitol were separated by chromatography on PEl-cellulose thin layers (Fig. 1) and assayed by scintillation counting. Using the fibre volume determined from mannitol radioactivity the concentrations of ATP and ADP within the fibres were calculated and compared with the concentrations in the incubating solution to determine the excess concentration (Marston, 1973).

Steady-State Intermediate in Relaxed Muscle

Fig. 1 shows that the relaxed fibres contained both excess ADP and ATP; Fig. 2 illustrates the concentration dependence of the excess ADP and ATP. The excess ADP exhibits all the characteristics predicted for the steady-state intermediate.

1) ADP is half bound in a similar concentration range to the Km for ATP hydrolysis (Table 1).

2) The maximum amount bound, 200 μM, corresponds to the estimated concentration of active sites in the fibre (Chaplain and Tregear, 1966; Tregear and Squire, 1973; Table 1).

3) The binding of ADP is accompanied by a corresponding loss of muscle stiffness. The correspondence has been clearly demonstrated over a range of ATP concentrations in rabbit fibres (Marston, 1973).

4) The rate of bound ADP release from the fibres equals the ATPase turnover rate in relaxed muscle (Fig. 3) and both rates are increased in parallel as the muscle is activated by increasing the free calcium concentration (Fig. 3). This point is not so well illustrated in *Lethocerus* fibres as it was in rabbit fibres (Marston, 1973) because the rate of diffusion of free nucleotide from the fibres is only twice the rate of bound ADP dissociation.

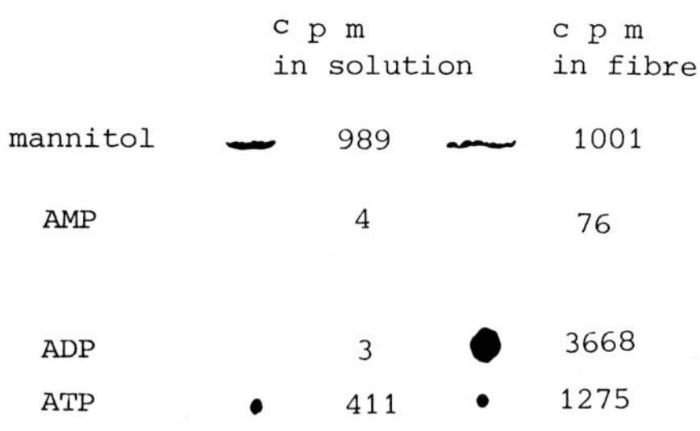

Fig. 1. Autoradiographs of ^{14}C mannitol, AMP, ADP and ATP separated by PEI cellulose chromatography.
left: 30 nlitre aliquot of incubation solution; ATP concentration is 15 μM.

right: TCA extract from approx. 1 cm. of muscle fibre, showing the excess ADP and ATP; fibre volume is 30.4 nlitres. Estimated ATP concentration in fibre is 46 μM, ADP concentration is 134 μM.

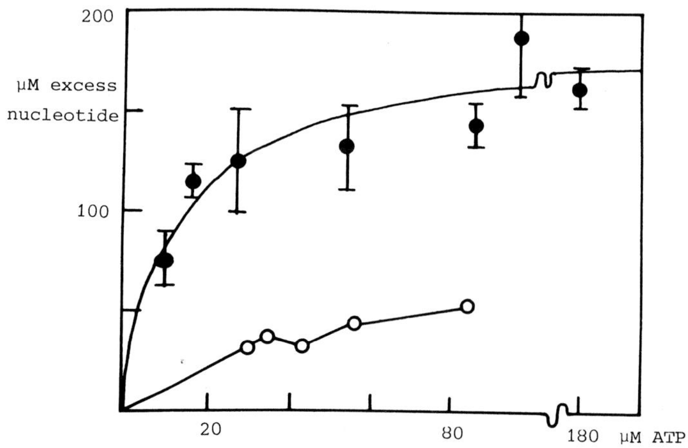

Fig. 2. Bound ADP (●) and ATP (O) in glycerinated *Lethocerus* fibres as a function of ATP concentration pH 7.1, 1°C, basic medium (see Table I) plus 10 mg/ml creatine kinase, 3 mM creatine phosphate. Solid line is best fit of ADP bound to simple binding equation (Table I) bars represent ± 5% fiducial limits.

5) The γ phosphate of ATP was also shown to bind at the active site in rabbit fibres (Marston, 1973) but has not been investigated in *Lethocerus* fibres.

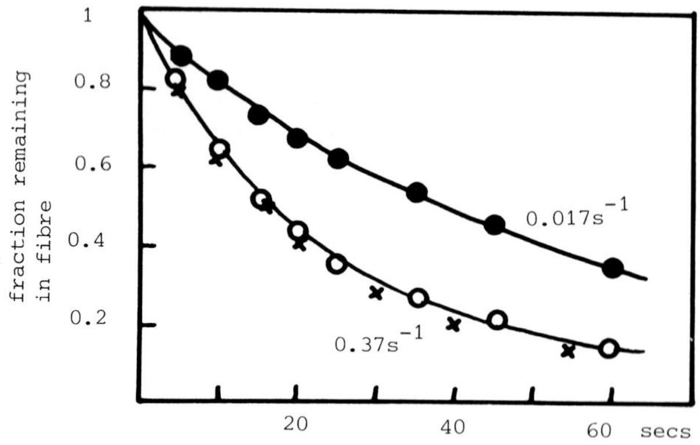

Fig. 3. Time course of release of bound ADP from relaxed (●) and activated (O) *Lethocerus* muscle. (x) represents release of ATP in relaxed muscle. The rate constants were estimated from semilog plots. Conditions same as Fig.2 except 4 mM Ca Cl_2 present in activated experiments, ATP concentration in medium, 57 μM.

TABLE I

NUCLEOTIDE BINDING TO GLYCEROL-EXTRACTED *LETHOCERUS* MUSCLE FIBRES

Nucleotide in incubating medium	Species bound	Maximum Amount bound μM	Dissociation constant μM
ATP	ADP	183 ± 14	12 ± 4
ADP	ADP	204 ± 8	11 ± 1.5
βγ(NH)ATP	βγ(NH)ATP	204 ± 15	111 ± 23

Basic incubating medium : 5 mM PIPES[+], pH 7.1, 5 mM Mg Cl_2, 5 mM EGTA, 50 mM KCl, $1^{o}C$.

Dissociation constant and maximum amount bound were determined by non-linear regression analysis of the data, assuming simple binding to one class of sites (Wilkinson, 1961).

[+]PIPES: piperazine, N:N' (ethanesulphonic acid).

The excess ATP in the fibre cannot be bound at the active site. This was demonstrated by comparing excess ATP and ADP in the fibre after TCA treatment and after chasing with a large excess of cold ATP. In the latter case labelled ATP at the active site would be released as ADP rather than ATP. In practice the excess ADP and ATP were the same using either treatment, indicating that the ATP bound elsewhere; the almost linear increase of bound ATP with increasing added ATP suggests non-specific binding sites.

Activated Muscle

Increasing the free calcium concentration from 10^{-9}M to 10^{-6}M activates the fibre ATPase about six-fold and the muscle produces tension. Solution kinetics predict that the predominant steady-state intermediates would be $M.ADP.P_i^{**}$ and $AM.ADP.P_i^{**}$. Binding studies are in agreement with this: when the ATP regenerating system is sufficiently active to remove free ADP, the amount of bound ADP in the fibre remains the same as it was in relaxed muscle, whilst the rate of ADP dissociation is increased so that the ADP release becomes limited by the rate of diffusion from the fibre (Fig. 3; cf. Marston and Tregear, 1974).

DISCUSSION

In summary, then, the binding experiments have shown that the dominant steady-state intermediate of relaxed muscle is M.ADP.Pi and in activated muscle it is M.ADP.Pi and AM.ADP.Pi. As far as it is possible to measure them, their properties are just as solution kinetics could predict.

This conclusion has been criticised by several workers who do not accept the Lymn-Taylor scheme. Inoue and Tonomura, (1975) suggested that the excess ADP in the fibre was not bound, but simply product which had not been reconverted to ATP and was trapped within the fibre because of slow diffusion out. This aspect of our procedure has been studied most carefully; Fig. 4 shows that there is indeed extra excess ADP at low creatine kinase concentrations, but the amount actually used was well in excess of that needed for complete removal of ADP. Barany and Barany, (1972) have pointed out that the ADP binding result is inconsistent with the amount of ADP they found in the whole muscle after freeze clamping. However, the amount of ADP extracted into perchloric acid is commonly less even than the known amount of non-exchangeable ADP bound on the actin filaments (Szent-Gyorgyi and Prior, 1966) so the reliability of the method is in doubt, possibly because myokinase can survive perchloric acid treatment.

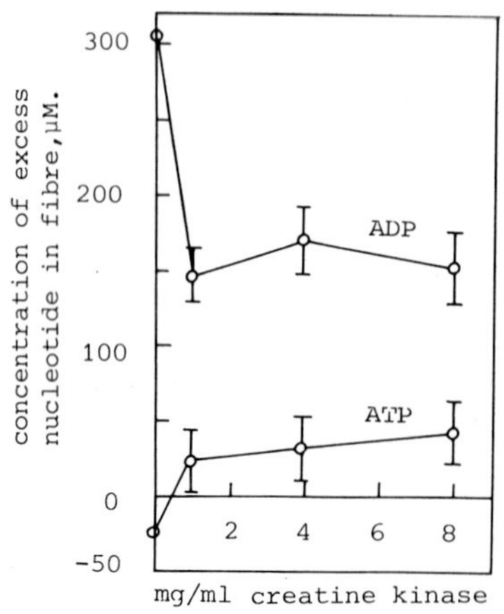

Fig. 4. Excess ADP and ATP in relaxed muscle fibres as a function of the concentration of creatine kinase. Basic medium 100 µM ATP, 3 mM creatine phosphate.

PART II: BINDING OF THE PRODUCTS AND ANALOGUES

The equilibrium complexes of ADP, Pi and non-hydrolysable ATP analogues with myosin in muscle fibres are of considerable interest to structural investigators since they represent static states of the actin-myosin interaction which are relatively easy to study. Equally the effects of the formation of such complexes on the mechanical kinetics of contraction can provide useful information for the interpretation of mechanical data.

Before any of these experiments can be usefully performed, it is necessary to determine whether the nucleotides bind to the active site, and with what affinity, since only those structural and mechanical changes which occur as the nucleotide is bound can be relevant to the properties of the active site during contraction.

PRODUCT BINDING

ADP binds tightly to *Lethocerus* muscle fibres (Marston, 1973; Fig. 5, Table I). The maximum amount bound corresponds to the known concentration of myosin sites in the fibre and the measured affinity is quite similar to that measured for acto-S-1 in solution (Beinfeld and Martinosi, 1975). It has not been possible to directly show that ADP binds at the active site, since competition between direct ADP binding and ATP-generated ADP binding cannot be measured but it has been shown that β-γ(NH)ATP competes directly with ADP binding and, as described below, β-γ(NH)ATP does bind at the active site.

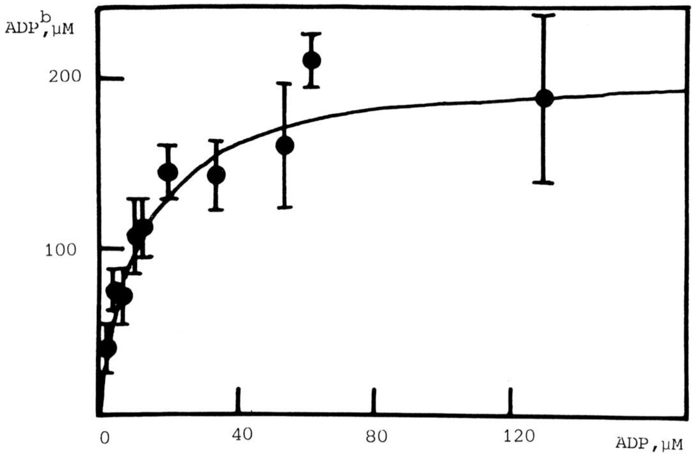

Fig. 5. ADP binding to *Lethocerus* muscle fibres as a function of ADP. Basic medium, pH 7.1, 1°C. Bars are 5% fiducial limits, line is best fit to simple binding equation (see Table I).

It was shown in rabbit muscle (Marston, 1973) and later in *Lethocerus* muscle (Rodger and Tregear, 1974) that ADP binding does not dissociate actin-myosin links, hence ADP forms a ternary complex with myosin and actin in muscle. Structural studies of this complex (Rodger and Tregear, 1974; Marston, Rodger and Tregear, 1976) showed that actin and myosin configurations were indistinguishable from rigor.

The ADP analogues, inosine diphosphate and uridine diphosphate, were found to bind to rabbit muscle fibres but with a much lower affinity than ADP (Marston, 1973). Phosphate binds only weakly to muscle fibres (Marston, 1973a); in rabbit muscle the dissociation constant was of the order of 1 mM and was independent of the presence of ADP at the active site. This compares with an estimate of 1.5 mM for pure myosin in solution (Bagshaw and Trentham, 1974). Phosphate does not relax muscle, even at a concentration of 20 mM (Marston,

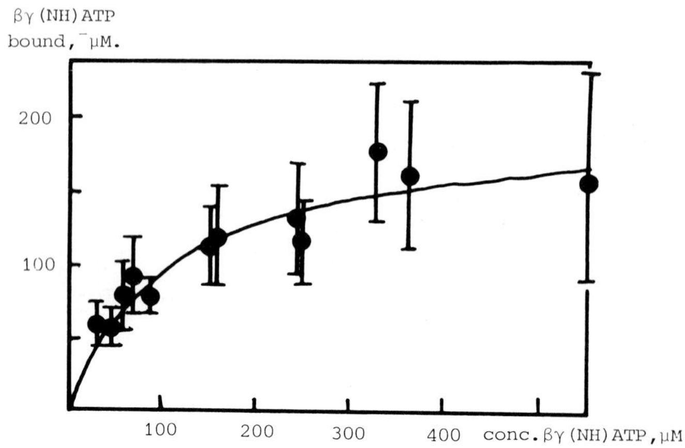

Fig. 6. βγ(NH)ATP binding to *Lethocerus* muscle fibres as a function of βγ(NH)ATP concentration. Basic medium, pH 7.1, 1°C plus 10 mM glucose, 2 mg/ml hexokinase.

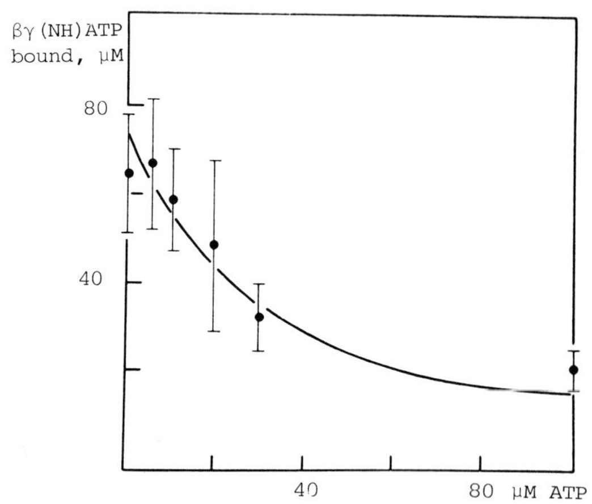

Fig. 7. Inhibition of |^{3}H|βγ(NH)ATP binding by non-radioactive ATP. Basic medium plus 10 mg/ml creatine kinase, 3 mM creatine phosphate. The curve was calculated on the assumption of competitive inhibition with K_d βγ(NH)ATP = 110 μM, B_m = 204 μM, K_i ATP = 19 μM.

1973a).

Both ADP and Pi have been reported to affect the mechanical performance of muscle contracting in ATP (Abbott and Mannherz, 1970; White and Thorson, 1972; Abbott and Steiger, 1977). The results of binding studies indicate that high levels of ADP and Pi would result in the formation of ternary complexes at the expense of M.ADP.Pi** and AM.ADP.Pi**. The predicted effect depends on the mechanism of contraction, but it is probable that the tension would become greater, the cycling time longer and the rate of some transients would be reduced.

β-γ imido ATP

This analogue is of great current interest because of the structural changes induced in the crossbridges when it binds to muscle fibres. Binding of $|^{3}H|$β-γ(NH)ATP may be studied by the same techniques used for ADP binding, but it is generally necessary to include an enzyme system, such as hexokinase plus glucose, to remove ATP which is a common contaminant of commercial preparations and which cannot be satisfactorily distinguished from β-γ(NH)ATP by chromatography.

β-γ(NH)ATP binds to *Lethocerus* fibres with a dissociation constant of the order of 100 μM (Fig. 6, Table I) (Marston, Rodger and Tregear, 1976). There is strong evidence that it binds at the myosin active site since the maximum amount bound (Table I) corresponds to the known concentration of sites and the binding is inhibited by ATP in a manner consistent with direct competition between the nucleotides (Fig. 7). Since pure β-γ(NH)ATP does not reduce the stiffness of rabbit (Marston, 1973a) or *Lethocerus* muscle fibres (Marston, Rodger and Tregear, 1976), it is likely that it binds without dissociating the crossbridges.

The binding of β-γ(NH)ATP induces a conformational change which has been detected by mechanical, electron microscope and X-ray diffraction techniques (Kuhn, 1973; Marston, Rodger and Tregear, 1976; Goody, Holmes, Mannherz, Barrington-Leigh, and Rosenbaum, 1975). The nature of this conformational change is the subject of several contributions to this symposium.

The correlation of binding with the conformational change has been unequivocally established by determinations of the concentration dependence of the change in muscle rest length (Marston, Rodger and Tregear, 1976) and the X-ray diffraction pattern (Goody, Barrington-Leigh, Mannherz, Tregear and Rosenbaum, 1976). However, the time dependence of dissociation of β-γ(NH)ATP from single muscle fibres (>80% in 40 secs) may not correlate with the time required for restoration of the rigor configuration, since it appears to take longer. This may be a result of diffusional delay in the removal of β-γ(NH)ATP from large fibre bundles; if not it indicates that there is not necessarily 1 to 1 correspondence between the state of the myosin substrate binding site and the configuration of the actin-myosin interaction.

DISCUSSION

The products of ATP hydrolysis and substrate analogues bind to the active site of myosin with affinities of the same order as their binding to actomyosin in solution. Those nucleotides that are not hydrolysed usually do not dissociate actin-myosin links when they bind, but they can alter the conformation of the actin-myosin links.

It is tempting to postulate these complexes (e.g. AM.ADP, AM.βγ(NH)ATP) as representing "frozen" stages of the crossbridge cycle; but studies of nucleotide binding cannot prove this hypothesis in fibres. Nevertheless, the mere observation that ternary nucleotide-myosin-actin complexes can be formed with different conformations is in agreement with the hypothetical crossbridge cycles of Huxley, (1969) and Lymn and Taylor, (1971) and encourages the belief that the basic cycle, as proposed, is true.

REFERENCES

Abbott, R.H. and Mannherz, H.G. (1970), *Pflügers Arch.* 321, 223.

Bagshaw, C.R., *et al.*(1974)*Biochem. J.* 141, 351.

Bagshaw, C.R. and Trentham, D.R. (1974), *Biochem. J.* 141, 331.

Barany, M. and Barany, K. (1972), *Cold Spring Harb. Symp.*, 37, 157.

Barrington-Leigh, *et al.* (1972),*Cold Spring Harb. Symp.*, 37, 443.

Beinfeld, M.C. and Martinosi, A.N. (1975), *J. Biol. Chem.*, 250, 7871.

Bullard, B., *et al.*, (1973),*Biochem. J.* 126

Canfield, S.P. (1971), *J. Physiol.* 219, 281.

Chaplain, R.A. and Tregear, R.T. (1966). *J. Mol. Biol.*, 22, 275.

Goody, R.S. , *et al.* (1976) *Nature*, 262, 613.

Goody, R.S., *et al*(1975), *Biophys. J.*, 15, 687.

Huxley, H.E. (1969), *Science*, 164, 1,356.

Inoue, A. and Tonomura, Y. (1975), *J. Biochem.(Tokyo)*, 79, 419.

Kuhn, H.J. (1973), *Experienta*, 29, 1086.

Lymn, R.W. and Taylor, E.W. (1971), *Biochemistry* 10, 4,617.

Marston, S.B. (1973), *Biochim. Biophys. Acta* 305, 397.

Marston, S.B. (1973a), *D.Phil. thesis, Oxford.*

Marston, S.B., Rodger, C.D. and Tregear, R.T. (1976), *J. Mol. Biol.,* 104, 262.

Marston, S.B. and Lehman, W.J. (1974), *Nature,* 252, 38.

Marston, S.B. and Taylor, E.W. (1977) in preparation.

Marston, S.B. and Tregear, R.T. (1974), *Bioc. Biophys. Acta* 333, 581.

Marston, S.B. and Tregear, R.T. (1974), *Bioc. Biophys. Acta* 347, 311.

Rodger, C.D. and Tregear, R.T. (1974), *J. Mol. Biol.,* 86, 495.

Rüegg, J.C. and Tregear, R.T. (1966), *Proc. R. Soc. B.,* 165, 497.

Sleep, J.A. and Taylor, E.W. (1976), *Biochemistry* 15, 5813.

Szent-Gyorgyi, A.G. and Prior, G. (1966), *J. Mol. Biol.,* 15, 515.

Tregear, R.T. and Squire, J.M. (1973), *J. Mol. Biol.,* 77, 279.

White, H. and Taylor, E.W. (1976), *Biochemistry* 15, 5818.

White, D.C.S. and Thorson, J. (1972), *J. Gen. Physiol.,* 60, 307.

Wilkinson, G.N. (1961), *Biochem. J.,* 80, 324.

REVERSIBLE TRANSFORMATION OF MECHANICAL WORK INTO CHEMICAL FREE ENERGY BY STRETCH DEPENDENT BINDING OF AMP-PNP IN GLYCERINATED FIBRILLAR MUSCLE FIBRES

H.J. Kuhn.

Dept. Physiol. II, University of Heidelberg, Im Neuenheimer Feld 326, D-6900 Heidelberg, FRG.

INTRODUCTION

A completely quantitative transformation of chemical free energy into mechanical work by contractile systems is possible provided that the system can be reversibly altered by length changes and by the addition of chemical reagents. Such energy transformations are coupled with a teinochemical effect: reagents which induce isometric relaxation are absorbed by the contractile system when it is stretched at constant thermodynamic activity of that reagent i.e. the teinochemical principle (Kuhn *et al.*, 1962).

A transformation of chemical into mechanical energy has been recently proposed (Kuhn, 1973) for glycerinated insect muscle fibres irrigated with the non-hydrolyzable ATP analogue AMP-PNP (Yount, 1971). It will now be shown using the same experimental model that mechanical energy is transformed into chemical energy, and that this energy transformation is associated with a stretch dependent binding of AMP-PNP.

METHODS

Glycerol-extracted dorsal longitudinal muscle fibres from *Lethocerus maximus* were glued between two glass rods of a tension recording apparatus. One glass rod was connected to an RCA5734 force transducer mounted on a micrometer drive, the position of which could be monitored by the displacement of a Grass force transducer FTO03C. The fibres were relaxed in ATP-salt solution with a low Ca^{++} concentration and contracted in ATP-salt solution with a high Ca^{++} concentration. Rigor contraction was generated following the procedure of White (1970): the bundles (5 - 12 fibres) were first incubated in relaxing solution and then immersed in an ATP-free rigor solution (20 mM imidazole, 10 mM Na-azide, 5 mM EGTA, 10 mM KCl, 5 mM $MgCL_2$). The ionic strength of all solutions was the same (I = 0.06). Mechanical measurements were performed as previously described (Kuhn, 1973). AMP-PNP concentration was measured spectroscopically at wavelength 260 nm using a Beckmann Acta MVI spectrophotometer.

RESULTS

Previous studies have shown that fibrillar insect flight muscle performs repeatable isometric tension cycles (Barrington-Leigh *et al.*, 1973; Kuhn, 1973) at constant immediate stiffness (Beinbrech *et al.*, 1976) after addition and removal of the ATP analogue AMP-PNP at concentrations up to 0.5 mM. Thereby the preparation was conditioned by the following procedure: 1) relaxation in ATP-salt solution containing EGTA; 2) subsequent removal of ATP which induced a rigor contraction (White, 1970); 3) the fibre bundles were then partially relaxed by adding AMP-PNP without showing an appreciable decrease in immediate stiffness (Beinbrech *et al.*, 1976); rigor tension was partially restored to a level which now remained constant when AMP-PNP was repeatedly added and removed from the fibres. This constant level of rigor tension was independent of the AMP-PNP concentration used (up to 1 mM). Figure 1 shows the monotonic decrease of isometric tension with increasing AMP-PNP concentration. The slope of the log concentration-tension curve seems to be steepest at an AMP-PNP concentration of about 0.1 mM reflecting an affinity constant of AMP-PNP binding of about 10 mM^{-1}. This confirmed earlier estimates of the affinity constant of AMP-PNP binding to the enzymatic sites of the myosin heads from chemical studies (Marston *et al.*, 1976) and from structural studies (Goody *et al.*, 1976).

As shown in Figure 2 a fibre bundle could be forced to produce positive mechanical work by stretching it in a solution containing 0.1 mM AMP-PNP and releasing it in a AMP-PNP free solution. A state of rigor contraction was achieved by immersing ATP-relaxed fibres into ATP-free rigor solution. Addition of AMP-PNP (0.1 mM) induced isometric relaxation (point A). A new rigor contraction was now induced by washing out the AMP-PNP-salt solution (point B). The fibres were now slowly released by 60 μm from B to C. The fibres were then relaxed from C to D after addition of AMP-PNP (0.1 mM). The initial state could be reached again by restretching the AMP-PNP relaxed fibre to A and subsequently producing a rigor contraction from A to B by immersing the fibres into rigor solution. During this sequence of events (B-C-D-A-B) the preparation does positive work corresponding to the area enclosed in the force-length diagram. The cycle could also be performed in the opposite direction (B-A-D-C-B); mechanical work was then performed on the fibre. At the end of the negative cycle the fibre bundle has returned to the same state as at the end of the positive cycle. The same mechanochemical state (same tension at the same length and at the same concentration of AMP-PNP) is thus attained after positive and after negative work cycles. This suggests that the chemical reaction between AMP-PNP and the fibre bundle as well as the response to lengthening are reversible. It therefore fulfills the prequisites of the

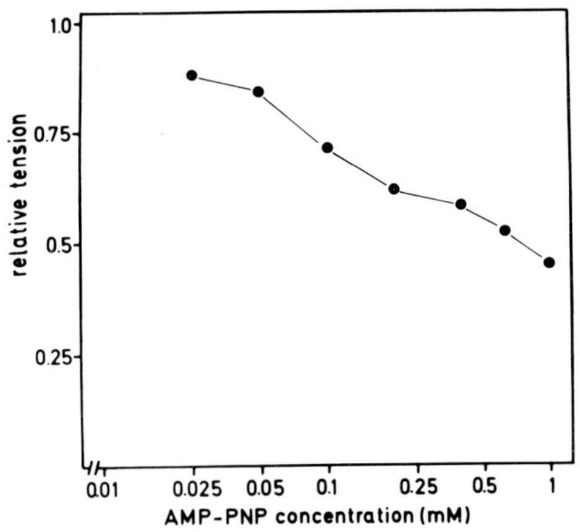

Fig. 1. Effect of adding AMP-PNP to fibres maintaining isometric rigor contraction. Mean values from 12 experiments. Diameter (●) corresponds to s.e. Relative tension values were obtained by dividing tension in AMP-PNP by the rigor tension (mean value 12.5 dynes per fibre). Fibre bundles isometrically contracted in rigor solution (10 mM imidazole, 10 mM Na-azide, 5 mM EGTA, 10 mM KCl, 5 mM $MgCl_2$) and subsequently relaxed by adding AMP-PNP in contractions indicated. Tension determined 20 min after change of incubation solution.
Conditions: pH 6.7, I = 0.06, $12^{o}C$,
bundles of 5 - 8 glycerol-extracted fibres from *Lethocerus maximus*.

teinochemical principle (Kuhn *et al.*, 1962) a quantitative transformation of chemical free energy into mechanical work and vice versa. The theory of teinochemical systems predicts a teinochemical effect: that an additional amount of substance is bound to the fibre when it is stretched in a solution containing AMP-PNP. Conversely, bound AMP-PNP should be released into the solution when the fibre bundle is released.

This prediction could now be verified directly. Fibre bundles were stretched in a compartment (S) containing AMP-PNP-salt solution and then released in a compartment (R) initially containing the same concentration of AMP-PNP (0.1 mM).

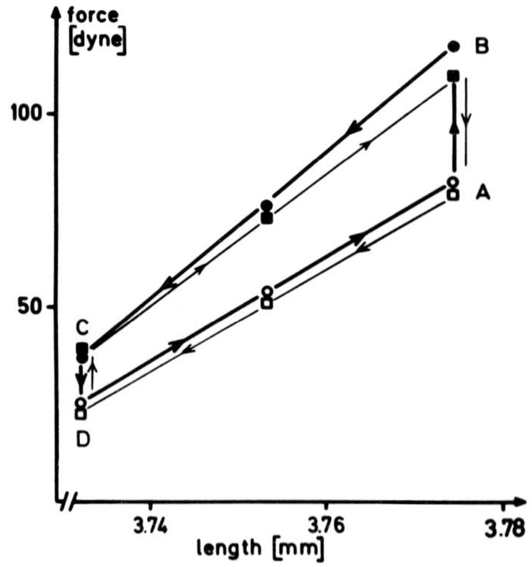

Fig. 2. Tension-length diagram obtained by stretching preparation in AMP-PNP-salt solution (O; D - A) and releasing in rigor solution (●; B - C) during a positive work cycle. A negative work cycle is obtained by stretching preparation in rigor solution (■; C - B) and releasing it in 0.1 mM AMP-PNP-salt solution (□; A - D). See text for further explanation.

Conditions: bundle of 8 fibres from *Lethocerus maximus*; pH 6.7; 10°C; I = 0.06.

A certain amount of AMP-PNP should have been transferred from bath to compartment R. However, this amount was so small that the procedure of stretching and releasing in compartments S and R respectively had to be repeated many times in order to accumulate the transferred AMP-PNP in quantities sufficient to produce a measurable concentration difference between compartments S and R. The results obtained are summarized in Table 1. Figure 3 shows a typical experiment demonstrating that the concentration difference between the compartments R and S increases with the number of transfers. The figure also shows that a measurable amount of AMP-PNP was transported against a concentration gradient of 2.5 μM after 100 cycles. In a control study it could be demonstrated that no transfer of AMP-PNP occurs if the bound AMP-PNP is displaced by 2 mM pyrophosphate added to the incubation baths R and S. In experiments

TABLE 1

Evidence for transfer of AMP-PNP after releasing fibres in compartment R and restretching them in compartment S.

	I: ΔC (μM)	II: $\partial N/\partial L$ (pmole cm^{-1} $fibre^{-1}$	III: $-\partial f/\partial \ln c$ (dynes/fibre)	IV: $\frac{-\partial f/\partial(\ln c)}{RT}$ (pmole cm^{-1} $fibre^{-1}$)
A	4.4 ± 0.5	61 ± 10	1.57 ± 0.14	66 ± 6
B	2.8 ± 0.4	71 ± 14	1.73 ± 0.15	72 ± 6
C	0.3 ± 0.5	8 ± 10	1.65 ± 0.10	------
D	0.1 ± 0.3	3 ± 11	0.02 ± 0.02	1 ± 1

I) Concentration difference between R and S determined after 100 cycles.

II) Amount of AMP-PNP transferred into compartment R per cm length change and per fibre. (calculated from the observed concentration gradients between compartments R and S assuming that the teinochemical effect is proportional to the change of fibre length).

III) Slope of the log concentration-tension curve.

IV) Amount of AMP-PNP transferred to R calculated from equation 1.

Initial volume was 0.295 to 0.302 ml in experiment A and 0.392 to 0.407 ml in experiments B. C and D. In experiment C the amplitude of length change was zero; in all other experiments, it was 60 μm. In experiment D, 2 mM pyrophosphate was added to the solution. All figures are mean values ±S.D. (n = 4). Initial AMP-PNP concentration 0.1 mM.

Conditions: pH 6.7; I = 0.06; 13°C

14 - 20 fibres per bundle of length 5 mm.

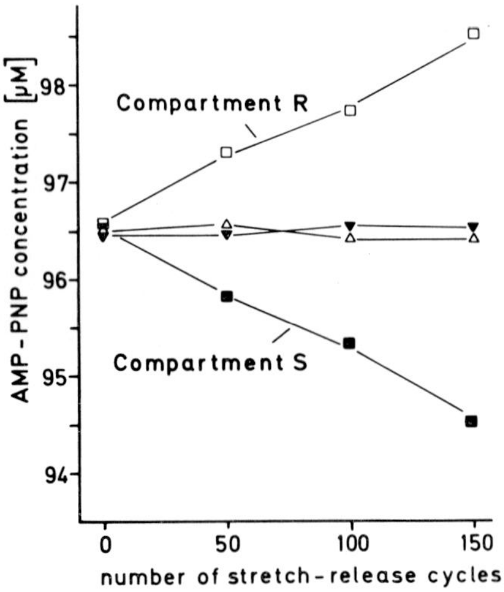

Fig. 3. Teinochemical effect of release (open symbols) and re-stretch (full symbols) of a bundle of 15 glycerol-extracted fibres from *Lethocerus maximus*. Stretch amplitude 60 μm. Duration of a cycle 3 min. □ , ■ 0.1 mM AMP-PNP; △ , ▲ 0.1 mM AMP-PNP + 2 mM pyrophosphate. Conditions: pH 6.7; $13^{\circ}C$; I = 0.06; AMP-PNP was determined spectroscopically. Fibre bundles in rigor were partially relaxed by addition of AMP-PNP, released in compartment R and re-stretched in compartment S.

such as these it is important that the state of a fibre is the same at the beginning and at the end of the stretch-release cycles. This could be achieved by ensuring that the tension never exceeded the yield point, above which plastic flow would occur (Kuhn, 1973). Since the cycles of length changes are started first by releasing isometrically contracted fibres, and subsequently restretching them to the initial length, the critical tension of the yield point was not exceeded. In a typical experiment, the tension at 0.1 mM AMP-PNP was 12.8 dynes per fibre at the beginning. After 150 cycles it was 12.0 dynes per fibre and recovered to 18.2 dynes per fibre after washing out the AMP-PNP. The latter tension was comparable to rigor tension , 18.5 dynes per fibre at the start.

DISCUSSION

Figure 3 shows that AMP-PNP linearly increases in concentration from 96.5 μM to 98.5 μM in compartment R (where the bundle was released) when increasing the number of release-stretch cycles from zero to 150. This reflects an AMP-PNP transport of ΔN = 0.34 pmoles per fibre (15 fibres) and per stretch-release cycle into compartment R (volume 0.4 ml). At the end of the experiment the concentration ratio between compartment R (c_R = 98.5 μM) and S (c_S = 94.6 μM) was c_R/c_S = 1.041. After 150 release-stretch cycles the molar free energy of dilution Δg of the compartments R and S hence increased from the initial value (Δg = 0) to $\Delta g = RT \times \ln (c_R/c_S) = 9.84 \times 10^8$ erg. Thus the free energy increase per cycle ($\Delta G = \Delta g \Delta N$) is ΔG = 325 μerg.

According to the Helmholtz law of conservation of free energy for isothermal processes the increase of free energy in the compartments must have been at least compensated by mechanical work. During the 150th cycle of Figure 3 the fibre bundle performed negative mechanical work when it was stretched by ΔL = 60 μm at the high force at "low" AMP-PNP concentration (94.5 μM) and when it was released by 60 μm at the low force and "high" AMP-PNP concentration (98.5 μM). The reversible mechanical work done by the fibres can be estimated as follows: The mean difference in equilibrium fibre tension (Δf) between compartment S and R is calculated from the slope of the log concentration-tension curve (cf. Figure 1 and Table 1, column III):

$$\Delta f = \frac{\partial f}{\partial (\ln c)} \times \frac{\Delta c}{c} = 6.2 \text{ mdyne.}$$

The reversible mechanical work ($W = \Delta f \Delta L$) corresponding to the area enclosed in a length-tension diagram (cf. also Figure 2) is then calculated to be W = 370 μerg. Note that the actual mechanical work done to produce the concentration gradients (cf. Figure 3) was much greater since it is also due to frictional, viscous resistances in the fibres etc. which were several times the reversible mechanical work calculated above. The reversible mechanical work (370 μerg) was thus completely converted into free energy of dilution of the AMP-PNP (325 μerg).

Based on thermodynamic cross relations given already by Gibbs (1878), Kuhn *et al.* (1960) and Katchalsky *et al.* (1960) have described such quantitative energy transformation of reversible mechanical work into free chemical energy by the equation:

$$\frac{\partial N}{\partial L} = -\frac{1}{RT} \times \frac{\partial f}{\partial (\ln c)} \quad (1)$$

Equation (1) could be experimentally verified for artificial contractile systems

(Kuhn *et al.*, 1962; Kuhn *et al.*, 1960; Pohl *et al.*, 1966) and may now be experimentally tested in the case of AMP-PNP induced relaxation. The expression $(\partial N/\partial L)_c$, i.e. the number (N) of moles AMP-PNP bound by one fibre per unit length change (L) at constant free concentration (c) of AMP-PNP, was calculated from the concentration gradients observed in transport experiments of the type shown in Figure 3 and summarized in Table 1. The values of $(\partial N/\partial L)_c$ listed in Table 1 (lines A,B; column II) indicate that about 0.7 pmoles AMP-PNP are bound to a fibre of length 1 cm when it is stretched by 1% L_i. When compared with the molar amount of enzymatic sites (10 pmole/cm; cf. Tregear and Squire, 1973) present in flight muscle of *Lethocerus*, this value reveals that about 7% of the sites are additionally loaded with AMP-PNP by a 1% L_i stretch (at AMP-PNP = 0.1 mM). The right hand expression in equation (1), $\partial f/\partial(\ln c)$, is the slope of the $\log_e$-concentration-tension curve (cf. Figure 1) of the isometric relaxation induced by 0.1 mM AMP-PNP; it may be roughly evaluated from the isometric relaxation induced by adding 0.1 mM AMP-PNP to fibres in rigor, cf. Table 1, column III. By multiplying these values by 1/RT (RT = 2.4×10^{10} erg. at 13°C) one obtains the amount of ligand additionally bound to the fibre per cm stretch predicted by the teinochemical equation 1 (cf. Table 1, column IV). It is seen from Table 1 that the independently determined expressions $(\partial N/\partial L)$ and $(-\partial f/\partial(\ln c) \times 1/RT)$ listed in columns II and IV do coincide fairly well. This provides for the first time full experimental proof that the teinochemical principle can be applied to muscle fibres still capable of active contraction.

AMP-PNP is bound by the fibre during stretch (note the decrease of AMP-PNP concentration in compartment S, figure 3) and AMP-PNP is released again when the bundle shortens. This is taken to indicate that AMP-PNP binding enables free energy to be stored during stretch. This free energy of binding is converted into free energy of dilution during the process of release. It is tempting to relate the described teinochemical effect to the known structural effects of AMP-PNP (Marston *et al.*, 1976; Beinbrech *et al.*, 1976; Goody *et al.*, 1975; Lymn and Huxley, 1973) and to the theory of Huxley and Simmons (1971). When a fibre is relaxed by AMP-PNP under isometric conditions, the cross bridges are probably rotated from an acute angle (Reedy *et al.*, 1965) to a perpendicular position (Huxley, 1969) which is stabilized upon binding of the ligand. When AMP-PNP is washed out with maintenance of isometric conditions the AMP-PNP dissociates from the myosin heads. The cross bridges return to an acute angled position while remaining attached to the actin, and generate tension. Conversely when a fibre is stretched, the model of Huxley and Simmons (1971) predicts that the cross bridges have a tendency to rotate into a more perpendicular position in which AMP-PNP binds to them. A subsequent release of the fibre favours the acute angled cross bridge conformation and the AMP-PNP is partially released from the myosin heads.

ACKNOWLEDGEMENTS

Supported by the Deutsche Forschungsgesellschaft, Grant RU 154/11-12.

I should like to thank Prof. Dr. J.C. Rüegg for his stimulating discussion. The technical assistance of Mrs. M. Reuther is gratefully acknowledged.

REFERENCES

Barrington-Leigh, J., Holmes, K.C., Mannherz, H.G., Rosenbaum, G., Eckstein, G., and Goody, R. (1973). *Cold Spring Harb. Symp. Quant. Biol.* 37, 443.

Beinbrech, G., Kuhn, H.J., Herzig, J.W., and Rüegg, J.C. (1976). *Cytobiology*, 12, 385-396.

Gibbs, J.W. (1878). *Trans. Com. Acad.* 3, 108-245, 343-524.

Goody, R.S., Holmes, K.C., Mannherz, H.G., Barrington-Leigh, J., and Rosenbaum, G. (1975). *Biophys. J.* 15, 687-704.

Goody, R.S., Barrington-Leigh, J., Mannherz, H.G., Tregear, R.T., Rosenbaum, G. (1976). *Nature* 262, 613-614.

Huxley, H.E. (1969). *Science* 164, 1356-1366.

Huxley, A.F. and Simmons, R.M. (1971). *Nature* 233, 533.

Katchalsky, A., Lifson, S., Michaeli, I., and Zwick, M. (1960). Size and shape of Contractile Polymers, Ed. A. Wasserman *Pergamon Press,* 1 .

Kuhn, W., Ramel, A., Walters, D.H., Ebner, G. and Kuhn, H.J. (1960). *Fortschritte der Hochpolymeren-Forschung* 1, 540-592.

Kuhn, W., Toth, I., and Kuhn, H.J. (1962). *Hel. Chim. Acta* 45, 2325-2342.

Kuhn, H.J. (1973). *Experienta* 29, 1086-1088.

Lymn, R.W. and Huxley, H.E. (1973). *Cold Spring Harb. Symp. quart. Biol.* 37, 449-453.

Marston, S.B., Rodger, C.D. and Tregear, R.T. (1976). *J. Mol. Biol.* 104, 263-276.

Pohl, W.G., Kuhn, H.J. and Kuhn, W. (1966). *Z. Nat. f.* 21a, 756-763.

Reedy, M.K., Holmes, K.C. and Tregear, R.T. (1965). *Nature* 207, 1276.

Tregear, R.T. and Squire, J.M. (1976). *J. Mol. Biol.* 77, 263-276.

White, D.C.S. (1970). *J. Physiol.* 208, 583 .

Yount, R.G., Babcock,D.,Ballantyne,W., Ojala, D. (1971). *Biochemistry* 10, 2484-2489.

MECHANICAL FACTORS AFFECTING THE ATP-PHOSPHATE EXCHANGE REACTION OF GLYCERINATED INSECT FIBRILLAR MUSCLE

M. Ulbrich* and J.C. Rüegg
Dept. Physiology II, University of Heidelberg, Im Neuenheimer Feld 326, D-6900 Heidelberg, FRG

*Present address: *Physiologisches Institut, Universität Würzburg.*

INTRODUCTION

Over two decades ago, Pringle (1957) showed that during flight fibrillar insect muscle oscillates according to a myogenic rhythm. This oscillation must be attributed to an "automatism" in the myofibrillar contractile apparatus, since myogenic oscillation can also be obtained with glycerol extracted fibre bundles from insect fibrillar muscle when they are attached to a suitable resonant lever system and suspended in a solution containing ATP as energy source and calcium ions as activator. Oscillation occurs because elongation of the preparation brings about a delayed activation of myofibrillar bundles, while release (shortening) causes a delayed deactivation. Accordingly glycerol extracted fibres can be made to oscillate and to do work when they are sinusoidally stretched and released at an appropriate frequency and amplitude in the range of o.5 to 3% of the fibre length (Jewell and Rüegg, 1966). Under these conditions the rate of ATP consumption greatly increases in order to provide the energy for the oscillatory work done (Rüegg and Tregear, 1966; Rüegg and Stumpf, 1969a; Mannherz, 1968; Steiger and Rüegg, 1969; Pybus and Tregear, 1973). In addition it was found that a quick stretch of glycerinated insect muscle not only activates the myofibrillar bundles from insect fibrillar muscle to generate a delayed tension (stretch activation), but also increases the rate of ATP splitting (Rüegg and Tregear, 1966; Rüegg and Stumpf, 1969b; Breull, 1971; Breull, Steiger and Rüegg, 1973).

In recent years the ATP splitting mechanism of the contractile apparatus has been elucidated and described in terms of a sequence of several reactions involving at least seven steps (Bagshaw *et al.*, 1972). Most if not all of these reactions are now considered to be reversible and it is even possible to synthesize ATP, bound to myosin, from added ADP and phosphate (Mannherz *et al.*, 1974). The first indication for a partial reversibility of the ATP splitting reaction came from studies of $^{32}P_i$ incorporation into ATP, using glycerinated insect fibrillar muscle (Mannherz, 1970; Ulbrich and Rüegg, 1971, 1976), psoas muscle (Gillis and Maréchal, 1971, 1974), and isolated actomyosin preparations (Hotta and Fujita, 1971; Ulbrich and Paulsen, 1976; Wolcott and Boyer, 1974; Ulbrich, 1975. This reaction is now ascribed to an ATP-phosphate exchange mechanism. Some previous investigators were unable to demonstrate an incorporation of labelled phosphate into ATP (Ulbrecht, M., 1962; Dempsey *et al.*, 1963; Inoué, 1973).

The question of the reversibility of the muscle mechanism had intrigued muscle physiologists ever since A.V.Hill and Howarth (1959) suggested that when contracting muscle is stretched the work done on the muscle might be partly transformed into chemical energy. Though a synthesis of ATP could not be demonstrated under these conditions, there is nevertheless a reduction of ATP breakdown during stretch (Curtin and Davies, 1973). In any case, estimates of *net* ATP production or consumption in muscle cannot exclude the possibility of a partial reversal of the ATP splitting reaction. As pointed out already, glycerinated insect fibrillar muscle from *Lethocerus maximus* proved to be very suitable for studying the relationship between the rate of ATP consumption and the mechanical processes. We have therefore chosen this preparation to examine in more detail the question of how the reincorporation of phosphate into ATP depends on calcium ion concentration and on mechanical events within the contractile system.

METHODS

Solutions and experimental details

Dorsal longitudinal muscles of *Lethocerus maximus* were extracted in a glycerol-water mixture (50/50 vol. %) buffered to pH 6.5 with histidine and stored at $-20^{o}C$ for 1-12 months up to use. Bundles of 6 fibres, 5 mm long, were mounted horizontally and glued with their ends on the two glass rods of a driven-oscillation apparatus described elsewhere (Steiger and Rüegg, 1969). One glass rod was connected to an electro-magnetic vibrator Ling Dynamics Model 101, the other glass rod to the anode pin of an RCA 5734 force transducer. Length changes produced by the vibrator were monitored by field-plates attached to the arm of the vibrator, and moving in and out of magnetic field. Length and tension signals were displayed on a Tektronix dual beam oscilloscope.

The fibre bundles were incubated in temperature controlled ($20^{o}C$) Perspex baths of 0.3 ml volume. Incubation solution contained 15 mM ATP, 15 mM $MgCl_2$, 20 mM imidazole/HCl buffer, pH 6.7, 4 mM Ca-EGTA (pCa = 2×10^{-5}M), 10 mM Na-azide, 1 mM ^{32}P-Phosphate of specific activity up to 50,000 cpm/μl (Amersham-Buchler). Relaxing solution contained 4 mM EGTA instead of Ca-EGTA. The radioactive phosphate was purified by paper chromatography (solvents according to Bandurski and Axelrod, 1951) and subsequent elution.

Prior to the experiment, the fibres were extended in relaxing solution (4 mM EGTA) until an elastic tension just appeared (L_o). In each experiment the same fibre bundle was transferred successively to up to 10 different baths while varying the experimental conditions. Each experimental parameter was tested twice in one fibre bundle. Incubation time was 40 min - 1 hr per bath.

Estimation of incorporation of $^{32}P_i$ into ATP-γ-^{32}P

Detection of the ATP-γ-^{32}P formed was done by transferring the δ-^{32}P quantitatively into glucose-6-P and separating the glucose-6-P from P_i by paper chromatography, as described by Mannherz (1970). At the end of the experiment, after removing the fibre bundle from the bath, 50 μl of a triethanolamine buffer solution (0.5 M, pH 8) containing 0.5 M glucose, and 10 μg hexokinase (Boehringer) were added to the incubation solution. After an incubation time of 30 minutes aliquots of 20 μl from each bath were applied to Whatman MM 3 paper and run in the solvent of Bandurski and Axelrod containing in addition 0.001 M EGTA (cf. Mannherz, 1970). The chromatogram was cut into strips of 3 cm x 1.5 cm which were counted in a scintillation counter (Beckmann LS 100) using a toluene cocktail (5.4 g PPO and 0.1 g POPOP per l). By this method a complete separation of the glucose-6-P from P_i was achieved and interference of polyphosphates with ATP was avoided.

In control experiments over 95% of the activity of added ATP-γ-^{32}P could be recovered in glucose-6-P after 30 minutes incubation with hexokinase. The relative rate of phosphate incorporation into ATP was therefore expressed as the ratio of the CPM rate of the glucose-6-P fraction over that of the P_i fraction. It was usually 0.05-0.2%.

RESULTS

Establishment of optimal conditions

As found previously, an increase in calcium ion concentration or repetitive stretching of insect fibrillar muscle caused a large increase in the rate of ^{32}P-phosphate incorporation into ATP (Figure 1). In order to evaluate the increase in exchange rate quantitatively, it is necessary to exclude non-actomyosin exchange reactions as well as artifacts caused by diffusional limitations, especially changes in the intrafibrillar levels of ADP and phosphates and depletion of ATP.

Exclusion of non-actomyosin exchange reactions

Sodium azide (10 mM) is known to be an inhibitor of mitochondrial ATPase

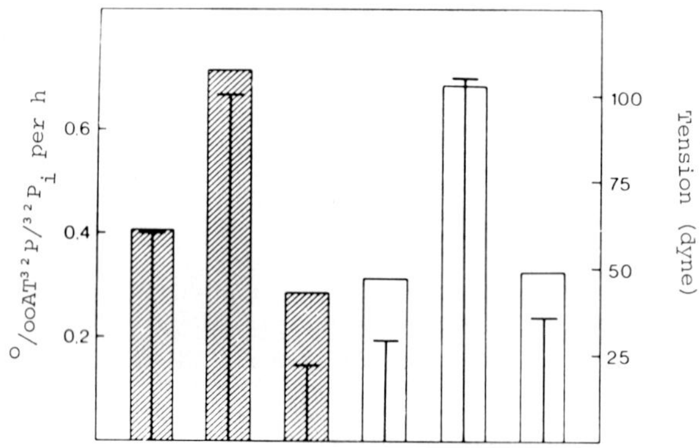

Fig. 1. Effects of prenylamine on incorporation rate and tension. Two fibre bundles in 6 successive incubation periods; in 2) and 5) with application of square wave stretch and release cycles (1 Hz, 1 % L_o), the others at resting length.

▨ incorporation rate in presence of prenylamine at 1.5×10^{-4} M.

□ in absence of prenylamine. Respective tension values are represented by vertical bars.

and of the mitochondrial ATP-P_i exchange reaction (vom Brocke, 1966; Bogucka and Wojtczak, 1966). Its effect on the ATP-P_i exchange reaction was found to be negligible and could not be enhanced by further addition of oligomycin (0.3 mg/ml), the classical inhibitor of mitochondrial exchange reactions (cf. Lardy *et al.*, 1958). The involvement of residues of the sarcoplasmic reticulum causing formation of ATP-^{32}P after addition of $^{32}P_i$ is unlikely under our experimental conditions of high pCa. Moreover, addition of 0.15 mM prenylamine which strongly inhibits ATP-^{32}P formation (Makinose, 1971) barely affects the basal exchange rate induced by stretch in glycerinated insect fibrillar muscle (Figure 1). Recently Morgan *et al*, (1976) referred to the possibility of an exchange reaction coupled to a system consisting of myosin light chain kinase, myosin and phosphatase. This system, while also dependent on calcium ion concentration, is improbable since its activity would not be expected to depend on actin-myosin interaction and on the extent of overlap between actin and myosin filaments. As found by Reiermann *et al*, (1977), the rate of calcium dependent ATP-P_i exchange observed with glycerinated psoas fibre bundles stretched to a sarcomere length of 4 microns is several times smaller than in fibres of 2.2 µm sarcomere length. This suggests that the Ca^{++}-activated exchange requires interaction of actin and myosin and must therefore be ascribed to an activity of the actomyosin contractile system. It had been shown previously that isolated actomyosin prepared by combining actin and myosin is capable of catalyzing the incorporation of phosphate into ATP, but neither actin nor myosin alone are. (Paulsen, 1976; Wolkott and Boyer, 1974; Hotta and Jujita, 1971).

Dependence on the concentration of ATP and ADP

The rate of ATP-P_i exchange in actomyosin preparations is independent of the concentration of Mg-ATP in the range 1-4 mM (Figure 1 in Paulsen, 1976) whereas in insect fibrillar muscle fibre bundles it is only constant in the range 10-20 mM. It is greatly reduced if the concentration is lowered below 10 mM (Figure 2) suggesting that below this concentration diffusional limitations lead to the development of an ATP-free core inside the fibres and hence to an inhibition of the exchange reaction. On the other hand, the concentration of Mg-ADP or the ratio of ATP to ADP does not seem to be critical. Less than 5% inhibition of the exchange rate is observed at both 10 and 15 mM ATP when the concentration of Mg ADP is increased to 5 or 10 mM.

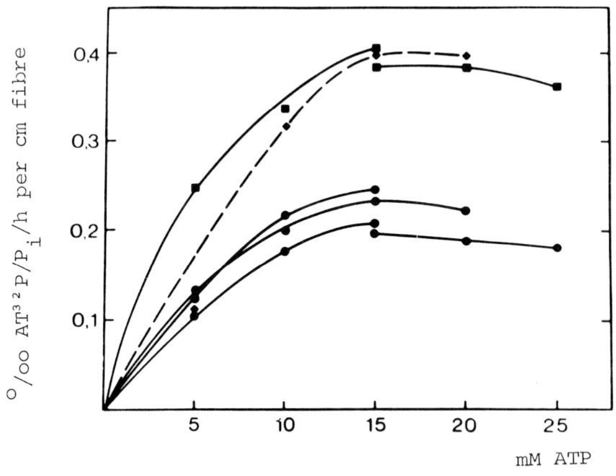

Fig. 2. Dependence of incorporation rate on Mg ATP concentration under isometric conditions (●——●), with application of rectangular stretch-release cycles (1 Hz) (■——■), and of sinusoidal stretch-release cycles (◆——◆). Each line represents an experiment with one fibre bundle.

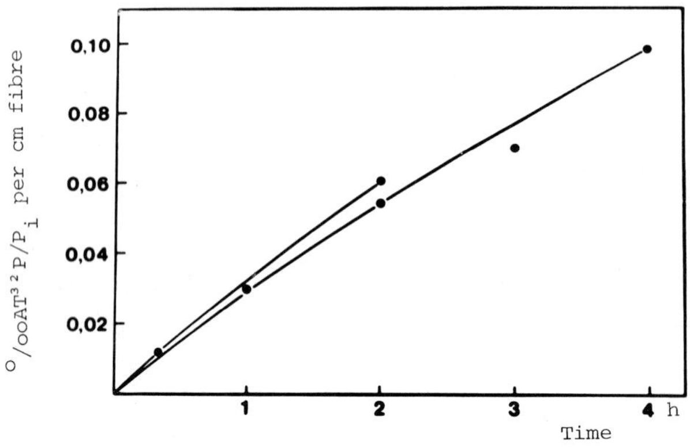

Fig. 3. Incorporation of $^{32}P_i$ into ATP-γ-^{32}P as a function of incubation time. Fibre bundles of *Lethocerus max.* incubated in 15 mM ATP, 15 mM $MgCl_2$, 20 mM imidazole/HCl/buffer (pH 6.7), 10 mM Na-azide, 4 mM Ca-EGTA, 1 mM ^{32}P-P_i.

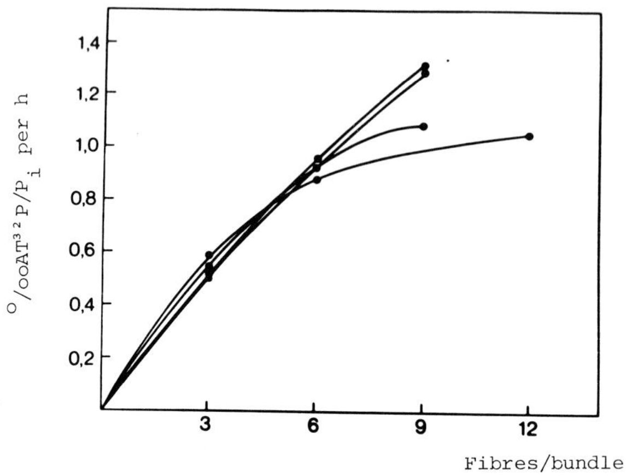

Fig. 4. Dependence of the incorporation rate on the number of fibres per bundle. Four series of experiments with increased number of fibres of the respective preparations incubated in 15 mM ATP-salt solution (see Fig. 3) and 1 mM ^{32}P-P_i.

Effect of incubation time, mass of tissue and temperature

Successive incubation periods give results usually consistent within about 10% of each other (Figure 7) and the amount of phosphate incorporated into ATP is found to be proportional to incubation time up to 4 hr (Figure 3). These results show that the 'age' of the preparation has little effect and that there is no progressive accumulation of inhibitory products.

The phosphate incorporation into ATP is proportional to the number of fibres in a fibre bundle (Figure 4) and comparable for bundles prepared from the same muscle. The incorporation rate in presence of 1 mM P_i is 0.05-0.1% per hour in a fibre of 1 cm length. This corresponds to an absolute rate of 1-2 pmoles .cm $fibre^{-1}$ min^{-1} or 0.5-1 nmoles per minute by 1 mg myosin. This figure compares reasonably well with a direct estimation of the phosphate incorporation into ATP catalyzed by isolated actomyosin: 1.3 nmoles.mg $myosin^{-1}$ min^{-1} (Paulsen, 1976).

A graph of the log of the incorporation rate plotted against the reciprocal value of absolute temperature (Figure 5) gives a straight line corresponding to a Q_{10} of about 4 and an activation energy of 24 Kcal/mole. This high Q_{10} is similar to that of the actomyosin ATPase in glycerinated insect fibrillar muscle (Schädler *et al.*, 1971) and in acto-heavy meromyosin (Barouch and Moos, 1971). To what extent are these temperature effects limited and distorted by diffusional processes?

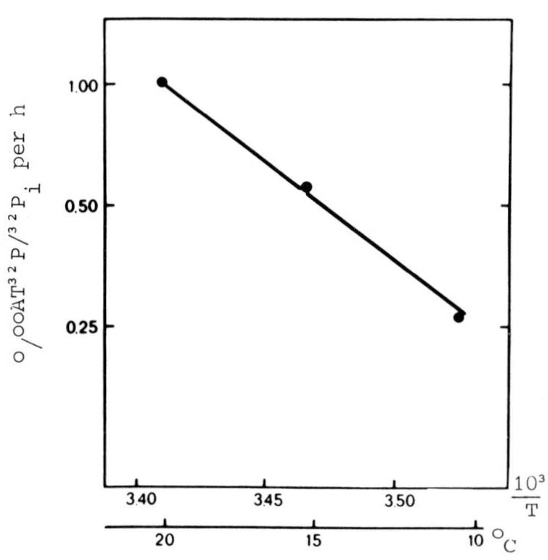

Fig. 5. Dependence of the incorporation rate on temperature. Mean values from two experiments.

A temperature-induced increase in ATPase activity would also be expected to increase the concentration of phosphate inside the fibre according to the relation

$$\Delta C = A/4\pi D$$

Where ΔC is the concentration difference between the phosphate concentration inside and outside the fibre, A is the rate of ATP splitting and phosphate production in 1 cm length of fibre and D is the diffusion constant. However an increase in phosphate concentration *per se* would be expected to decrease rather than increase the fraction of labelled phosphate incorporated into ATP. Furthermore, this fraction is little (i.e. less than 15%) affected by increasing the outside concentration of phosphate to 1-2 mM. Taken together, the experiments suggest that diffusional limitations caused either by substrate depletion or ATPase dependent accumulation of ADP and phosphate within the fibre bundle are not seriously interfering with the determination of the exchange rate, and that we may be justified in expressing results in terms of rates per unit length of fibre.

Ca^{++}-activation of ATP-P_i exchanges

There is a similar dependence on Ca-concentration for both phosphate incorporation rate and ATPase activity (Figure 6). At the highest Ca-concentration attainable with the EGTA-buffer system, P_i incorporation is 8 times and ATPase 5 times as high as in the Ca-free solution, in agreement with previous investigations of the Ca-dependent ATPase activity of glycerinated fibres from *Lethocerus maximus* (Rüegg and Tregear, 1966). The calcium ion dependence of immediate fibre stiffness is quite similar (Ulbrich and Rüegg, 1976), stiffness being a measure of the number of cross bridges attached to actin at any one moment (Huxley, 1974), i.e. for the concentration of actomyosin complexes. The parallelism of stiffness and rate of ATP exchange reaction under a variety of conditions of Ca^{++} activation may therefore be further evidence showing that the exchange is dependent on actin-myosin interaction, presumably on the formation of a "high energy" actomyosin ADP intermediate.

Stretch-activation

Even at optimal calcium ion concentration (greater than 10^{-5}M) repetitive square wave stretches of 1 Hz frequency and 1% L_o amplitude caused as shown in Figure 7 a large increase in both ATPase activity and rate of phosphate incorporation into ATP. The latter was on the average 1.64 times greater during the application of square wave stretches than under isometric conditions. In other experiments the effect of different amplitudes of stretch was compared

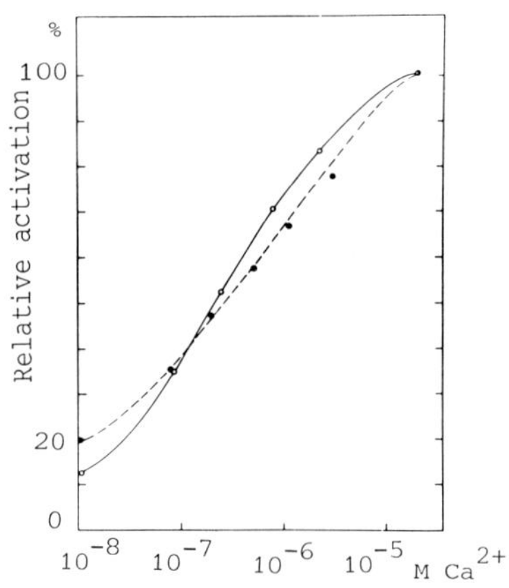

Fig. 6. Relative activation of the incorporation rate (O——O) and of ATPase (●---●) by Ca^{2+} ions.

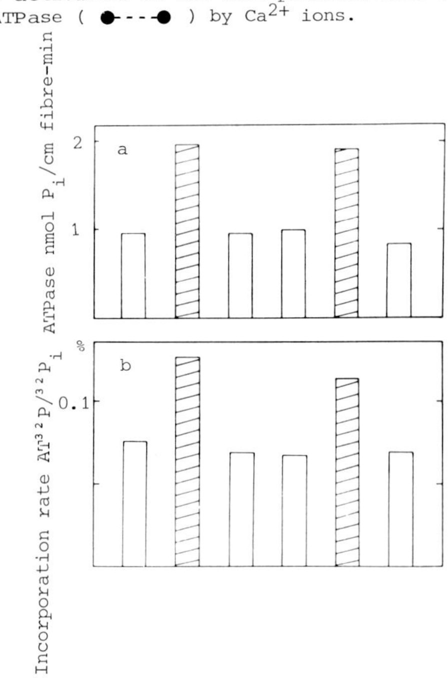

Fig. 7. Increase in the ATPase activity (a) and the rate of $^{32}P_i$ incorporation during rectangular stretch and release cycles.(b) Six consecutive incubation periods of 1 fibre bundle.

□ = isometric contraction ("rest")

▨ = stretch cycles (1 Hz, 1% L_o).

to the effect of 1% stretches used as reference value.

Dependence on amplitude of rectangular stretch cycles

When amplitude of rectangular stretch-release cycles is increased from 0.25% L_o up to 3% L_o, an increase in stretch-induced incorporation rate and ATPase as well as in delayed tension occurs (Figure 8). There is no linearity over the total range of amplitudes, but the largest relative effects are achieved at lengths up to 1%L_o. At higher amplitudes up to 3% L_o, there is little additional activation.

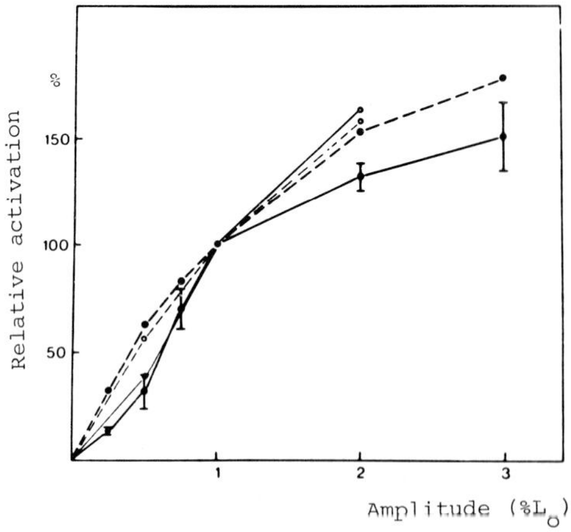

Fig. 8. Increase in activation of the incorporation reaction (●——●), and stretch-induced delayed tension (●- - -●) in one set of fibre bundles and ATPase activity (○- - -○) and stretch-induced delayed tension (○——○) in a second set, due to increased stretch amplitude; Relative to 1 % L_o amplitude at 1 Hz frequency.

Effect of stretch duration

In contrast to repetitive stretches produced by square wave length changes of 1% performed at a frequency of 1 Hz (cf. Figure 7), a static extension by 1% had little activating effect. Subsequent experiments showed that following a quick stretch, extra incorporation induced by the length step occurs almost entirely within the first 100 ms after the step. Thereafter i.e. 100-500 ms after the step the rate of phosphate incorporation is only a little higher than before stretching. In these experiments 45 fibre bundles were subjected to rectangular stretch-release cycles of an amplitude of 1% L_o and applied with a frequency of 1 Hz during different test periods of 1 hr each. The relative amount of phosphate incorporated was then determined for each test period. The incorporation obtained during test periods at rest length was then subtracted from these data in order to obtain the 'extra incorporation' induced by stretch for a test period of 1 hr. Dividing this value by the number of stretches during the test period one obtains the stretch induced extra incorporation per stretch or release. In experiments with each fibre bundle, test periods at rest length alternated with test periods during which the fibres were subjected either to repetitive (1 Hz) reference square wave stretches of 100 ms or to test square wave stretches of 5, 10, 25, 50, 200 or 500 ms duration. The stretch induced incorporation obtained with repetitive 100 ms stretches is then compared with the effect obtained with stretches of a different duration, but applied with the same frequency (1 Hz). In Figure 9 the relative amount of phosphate incorporated is plotted as a function of stretch duration, and this exponentially delayed curve is taken to represent the time course of the stretch induced formation of ATP-^{32}P. This time course is remarkably similar to that of the stretch-induced delayed rise of tension (Figure 9), and to that of stretch induced extra-ATP splitting (Breull *et al.*, 1973). Figure 9 shows that most of the stretch induced incorporation of phosphate into ATP occurs in a transient phase ("early burst") early in the phase of delayed tension rise and very little further stretch induced incorporation occurs later on, in the "steady state", i.e. in the phase of tension maintenance about 100-500 ms after stretch. The slope of the curve between 100 and 500 ms presumably reflects the steady state rate of stretch-induced phosphate incorporation caused by a static extension effect.

Further analysis of the time course shown in Figure 9 is obtained by subtracting the experimentally determined incorporation (I_e) obtained during a given time from the corresponding extrapolated value of the steady state phase (I_s). If the difference ($I_s - I_e$) is plotted on the ordinate of a logarithmic scale versus time. (Figure 10), a straight line is obtained indicating first order reaction kinetics with a rate constant of 27 sec^{-1}. This compares favorably with the rate constant of the delayed tension rise (31 sec^{-1}).

ATP-phosphate exchange during myogenic oscillation of fibre length

As the incorporation of $^{32}P_i$ predominantly occurs briefly after stretching (cf. Figure 9) it is obvious that in a given time period many brief stretches would produce a far larger increase of incorporation than fewer stretches of longer duration, even if the total times of extension were identical. In the experiments shown in Figure 11, fibre bundles were subjected to square-wave stretch

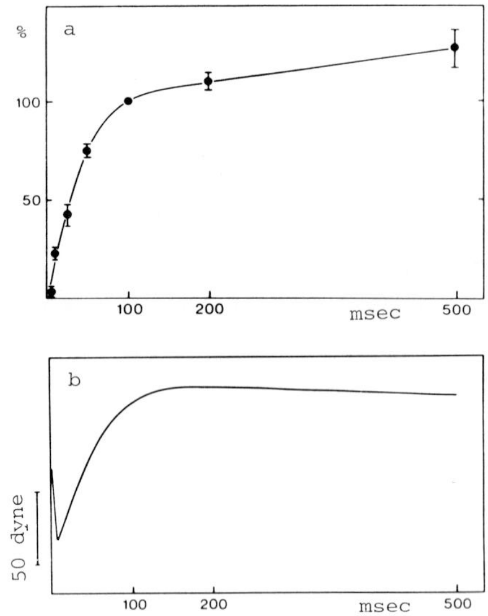

Fig. 9. a) stretch-induced exchange rate (relative) as a function of stretch duration of square wave stretch-release cycles (Frequency: 1 Hz, Amplitude: 1% L_o).
Reference value (100%): 100 msec.

b) time course of stretch-induced tension development

and release cycles of constant amplitude (1% L_o), but different frequency. In all test periods (1 hr) the total time of extension equalled the total time at the released length. In Figure 11 the amount of stretch-induced incorporation obtained in test periods of 1 hr with 32 fibre bundles stretched at 1 Hz (Reference effect = 100%) was compared with the effect produced at different frequencies of the length change. The incorporation effect increases up to 5 Hz and then decreases again because of the small incorporation obtained with very brief stretches (cf. Figure 9).

The frequency dependence of the ATP-P_i exchange in 24 fibre bundles caused by sinusoidal stretches and releases is quite similar (Figure 12), and it is also similar to the frequency dependence or ATPase (Steiger and Rüegg, 1969). Furthermore, the absolute increase in incorporation rate observed with 1% length change and at the optimum frequency was less than 10% different in experiments

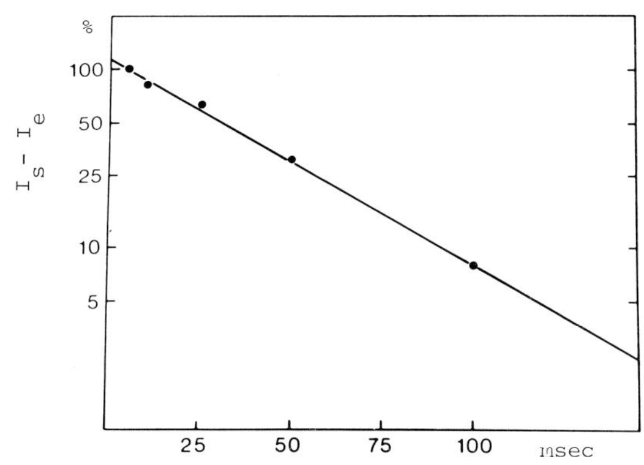

Fig. 10. Semi-logarithmic plot of the differences (I_s-I_e) found for the various stretch durations (see text and Fig. 9).

I_e = exchange rate of the early phase.
I_s = extrapolated exchange rate of the steady state phase.

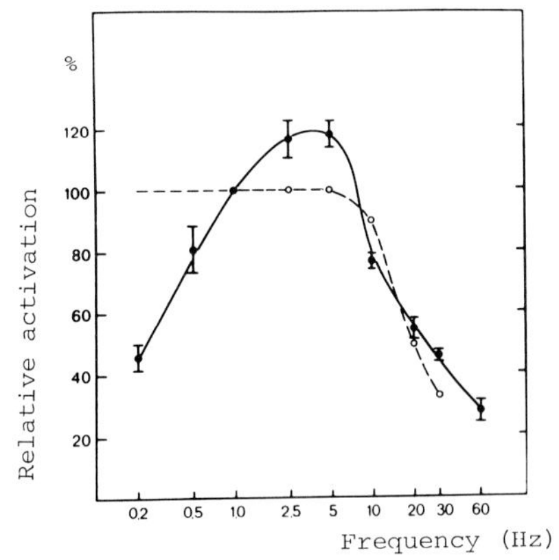

Fig. 11. The dependence of the relative activity of the stretch-induced exchange rate (●---●) and the size of delayed tension (O---O) on the frequency of square wave stretch and release cycles. The times in stretch and release are equal.

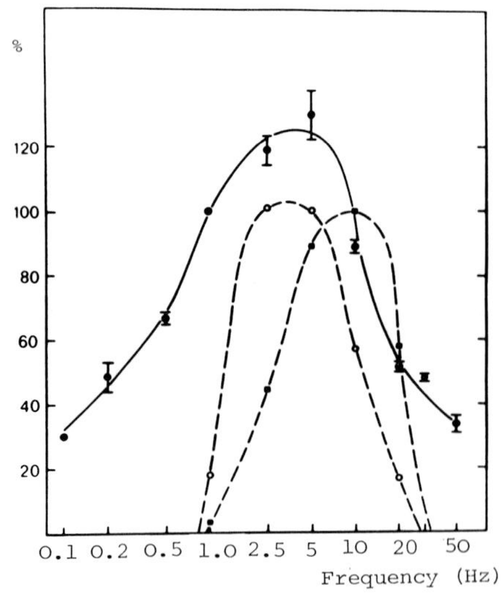

Fig. 12. The dependence of stretch-induced exchange rate (●——●), the work per cycle (O----O), and the power (■----■) on the frequency of sinusoidal stretches (Amplitude 1% L_o). Exchange rate is expressed in % of the reference value at 1 Hz (= 100%), work and power in % of their respective maximal values.

with sinusoidal and square wave stretches. The observed P_i-incorporation amounted to 0.2% per min in 1 cm fibre or 2.7 pmoles.cm fibre^{-1}min^{-1}. This is two times higher than the incorporation obtained at fixed length and about 0.2% of the ATP splitting rate under these conditions.

When fibres were stretched and released in a sinusoidal fashion at an amplitude of 1% and with a frequency of 1-20 Hz they were driven to oscillate (cf. Jewell and Rüegg, 1966) to perform oscillatory work and to put out oscillatory power. The latter was of the order of 5 ncal.sec^{-1} cm fibre^{-1} at the optimum frequency of 10 Hz. At frequencies higher than 20 Hz and lower than 1 Hz work was done *on* the fibres, i.e. the bundles performed negative work. Under these conditions the phosphate incorporation is larger than in unstretched preparations as first shown by Mannherz (1970) but it is clearly smaller than under conditions of optimal power output. These results do not favour the idea of a reversal of chemomechanical energy transformation caused by the performance of negative work. Rather, there may be a correlation between positive oscillatory work per cycle and rate of phosphate incorporation into ATP. The similarity of the frequency dependence of the phosphate incorporation during repetitive sinusoidal and rectangular stretches is so striking that both processes ought to be described by the same mechanism. At each frequency the amount of stretch induced incorporation obtained in a given test period (cf. Figure 11) ought to correspond to the product of the number of stretches times the induced incorporation per stretch or oscillatory cycle. The latter depends on stretch duration, as shown (Figure 9). At frequencies above 5 Hz, corresponding to a stretch duration of 100 ms, the predicted correspondence is remarkably good (Ulbrich, 1975). At lower frequencies howëver an additional contribution by the static extension, causing a steady state rate of incorporation, has to be considered as well. Interestingly, the frequency dependence of the ATP-phosphate exchange reaction and of the output of oscillatory power or of the work cycle are quite similar (Figure 12). This need not indicate a causal relationship but it may simply reflect the relation of both the mechanical output and of the exchange activity to the rate of ATP splitting.

DISCUSSION

Possible mechanism of ATP-P_i exchange

Many of the results described in this paper have clearly demonstrated that an incorporation of $^{32}P_i$ into ATP is catalyzed by myosin only under conditions of actin-myosin interaction. Since the formation of myosin-bound ^{32}P-ATP occurs easily with myosin alone because of the ready reversibility of the hydrolytic step of ATP splitting (Mannherz *et al.*,1974; Wolcott and Boyer, 1974), the interaction with actin is presumably required to facilitate the dissociation of the myosin-^{32}P-ATP complex into myosin and labelled ATP. While the incorporation of labelled phosphate into protein-bound labelled ATP *may* occur in the myosin product state, i.e. in cross bridges detached from actin, an exchange reaction occurring after combination of the myosin product with actin must be considered as well. For, according to the mechanism proposed by Lymn and Taylor (1971), it is the myosin-product complex which combines with actin to form an actomyosin-ADP intermediate which then dissociates. Alternatively, it might react with added ^{32}P-phosphate, which becomes incorporated into ATP. The nucleotide triphosphate bound to the attached cross bridge would be expected to prevent the power stroke of the attached bridge; and it would presumably cause the bridge to detach without the intervention of new ATP. This detached bridge would reattach anew after the bound ATP is hydrolyzed. It is tempting to speculate that an ATP-P_i exchange reaction in cross bridges attached to actin might give rise without *net* ATP splitting to the breaking and reformation of actin-myosin cross linkages, which are capable of bearing (but not generating) tension during stretch, as postulated by Curtin and Davies (1973).

Possible nature of coupling of phosphate incorporation into ATP and tension development

In view of the similarity in the rate constants of stretch induced phosphate incorporation and tension development it is tempting to speculate that both processes are interdependent. Possibly, the mechanical as well as the chemical processes reflect the formation of a "high-energy" intermediate (actomyosin-ADP?) which presumably accumulates after stretch. Evidence obtained by Herzig (1977) suggests that a quick stretch may also give rise to an increase of the population of "vertically" attached cross bridges. The possibility may be considered that 'energy rich' actomyosin-ADP complexes causing the exchange reaction do in fact correspond to the population of myosin cross bridges vertically attached to actin. In this respect it is interesting to note that ADP (e.g. Beinbrech, 1977) or certain ADP analogs like AMP-PNP (Beinbrech *et al.* 1976; Marston *et al.*, 1976; Holmes *et al.*, 1976) do seem to stabilize the cross bridges in a predominantly vertical position. It seems possible however that such cross bridges may be in equilibrium (cf. Lymn and Huxley, 1973) with nucleotide-loaded bridges attached to actin at an acute angle.

ACKNOWLEDGEMENTS

This work was supported by the Deutsche Forschungsgemeinschaft Grant Ru 154/9-12.

REFERENCES

Bagshaw, C.R., Eccleston, J.F., Trentham, D.R., Yates, P.W., and Goody, R.S. (1972). *Cold Spring Harb. Symp. quant. Biol.* 37, 127.

Bandurski, S. and Axelrod, B. (1951). *J. Biol. Chem.* 193, 405.

Barouch, W.W. and Moos, C. (1971). *Biochim. Biophys. Acta* 234, 183.

Beinbrech, G., Kuhn, H.J., Herzig, J.W., and Rüegg, J.C. (1976). *Cytobiologie* 12, 385-396.

Beinbrech, G. this volume.

Bogucka, K. and Wojtczak, L. (1966). *Biochim. Biophys. Acta* 122, 381.

Breull, W., Steiger, G., and Rüegg, J.C. (1973). *J. Mechanochem. Cell Motil.* 2, 91.

Breull, W. (1971). *Experienta* 27, 779.

Brocke, vom, H.H. (1966). *Pflügers Arch.* 290, 70.

Curtin, N.A. and Davies, R.E. (1973). *Cold Spring Harb. Symp. quant. Biol.* 37, 619-626.

Lardy, H.A., Johnson, D., and McMurray, W.C. (1958). *Arch. Biochem. Biophys.* 78, 587.

Lymn, R.W. and Huxley, H.E. (1973). *Cold Spring Harb. Symp. quant. Biol.* 37, 449-453.

Lymn, R.W. and Taylor, E.W. (1971). *Biochem.* 10, 4617.

Mannherz, H.G. (1968). *Pflügers Arch.* 303, 230.

Mannherz, H.G. (1970). *Febs. Letters* 10, 233.

Mannherz, H.G., Schenk, H., and Goody, R.S. (1974). *Eur. J. Biochem.* 48, 287.

Makinose, M. (1971). *Febs. Letters* 12, 269.

Morgan, M., Perry, S.V., and Ottaway, J. (1976). *Biochem. J.* 157, 687.

Paulsen, G. (1976). *Eur. J. Biochem.* 61, 77.

Pringle, J.W.S. (1957). *Cambridge Univ. Press.*

Pybus, J. and Tregear, R.T. (1972). *Cold Spring Harb. Symp. quant. Biol.* 37, 655.

Dempsey, M.E., Boyer, P.D., and Benson, E.S. (1963). *J. Biol. Chem.* 238, 2708.

Gillis, J.M., and Maréchal, G. (1971). *J. Physiol.* 214, 41P.

Gillis, J.M. and Maréchal, G. (1974). *J. Mechanochem. Cell Motility* 3, 55.

Herzig, J.W. this volume.

Hill, A.V. and Howarth, J.V. (1959). *Proc. R. Soc. B,* 151, 169.

Holmes, K.C., Goody, R.S., Mannherz, H.G., Barrington-Leigh, J., and Rosenbaum, G. (1976). *26. Col. Mosbach 1975. Molecular Basis of Motility, Springer Verlag p. 26.*

Hotta, K. and Fujita, Y. (1971). *Physiol. Chem. & Physics* 3, 196.

Huxley, A.F. (1974). *J. Physiol.* 243, 1.

Inoue, A. (1973). *J. Biochem.* 73, 1311.

Jewell, B.R., and Rüegg, J.C. (1966). *Proc. R. Soc. B* 164, 428.

Reiermann, H.J., Herzig, J.W. and Rüegg, J.C. (1977). *Basic Res. Cardiol.,* 72,133.

Rüegg, J.C. and Tregear, R.T. (1966). *Proc. R. Soc. B* 165, 497.

Rüegg, J.C. and Stumpf, H. (1969). *Pflügers Arch.* 305, 21-33.

Rüegg, J.C. and Stumpf, H. (1969). *Pflügers Arch.* 305, 34-46.

Schädler, M., Stieger, G.J., and Rüegg, J.C. (1971). *Pflügers Arch.* 330, 217.

Steiger, G.J. and Rüegg, J.C. (1969). *Pflügers Arch.* 307, 1.

Marston, S.B., Rodger, D.C. and Tregear, R.T. (1976). *J. Mol. Biol.* 104, 263-276.

Ulbrich, M. (1975). *Dissertation, Bochum.*

Ulbrich, M., and Rüegg, J.C. (1971). *Experienta* 27, 45.

Ulbrich, M., and Rüegg, J.C. (1976). *Pflügers Arch.* 363, 219.

Ulbrecht, M. (1962). *Biochim. Biophys. Acta* 57, 438.

Wolcott, R.G. and Boyer, P.D. (1974). *Biochim. Biophys. Res. Commun.* 57, 709.

CAPTURE AND CULTURE

THE AVAILABILITY OF INSECT FIBRILLAR MUSCLE

J.W.S. Pringle.

A.R.C. Unit, Department of Zoology, South Parks Road, Oxford, U.K.

HISTORY

The occurrence in insects of a type of striated muscle with unusual properties was first suspected in 1947 during the course of experiments on the halteres of muscid Diptera. These dumb-bell shaped organs, which are modified hind wings, function as gyroscopic sense organs and are kept in motion during flight by means of a small muscle attached to their base. The contractions of this muscle are balanced by an elastic restoring force and the system as a whole is mechanically resonant. Two experiments were significant in the present context; the oscillation frequency could be changed reversibly by altering the inertia of the terminal knob of the haltere and the oscillation frequency remained constant at 120 Hz when the frequency of electrical stimulation was varied in the range 40-400 Hz (Pringle, 1948).

The first experiments on the physiology of fibrillar flight muscle were also done on flies (Pringle, 1949). By the simple technique of inserting wire electrodes into the thorax of a fly fixed so that it could perform normal wing movements, it could be shown that there is no temporal relationship between the muscle potentials signalling the arrival of motor nerve impulses and the mechanical oscillation. The hypothesis was put forward that contraction of these muscles is activated by stretch and that this property is responsible for maintenance of the oscillation.

It is not feasible to make a preparation of the isolated flight muscle of flies. This difficulty was overcome independently by Boettiger (1957), who succeeded in attaching the mesophragma of the dorsal longitudinal flight muscle of queen *Bombus* (Hymenoptera) to recording apparatus (Fig. 1A) and by Pringle (1954) who used the timbal sound-producing muscle of cicadas of the tropical genus *Platypleura*, at the dorsal end of which the fibres are conveniently inserted on a plate with a thread-like apodeme (Fig. 1B). Neither preparation is wholly convenient, the *Bombus* muscle because of its small size and the difficulty of achieving a non-distorting attachment of the recording apparatus and the cicada timbal muscle because suitable cicadas are not readily available in temperate countries. Not all cicada timbal muscles have the self-oscillatory mechanism (Hagiwara, 1953; Hagiwara, Uchiyama and Watanabe, 1954).

The next preparation to be used was the fibrillar basalar flight muscle

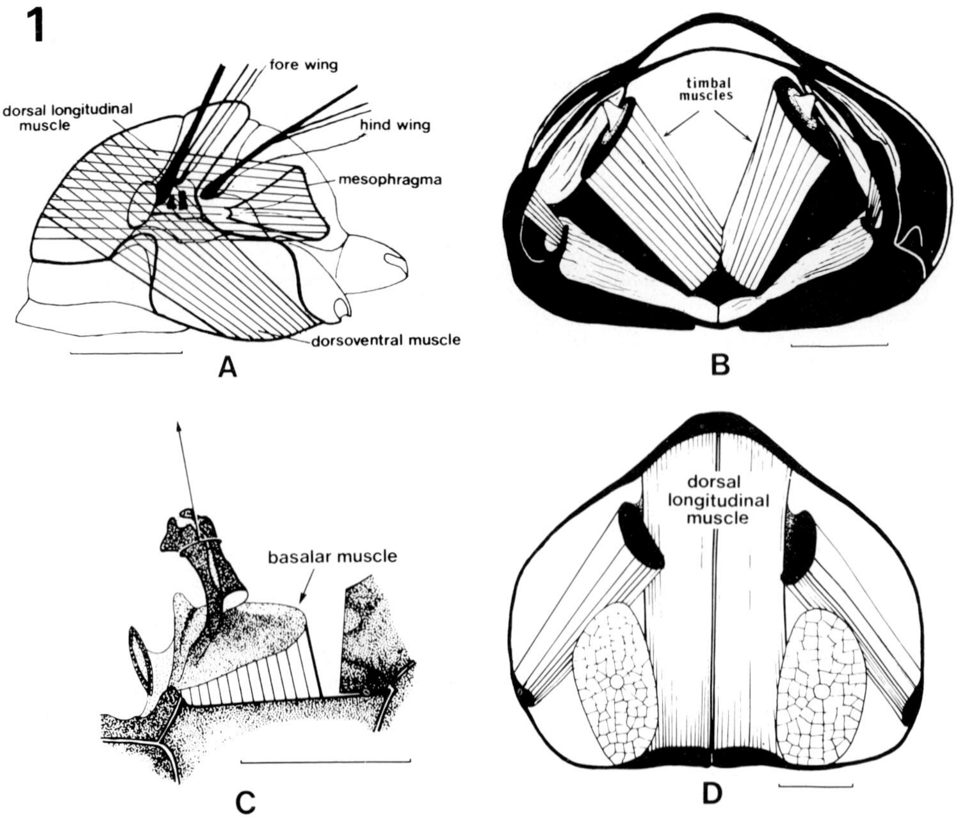

Fig. 1 Preparations of insect fibrillar muscle. A, thorax of *Bombus*, showing dorsal longitudinal muscle and its mesophragma. B, posterior view of 1st abdominal segment of a cicada, showing timbal muscles. C, part of pleuron of *Oryctes*, showing mode of attachment to apodeme of basalar muscle . D, internal view of upper part of thorax of *Lethocerus*, showing dorsal longitudinal muscle and the other fibrillar muscles. Scale 3 mm. (A, modified from Pringle, 1961; B, Pringle, 1957; C, Darwin and Pringle, 1959).

of the tropical beetle *Oryctes rhinoceros*,, obtained from Sri Lanka and from the Central Coconut Research Station, Kayangulam, Kerala, India (Darwin and Pringle, 1959). The dorsal insertion of this muscle is also on a disc-like apodemal plate and a wire attached to the detached pleural process pulls evenly on all the fibres (Fig. 1C). As with the cicada timbal muscle, the ventral insertion is diffuse and has to be left intact in the partly dissected insect. In spite of the fact that the muscle is thus not fully isolated, the basalar muscle of large beetles remains the best preparation for the study of the physiology of the living muscle. It has the advantage over indirect flight muscles that the nervous excitation can be varied over a wide range using the facilitation present at the neuromuscular junction.

Several species of large beetle can be bred successfully in the laboratory. *Oryctes spp.* have the typical grub-like larvae of lamellicorn beetles and survive well in damp soil if fed with bisected root vegetables placed on the surface; swedes (*Brassica napus)* are suitable and cheap. The colony may develop a rich culture of mites, but these appear to be merely inquiline and do not harm the beetles unless their numbers increase to the point when the spiracles become blocked. A tropical rose-chafer (*Pachnoda ephippiata,* Cetoniidae) can also be bred in large numbers in soil on a diet of tinned apricots; the adults eat the fresh fruit and the larvae eat the decayed remains. This beetle, although small (2.8 cm long) has an equally convenient anatomy (Ballantyne, 1976) and the culture does not become as infected with mites as that of *Oryctes*. The largest beetles to be used in the Oxford laboratory are scarabeid dung-beetles of the genus *Heliocopris* obtained from Kenya from elephant dung; these are up to 6 cm long, weigh 20 g and have about 2 g of extractable flight muscle. They cannot be bred in the laboratory, but can sometimes be obtained in large numbers at lights during rainy weather in areas where elephants are common.

Beetle fibrillar muscles, like those of flies, bees and wasps, have a small number of fibres of large diameter. The fibres are bound together by a dense network of air-filled tracheae which penetrate deeply into invaginations in the fibre surface. In the beetle basalar and cicada timbal muscles, the main tracheal trunks penetrate the muscle from outside, making it possible for the aeration to be preserved in the semi-isolated preparation. The anatomy of the tracheae, however, makes it impossible to separate the individual fibres without damage and these muscles are not suitable as starting material for experiments on single muscle fibres or even small bundles.

In the search for a still better fibrillar muscle, the dorsal longitudinal muscle of belostomatid bugs (Hemiptera, Heteroptera) (Fig. 1D) was first explored in 1961 in dead speciments brought to England from Uganda by Dr. P.L. Miller. Uniquely in insects of this Order, the indirect flight muscles have a large number (about 1,800) of small-diameter fibres (70 μm) and the main tracheal trunk runs down the middle of the muscle with branches radiating outwards and terminating in blind sacs (Ashhurst, 1967). There is little or no penetration of tracheae into the fibres and the individual fibres can therefore be separated without damage. At both ends the muscle insertions are diffuse and it is not easy to attach the whole muscle to recording apparatus without uneven loading of different parts of the cross-section. Adequate ventilation seems to depend more on abdominal pumping movements than in other insects with fibrillar flight muscles and preparations of the whole muscle rapidly become anoxic when active. This insect is therefore not suitable for experiments on living muscles, but is ideal for experiments using glycerol-extracted fibres, since the small fibre diameter makes for rapid diffusion. In *L. maximus* the fibres are 2 cm long.

Belostomatid bugs are aquatic and fly infrequently though well. In their feeding habits and reproduction they fall into two groups (Fig. 1 of Cullen, this volume). *Belostoma* and *Limnogeton* are snail eaters and their flight muscles seem to be softer than those of other genera; they are also smaller. *Hydrocyrius*, though taxonomically related to *Belostoma*, is carnivorous on fish, as also is the other group including the genus *Lethocerus*. In the first group, including *Hydrocyrius*, the female lays eggs on the back of the male, which aerates them by floating at the surface until the nymphs hatch. *Lethocerus* females lay eggs on vegetation just above the surface of the water. The mating procedure is unknown in both groups, but *Hydrocyrius* and *Limnogeton* have been bred successfully in the tropical climate room in the Oxford laboratory. The care of the nymphs is laborious, since they are highly cannibalistic. In spite of much effort, *Lethocerus* has never been bred successfully, though eggs collected in the wild have been reared to maturity.

All water-bugs have a tendency to degenerate their flight muscles if food is short, though the controlling factors have not been fully understood (Cullen, 1969). *Hydrocyrius* is significantly more prone to this effect than is *Lethocerus* and since it is impossible to determine by external examination what is the state of the flight muscles, insects collected in the water are liable to include a lot of specimens that are useless for muscle experiments. The test of ability to fly cannot be applied, since the living insects are induced to fly only with difficulty (Barber and Pringle, 1966). The only really satisfactory method of collecting suitable insects is to catch them at lights, to which they are strongly attracted, but since large numbers fly only in rainy weather, the timing of the operation is difficult to plan long in advance.

Biophysical experiments at Oxford have largely been conducted on single fibres or small bundles isolated from glycerol-extracted dorsal longitudinal muscles of *Lethocerus cordofanus*, *L. maximus* and *L. annulipes*. Because of shortage of material and the greater size of *Heliocopris* beetles, biochemical experiments have largely been done using these muscles as the starting material. Recently some biophysical experiments have been conducted on the dorsal longitudinal muscle fibres of carpenter bees of the tropical genus *Xylocopa,* which resemble *Bombus* in their anatomy but are twice as large. These insects were obtained alive from Kenya where they can be caught in the daytime on flowers or at their nesting sites in the wood of houses. They are active insects, difficult to maintain in the laboratory and have to be killed and used soon after arrival. Their one advantage over *Lethocerus* is that, as in other Hymenoptera, the muscle fibres can be extended in the relaxed state without much damage to their fine structure.

Maintenance of a regular supply of large insects with fibrillar flight muscles remains a chronic problem. Given sufficient semi-skilled assistance, it would be possible to obtain hectogram or even kilogram quantities of fibrillar muscle for biochemical analysis by dissection of very large numbers of small insects; beetles would probably be the Order of choice. Small fibre size gives Heteroptera such an advantage over other Orders for biophysical work that it would be worth putting more effort into the use of some of the smaller temperate genera such as *Notonecta*, from which fibres 6 mm long can be obtained. Dr. D.C.S. White at York University has explored this possibility. But there is now a considerable literature on *Lethocerus* and each change of species carries with it dangers of confusion owing to at least quantitative differences in parameters. Effort should therefore be concentrated on the maintenance of adequate supplies of living *Lethocerus* and, possibly, on development of the technology needed to breed either this genus or the easier *Hydrocyrius*, which appears to have no relative disadvantages apart from its greater tendency to muscle degeneration.

OLD WORLD SOURCES OF *LETHOCERUS SPP.*

The genus *Lethocerus* is world-wide in its distribution, but is restricted to warmer climates. Apart from one species which can survive the winter temperatures of United States of America, this is a tropical and sub-tropical genus, likely to be found wherever there are lakes or permanent, slow flowing water. The distribution of Old World species is shown in Fig. 2 (overleaf).

The most widely distributed species are *Lethocerus cordofanus* from Africa, *L. indicus* from India and the Far East and *L. insulanus* from Australasia. These are all of comparable size and their dorsal longitudinal flight muscles can provide fibres 1.2 - 1.5 cm long. The rainy seasons in these parts of the world differ and, given an adequate organisation, it should be possible to maintain supplies for at least six months of the year from these countries. In fact and in spite of extensive correspondence and a few personal visits, regular supplies have been obtained from only one country - *L. cordofanus* from Uganda.

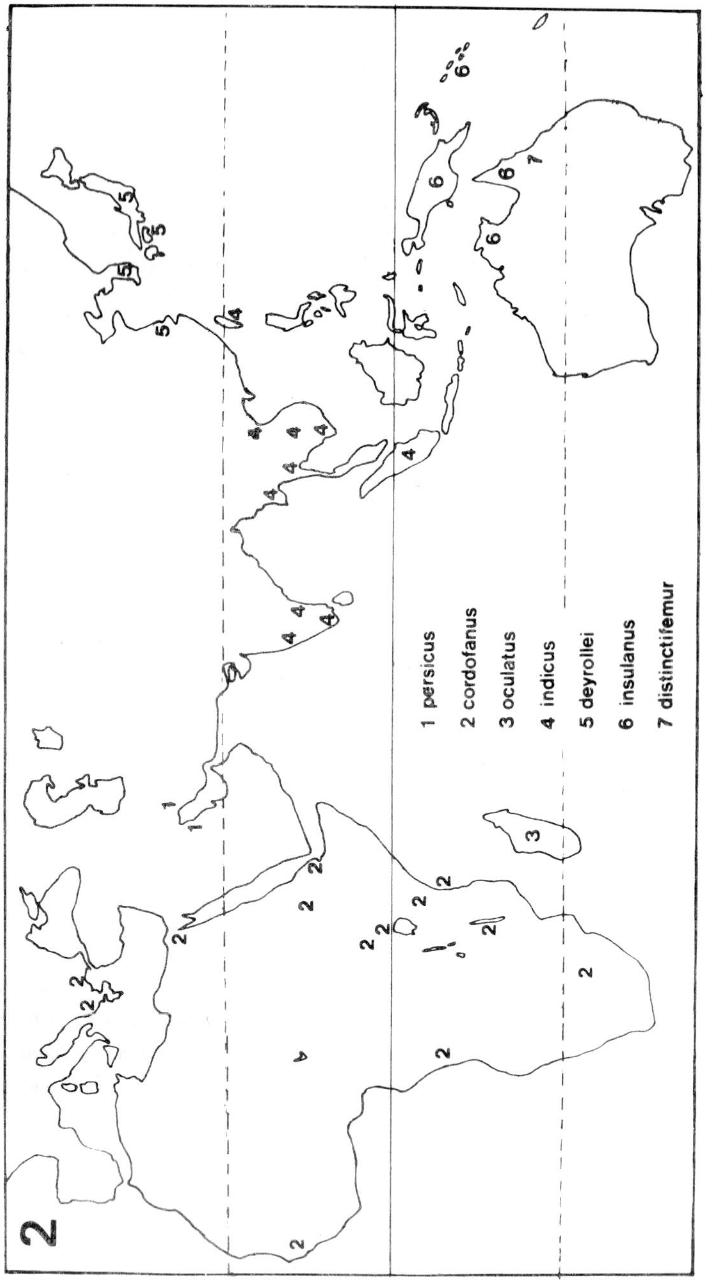

Fig. 2 Distribution of the seven Old World species of *Lethocerus* (based on Menke, 1960 and on the collection in the British Museum).

The problem is not mainly one of difficulty of collection, though the extent to which the three species come to lights varies. Nor is it one of longevity, since water bugs survive well in transit if properly packed. It is mainly a problem of the willingness of someone living in an appropriate area to go to the trouble of collecting the insect, keeping them in good condition until a sufficient number have been obtained, packing them suitably and arranging for their dispatch by air freight. Regular success has been achieved only when there is a direct flight from an airport near to the collecting area to the country where they are needed, but there is no obvious reason why one trans-shipment should cause difficulties.

The insects are best packed by inserting them individually into thin plastic sleeves, cut to appropriate length from a roll of tubing and perforated with scissors or a punch. After inserting the insect, the ends of the tube are closed with a stapler. It is important that the insects be restrained and they must be treated individually to avoid mutual fighting. Once enclosed in this way, fifty insects may be packed on top of each other in a perforated tin lined with damp paper and sent off without further protection. It is preferable to arrange for the package to be kept in the cabin of the aircraft, but casualties have not been too high in consignments subjected to freezing temperatures for a short time. On arrival (we collect from the airport) the insects should be freed and allowed to warm up before being put into water, since they may otherwise drown.

Supplies from Africa have been maintained due to the efforts of a technician, Mr. Joseph Ngirumwe. Willingness to collect water bugs has been expressed in writing by people in many other parts of the world, but few have ever arrived. The addresses of the most hopeful places can be supplied on request. There are still plenty of places where the insects must be common, though the virtual extermination of water-bugs from Japan shows that they are sensitive to insecticide run-off. Even with a considerable increase in cost, expenditure on living material still represents only a small fraction of the total budget of a typical biophysical or biochemical research group and it ought not to be too difficult to find new sources. Our own experience shows how difficult it is in practice.

ACKNOWLEDGEMENTS

It will be clear from this article how much the Unit in Oxford is indebted to Mr. Joseph Ngirumwe, without whose efforts much of the research work would have been impossible. Thanks are also due to Dr. Fred Bennett in Trinidad and to other people in the tropics who have taken the trouble to collect and send living material to us. Dr. Paul Freeman, British Museum (Natural History), kindly provided some of the information in Fig. 2.

REFERENCES

Ashhurst, D.E. (1967) *J. Cell Sci.*, 2, 435.

Ballantyne, D. (1976) Ph.D. Thesis, Oxford University.

Barber, S.B. and Pringle, J.W.S. (1966) *Proc. Roy. Soc. B*, 164, 21.

Boettiger, E.G. (1957) In *Recent Advances in Invertebrate Physiology*, ed. by B.T. Scheer, p. 117. University of Oregon Press, Eugene.

Cullen, M.J. (1969) *Proc. R. ent. Soc. Lond. (A)*, 44, 123.

Darwin, F.W. and Pringle, J.W.S. (1959) *Proc. Roy. Soc. B*, 151, 194.

Hagiwara, S. (1953) *Jap. J. Physiol.*, 3, 284.

Hagiwara, S., Uchiyama, H. and Watanabe, A. (1954) *Bull. Tokyo med. dent. Univ.*, 1, 113.

Menke, A.S. (1960) *Austral. J. Zool.*, 8, 285.

Pringle, J.W.S. (1948) *Phil. Trans. Roy. Soc. B*, 233, 347.

Pringle, J.W.S. (1949) *J. Physiol.*, 108, 226.

Pringle, J.W.S. (1957) *Insect Flight*, Cambridge University Press.

Pringle, J.W.S. (1961) *Natural History*, August-September 1961, American Museum of Natural History, New York.

Pringle, J.W.S. (1954) *J. Physiol.*, 124, 269.

NEW WORLD SOURCES OF LETHOCERUS

Michael K. Reedy

Duke University Medical Center, Durham, North Carolina, U.S.A.

Since the spring of 1973 I have tried to develop new sources of *Lethocerus* muscle which might make me independent of the Oxford connection to the African Supplies. My laboratory required large numbers to permit biochemical and immunocytochemical work with flight muscle myosin. Standing orders with Dr. Fred Bennett in Trinidad and Harold Williams in Dunellon, Florida, had not brought me any live bugs since 1969. Our experience since then can be briefly summarised. Collectors provided hundreds of bugs in Florida, in 1974 and 1975, and scores of bugs in 1973, 1976 and early 1977. A hundred bugs were easily collected but impossible to export from Mexico. A single, small, unrepeated shipment came from Brazil in 1974. So far efforts to recruit collectors by corresponding with students and university entomologists in Jamaica and Venezuela have not been productive. A dedicated student collector was unable to find any Belostomids during three wet weeks of July and August, 1975 in Trinidad. And though it comes from the old world and not the new, it seems worth noting that an enterprising citizen of Bangkok is prepared to guarantee about 500 bugs for an advance payment of $1,500, an offer so far untested.

Our level of need has varied from desperation, when one or two bugs of over 3 cm would be welcome, to luxury seeking, when it would be nice to have another hundred for an extra muscle protein preparation, or to have one or two very large bugs for X-ray diffraction. My laboratory recognises three levels of need, which we express to collectors in this way.

1. A few bugs, even small ones (2.5-6.5 cm) supply all the muscle we need for months of electron microscope and mechanical experiments, using 5 mm bundles of 2-10 fibres.
2. A few large bugs, 7cm - 11 cm, are necessary and sufficient to support X-ray diffraction studies which require a 10-18mm bundle of 50-100 fibres for each experiment, or for studies where the slow kinetics of large insect muscle are felt to be advantageous.
3. Scores or hundreds of medium to large bugs, 5.5-7.0 cm are used fresh to prepare bulk washed myofibrils (20-50 mg/bug) for contractile protein studies, or are held alive to guarantee fresh material for all types of experiments during 6-10 months of the year when live bugs are normally not found by collectors.

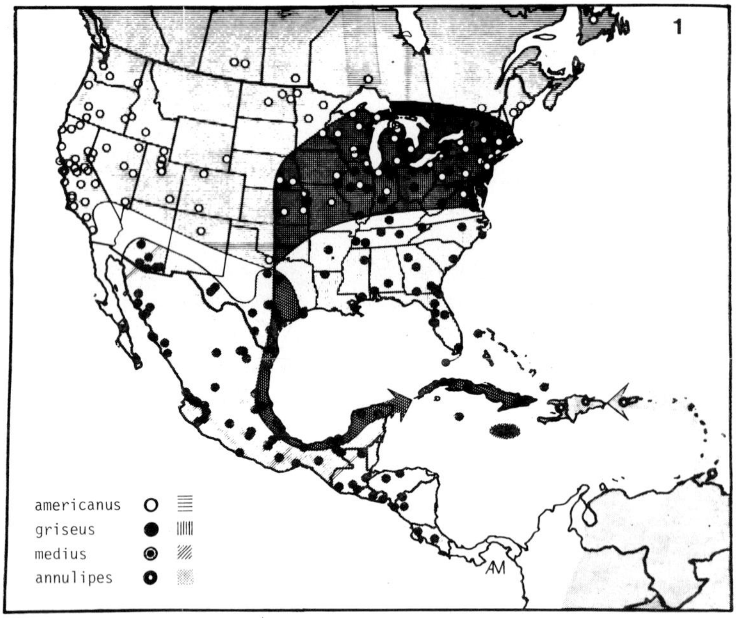

Distribution for *Lethocerus americanus, griseus, medius* and *annulipes*. (Menke, 1963)

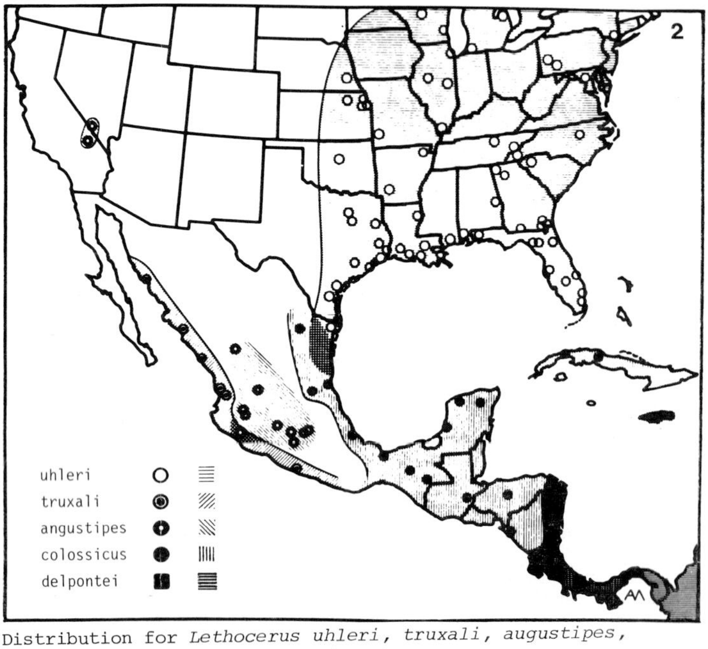

Distribution for *Lethocerus uhleri, truxali, augustipes, colossicus* and *delpontei*. (Menke, 1963)

4. If only small bugs are available, a dozen or two seems plenty, even in desperate times, because they would be useful only in small-bulk projects. On the other hand, if as many as a dozen or two 9-11 cm bugs were ever to reach us alive, even large-bulk biochemical projects might be supported with conveniently few dissections.

Three separate kinds of sources might supply the separate kinds of need, to the extent that one could safely assume that flight muscle from all *Lethocerus* species will be closely comparable. First, few and small bugs may be had by local collections in many parts of North America, according to the distribution maps 1 & 2 from Menke (1963) shown here. The very largest bugs will be available only from Trinidad and South America *(L. maximus)* and from Southeast Asia *(L. indicus)*. Now, X-ray diffraction chambers of present design (especially the latest Heidelberg model) permit firm mounting of somewhat shorter muscle bundles than thoes provided by *L. maximus,* but part of the X-ray beam is wasted if bundles are shorther than 10 mm. The largest 70-75 mm specimens of the species kept at Oxford, *L. cordofanus,* provide many usable bundles of 10-14 mm per carcass. The same maximum size (and presumably, muscle length) is reached by the South American *L. mazzai* (Cullen, 1969), and also by *L. truxali* (Mexico, west coast) and *L.colossicus* (Jamaica, Cuba, Central America and Eastern Mexico) according to Menke (personal communication), so that special efforts to obtain these might be justified for X-ray diffraction work. As for biochemical work, it must wait on the availability of large numbers of easily collected medium or medium to large sized bugs, such as the 50-70 mm *L.griseus* obtained for our work from Florida.

Now I shall mention our collection efforts, and follow with some discussion of the general problems of getting live giant water bugs.

Florida

Dr. Jon Hering at the Smithsonian Natural History Museum told me that the major centres of aquatic Hemipteran scholarship in the United States have been California, Kansas and Florida, with Florida showing great present-day activity in this field. My request for assistance in recruiting collectors, addressed to the Chairman of the Department of Entomology, University of Florida, Gainesville, was posted or announced and several graduate students in that Department have responded to my prices and instructions and sent me live water bugs since 1973. Of these, Ken Kopf was a particular mainstay during the 1974 and 1975 summers of greatest abundance, and he also went to Trinidad for me in 1975. Another student, David Mays, regaled me with recollection of his personal experience at one time in Southern Florida near Lake Okeechobee when a single rural intersection light attracted so many bugs that the intersecting roads were literally paved with them and rendered hazardous. (A student in my laboratory recalls a similar episode at Fayetteville here in North Carolina ten years ago when a swamp was drained on one side of the road, causing the road to be covered with water bugs moving across to an undrained swamp.) None of the Gainesville student collectors recalled finding the waterbugs in water habitat. They collected bugs off the pavement when the bugs swarmed to large mercury arc lights at a shopping mall on the edge of town or various automobile sales lots a mile or two within the outskirts of Gainsville (pop. 67,000). The swarming which made collection easy seemed to occur especially after a few days of heavy rainfall, roughly correlated with the bright phases of the moon. Both *L. griseus* and *L. uhleri* were collected together. *L. griseus,* with dark and light brown stipes running longitudinally on the ventral surface of the abdomen, was the larger and formed 70-90% of each shipment, while *L. uhleri,* whose unstriped ventral abdomen showed various light shades of yellow, green or rose, were frequently smaller than the 50 mm cutoff I had initially requested. About half the *L. griseus* were between 6 cm and 7 cm in length, while the larger *uhleri* reached about 6 cm late in the season. Collections and shipments were clustered in 2-8 days each month between late May and late September, but have already occurred by March 15 in 1977. The students recruited one another after my letter had first come to the attention of one of them, and I hope for continued collections from this group.

In April 1975 I spent several hours driving in the countryside around Orlando, Florida, showing photographs of Lethocerus sp while interrogating service station and shopping mall personnel as well as the citrus pest study crew at a government agricultural station. Few people had noticed the bugs in spite of the many nearby lakes. "Palmetto bugs", often mentioned, may be Lethocerus, but I can't be sure. One of the pest station crew was sure he had seen many bugs as big as the life-sized photograph I showed him of *L. maximus* (Cullen, 1969) while he was working in 1968 under night lights constructing the Orlando jet airport. Another witness recalled similar sized bugs being exterminated in a store room during his school days. Hopeful as this made me that very large bugs might be collected in Florida, I am equally prepared to see it as a trick played by memory, exaggerating something rather large and threatening into something still larger and more threatening. My student collectors assured me that no bug longer than about 7 cm was represented in the university museum collection of Florida Belostomatids.

Presently I am trying to recruit a more widespread group of collectors than previously, especially in south central Florida, where the report of roads slick with *Lethocerus* originated. My effort to recruit high school boilogy class members met a promising beginning in a class from Sebring, Florida, and my current effort is to attempt to recruit collectors through the local Boy Scout organisations in Florida. For transportation I would expect to continue using air mail special delivery parcels, which have served me well in past years.

Meanwhile, the Harold Williams firm in Dunellon, Florida, has again become active and willing to supply live *Lethocerus* through the efforts of its founder's widow, Mrs. Thelma Williams, Box 224, Dunellon, Florida 32630.

The distribution maps from Menke do not indicate a most favourable feature of Florida and coastal Louisiana. Menke told me that these are the only regions of the U.S. where swarming of *Lethocerus* to lights seems to reach nuisance levels nowadays. A less dramatic response, to a mercury arc light trap was recounted to me by Dingle concerning his experience in Lake Michigan collecting *L. americanus* in 1960. His summer-long collection was marked by two short periods of swarming when up to 20 bugs per night would be trapped, occurring in May and July, with the remaining nights producing 0-3 bugs each night

Mexico

Terry Taylor, an American collector whose firm* supplies preserved speciments of many kinds of insects, including the two large species of Mexican *Lethocerus,* told me that he had caught hundreds with a light trap in Vera Cruz in a few hours. When his collecting took him there again, in July 1974, he caught and packaged a hundred *L. colossicus*. However, the local postmaster cautiously invoked a law forbidding "The unauthorised export of Mexican treasures and antiquities" and refused to accept the airmail shipment. Later, the Mexican Embassy in Washington D.C. helped me to obtain official permission to export bugs, but the permit, issued for 6 months, only reached me three or four weeks before its expiration date, and was not helpful at that point. I believe a direct personal approach might still manage to clear the way for collection from Mexico.

Trinidad

I first requested *L. maximus* from Dr. Fred Bennett in April 1969, and received three bugs within six weeks. In subsequent years his collectors have brought him too few of these to allow more than 1-3 shipments per year to Ruegg, whose research requires the largest, slowest muscles, and perhaps one per year to Oxford, where the declining availability of *L. maximus* since about 1967 has

---

* Combined Scientific Supplies,
P.O. Box 1446,
Fort Davis,
Texas 79734,
United States of America.

shifted emphasis to use of *L. cordofanus*. Cullen's (1969) collection records show that even a full-time collection effort may be rewarded only sporadically and sparsely. Realising that Bennett had no time for such an effort, I arranged in 1975 to send Ken Knopf, an entomology graduate student who had sent me 300 live *Lethocerus* from Florida in 1974, so he could make the effort full-time for three weeks of July and August, including two weeks around the full moon, selected as prime collecting time from Cullen's indications. Driving all over Trinidad, night light trapping by various lakes, talking with automobile service station night operators and with individuals who had previously brought *L. maximus* to Bennett -- all failed to produce even a single small *Lethocerus*. The discrepancy with Cullen's experinece seemed difficult to explain, although it was locally explained as due to a rather dry rainy season. Neither pollution nor extermination efforts were identified. The general scarcity of *L. maximus* on Trinidad since 1970 is not easily correlated with rainfall; I reviewed monthly rainfall records from 1965-1974 for Piarco airport in Trinidad and could recognise no difference between the good old days and the lean recent days. Possibly a real population fluctuation, or a dislocation of the peak rainy season were major causes of Knopf's empty-handedness. In September 1976 and again in January 1977, Ruegg's laboratory has received three *L. maximus* from Bennett, so the bugs are still there, but presently in numbers too small and at times too unpredictable to justify any more collecting expeditions exclusively to Trinidad.

Venezuela

My correspondence in 1973 and 1974 with a university entomologist in Maracay and another in Maracaibo, men who regularly collected in or near aquatic habitats using light traps, met at first with reassurances that they could be reasonably sure of sizeable catches of *L. maximus and L. annulipes* which they would try to send me. However, none were sent; a small one was once caught and reported to me. Menke has suggested to me that the region south of Maracay would be more promising than the drier environs of Maracaibo. The Calabozo Reservoir and the more remote Orinoco flood plain, both south of Maracay, are regions hwere he has personally collected a sizeable series of *L. maximus* in one summer, although in another summer he took almost no Belostomatids at all with the light trap.

Jamaica

At Cullen's suggestion, I have tried to investigate by correspondence the reports of good numbers of large *Lethocerus* in the wetter regions of Jamaica. Various scientists and students in Jamaica have suggested that one of the better regions would be the coastal and near coastal swamps of the Black River region, about 100 miles west of Kingston and 40 miles south of Montego, the two nearest airports. A U.S. Peace Corps Volunteer was kind enough to announce my offer to some Jamaican High School students in the Black River area. I am told that the students recognised the bugs from Cullen's photographs, but were reluctant about collecting them. I still do not have any reports direct from Jamaica about the actual size, numbers, season or light-swarming behaviour of local *Lethocerus*, which should be *L. colossicus* according to Map 2. A mounted specimen from Jamaica in the University of Florida in Gainesville was as large as the *L. maximus* in Trinidad and much larger than Florida *Lethocerus*, according to students who examined them for me. I take this as less apocryphal than the reports which placed such large bugs in Florida itself.

Collections

With no first hand experience in collecting or observing *Lethocerus* in the wild, I have relied completely on the very useful account of his 1966-1967 study in Trinidad given by Cullen (1969), and on oral reports from successful collectors who have supplied me, in trying to inform potential new collectors where, when and how (rainy season, full moon, leather gloves etc.) they might best collect the bugs. When no bugs are sent to me, it has been practically impossible to know whether the difficulty rests upon local water bug behaviour, on the information I have offered, or on the interest and willingness of collectors. In the case of Trinidad, it was a great help to have the report of an experienced and elsewhere successful collector to point up how one region may vary in productivity from one year to another, and how variable the season may be. For these reasons, I am trying to assure a supply by recruiting collectors in several different Florida towns, and I will be trying to learn whether and where in Jamaica and Venezuela the larger bugs ever fly to lights in such numbers as Taylor has reported in Vera Cruz. Boy Scout groups and high school biology classes seem possible sources of collecting talent for my recruiting efforts in Florida. The flying to lights behaviour which is essential for trapping of flight worth bugs is not likely to be observed except where powerful vapour arc lights or strong UV sources are introduced. Night and rural police automobile patrols seem possible sources for such information. Another alternative would be to set up generators with mercury arc lamps as trapping and collecting devices in promising swamp or lake regions far from any other lights. Cost, placement, maintenance and security of such equipment would require consideration to make this effective. Mercury vapour arcs seem more effective for *Lethocerus* than traditional battery driven black lights used for insect light traps, according to my discussions with Florida students from Gainesville, with professional collector Terry Taylor, and with Hugh Dingle. Dingle pointed out to me a possible problem due to the efforts of arc lamp manufacturers to reduce th UV radiation from their lamps. Dingle's (1960) *Lethocerus* collecting on Lake Michigan was conducted with a lamp type whose efficacy he felt could not be matched with lamps he tried 5-7 years later in Iowa, lamps of distinctly lower specified UV radiating power. The ultimate effect of this development could require a protected supply of old style mercury arc lamps in order to assure their attractiveness to *Lethocerus*.

Compensation to collectors must be adjusted so it is not a limiting factor in getting their cooperation. My laboratory pays $0.50 per live bug over 6.0 com from Florida, with bonuses for shipments of large numbers, and token payment of $0.10 for up to 50 dead arrivals. I have offered $1-3 per bug for the larger specimens from Jamaica and Venezuala without result. Perhaps larger prices for an initial shipment could help to open up collecting efforts and prove supplies in these more difficult localities.

Certain and satisfactory holding procedures at the collection end are still not worked out. Covered plastic dishpans with 1 cm or less of water are the usual place to accumulate a catch. Large catches of 50-100 bugs have survived 8-30 hours at 20-25° room temperature and other batches have survived the same periods at 4° C. Still other catches have turned up with 30-70% dead after overnight holding at each of these temperatures. 20-25°C seems a little safer in principle for summer-adapted insects, unless or until heavy losses occur at this temperature. Aquarium facilities like those used for laboratory holding, but for 100 bugs or more, should provide ideal bug survival, but would overburden collectors who have limited time and space to spend on such facilities, even if I were to pay for the equipment itself and for food minnows.

Transportation

Our packaging and shipping instructions are now modelled after those of Pringle's Oxford laboratory. Early on, we lost about 300 bugs in various shipments over 10 days in 1974, because they drowned in pooled water inside large plastic bags full of soggy tissue paper and bugs. Since then, we urge polythylene sandwich bags, one bug per bag, stapled shut to partly immobilise bugs, with bottom corners cut off for air. "Whirl-Pak" wire-closed polythene bags are preferred as most convenient by the Florida collectors. Bagged bugs are layered with a few interleavings of damp paper towels in a multiply perforated carton and mailed Air Mail Special Delivery at a time previously identified as likely to give the shortest transit interval. Postage costs are $3-$7 per shipment. The shipping label, supplied by me, includes a photocopy of a life-sized *Lethocerus* "speaking" via cartoonist balloon to request RUSH! LIVE BUGS! treatment, no freezing or heating, and to give my lab and home telephone number with request that the town post office staff telephone me at once when bugs arrive, regardless of the hour, so I can go down and pick them up if I choose. (Personal discussions with the Post Office staff were useful in setting this up). Previous to this, the collector will have phoned me as soon as shipment is ready or posted, so my laboratory can be ready for whatever we have decided to do with the bugs. Usually we put 5-30 of the largest in an aquarium to keep live for glycerination over the next days, weeks or months. The rest are dissected, and the dorsal and longitudinal and oblique flight muscles converted to a washed myofibril preparation for protein work. Now, we are considering re-shipping some of the bugs to other laboratories, as Pringle's laboratory has done over the years.

International shipments require air freight, costing $18-$25 for a minimum parcel up to 10 lbs., and they require careful planning of the timing and connecting flights, so they don't spend 20-40 hours or a whole weekend in a hot tropical freight shed or agriculture department inspectors' office. Clearly productive collecting areas that are too far from convenient air terminals with good through flight connections will not contribute much to international research on *Lethocerus*. Collectors must telegram when shipments are made and planes must be met. Moreover, a special import permit with associated import labels for each shipment must be obtained at one or two chosen ports of entry, so the airline flights used must be chosen accordingly. Airmail special delivery to and from my Caribbean and Latin American correspondents spends 5-10 days in transit, so air freight is necessary, with personal drop off and pick-up at both ends. I try to have telephone contact with freight handling and Agriculture inspection personnel at ports of entry and airports where freight transfer to a connecting flight occurs.

Receipt and Holding

Upon receiving a parcel, we empty all bug bags into a few dishpans with 1 cm of water, sorting into lively, sluggish and suspected dead categories. Bugs who show signs of life but not liveliness are dissected at once for fibrils or glycerol extraction. The largest lively bugs are held for later use in a 20 gallon room temperature (23°C) aquarium with a screen top. They are not caged separately, as at Oxford; they huddle together but do not exhibit either lethal or reproductive consequences of all this body contact. The two Florida species *L.griseus* and *L.uhleri* show no emnity toward one another nor toward an occasional *L. cordofanus* from Oxford who shares their tank. Half inch square mesh standard galvanised wire fencing, formed into cylinders of 5 inch diameter, initially used as cages, are now offered open end up as perches to make it easy for the bugs to back their respiratory "tail pipes" to the surface for needed air. Metal from the unpainted wire which may go into the solution seems to harm the bugs not at all. A soft, fine mesh non-metallic window screening is clamped all around the aquarium rim with clothes pins, close spaced to prevent escapes, of which we have had several episodes with newly caught bugs, but never with aquarium-habituated bugs. The bugs are fed bait minnows, 2-3 inches long, called "crappieminnows" and "bass minnows" by the several fish bait dealers in Durham. Three dozen minnows cost $1.50-$2.00 and should be kept without feeding (it kills them) separately until offered to the bugs. The bug aquarium is filter-pump fitted and aerated for the sake of the minnows. Nevertheless, all the minnows sometimes die overnight as if asphyxiated or poisoned, with red gills but no wounds to suggest bug attack. We have wondered whether aquarium bound bugs, whose appetite is obviously only very occasional, might injure these fish by non-feeding strikes (not observed) or intoxicate them by discharging the *Lethocerus* salivary glands into the water at intervals. In any event, many more fish die than are eaten.

We immobilise by chilling the bugs in crushed ice before beheading and dissecting them. Even unchilled and lively they swim or scuttle only to escape, not to attack, until they are picked up. They are so armour bound by heavy chitin that any grip from behind or to the rear keeps fingers out of reach of the extended, telescoping piercing beak.

No sign of degeneration has been visible in flight muscles of the Florida bugs even after 7 months of aquarium life. It remains to be shown that flightless aqarium bound muscles of the Florida species are equivalent to those of *L. cordofanus* in producing mechanical and structural results of the higest quality. Flight testing before glycerination would be useful, if it were possible (*cf.* Dingle, 1961), but no one has yet succeeded in this. We have dropped and whirled tethered freshly caught *Lethocerus* without ever producing flight. Only once, in several accumulated hours of leaving bugs in open dish pans with sides too high to climb out, has one ever flown, heard but not seen, launching itself with a sudden buzz straight across the room into a window pane which stunned it. We have yet to properly test Peter Schneider's idea that controlled heating of a surface perch will facilitate and force flight of captive bugs.

Miscellany

In December 1975 eight bug eggs (position in aquarium unknown) hatched one night among hundreds dropped with no result. The nymphs lived only 3-5 days, since we never tried to find any specially appropriate food for them. No report of such hatching has come from Oxford yet.

Most dissectors are familiar with the fragrant spicy material from the abdominal glands of the males, analysed by Butenandt and Tam (1957) as ^{2}-Hexenol acetate. The last step of their synthetic procedure was easily carried out in my laboratory using 1:1 stoichometric amounts of the trans-2-hexen-1-ol & acetyl chloride available from Aldrich Chemical Company. About 50 ml was synthesised by adding acetyl chloride to the alcohol in a small filter flask, ice chilled in a fume hood, quickly stoppering the flask to force the rapidly evolving HCl gas through an NaOH charged drying tube on the side arm. The resulting ester can be neutralised by shaking over aqueous $NaHCO_3$ in a separatory funnel until the aqueous phase is approximately pH 7. The fluid now smells exactly like the bug secretion. However, there has been no response by aquarium-held bugs when soaked cotton-tipped applicators were placed over the water near climbing perches. 30 ml of this went to Trinidad for testing in the wild in 1975 but no bugs meant no opportunity. Its function, as sex attractant or whatever, is still not established to my knowledge, and I include the preparative method here to encourage others to identify any behavioural responses which might be relevant to collecting.

Acknowledgement

Supported in part by research grants from the U.S. National Institutes of Health and from the Muscular Dystrophy Association of America.

References

Buttenandt, A. and Tam, N.D. (1957) *Z.physiol. Chem., Hoppe Seyler's*, 308,277

Cullen, M.J. (1969) *Proc.Roy.Ent.Soc.London (A)*, 44, 123

Dingle, H. (1961) *Biol. Lab. Woods. Hole.*, 121, 117

Menke, A.S. (1963) *Ann.Ent.Soc.Am.*, 56, 261

THE BREEDING OF GIANT WATER BUGS IN THE LABORATORY

Michael J. Cullen

Muscular Dystrophy Research Laboratories, Regional Neurological Centre, Newcastle General Hospital, Westgate Road, Newcastle-upon-Tyne, U.K.

INTRODUCTION

The breeding of giant water bugs in the laboratory is difficult. This difficulty is the result of certain features of water bug biology, i.e. they are predacious, they require an abundant source of food and the timing of their reproductive phase is probably dependent on the timing of their dispersal (flying) phase. It is not accidental that the species of insects which are most often bred in the laboratory are species which are phytophagous, sarcophagous or omnivorous. Insects which depend on living prey present certain problems of their own.

The family Belostomatidae contains seven genera, and I have had experience of breeding four of these: *Belostoma*, *Hydrocyrius*, *Limnogeton* and *Lethocerus*. The genus *Lethocerus* differs from the others in that the females lay their eggs above the water surface on the stems of semi-aquatic vegetation. Female bugs of the other belostomatid genera lay their eggs on the hemelytra of the males which subsequently carry the eggs until they hatch. In my experience the species with this form of oviposition are less difficult to breed than is *Lethocerus*.

MATERIALS AND METHODS

The results section below will deal with my experiences of breeding giant water bugs (1) in captivity outdoors in the grounds of the Commonwealth Institute of Biological Control in Trinidad, (2) in the laboratories of the University of the West Indies in Trinidad, and (3) in the laboratory in Oxford. The circumstances and methods used in each case were different and will be explained at the appropriate point.

It has always been necessary to maintain a stock of prey species as food for the water bugs. In Trinidad, where fish, tadpoles and aquatic invertebrates were abundant, this was not a problem. In Britain tadpoles are only seasonally available and we relied largely on fish. Some of these were native British fish caught outside and brought into the laboratory, but mostly they were tropical fish bred in the laboratory. These were live bearing fish of the family Poeciliidae, in particular the molly *Poecilia latipinna*, the guppy *Poecilia reticulata* and the platy *Xiphophorus maculatus*. These species of fish bred easily and, given adequate attention, large stocks of them could be built up. It tended to happen, however, that when the April-June input of large numbers of water bugs from Uganda occurred, the fish stocks were reduced to very low numbers and required augmenting by purchases from tropical fish dealers.

RESULTS

The six species of belostomatid bug which have been kept in the laboratory are shown in Figure 1. These will be considered in turn.

Lethocerus maximus

In Trinidad *L. maximus* (Fig.2), caught in light traps or by net, were housed in cages built over concrete tanks in the grounds of the Commonwealth Institute of Biological Control. The tanks were planted with the water lily *Nympheae amazonum*, the water hyacinth *Eichhornia crassipes* and the sedge *Rhynospora corymbosa*. The tanks were kept stocked with fish *Poecilia* and *Rivulus*, tadpoles *Hyla* and water beetles *Hydrophilus*.

TABLE I

The number of eggs per batch, the height above the water surface and the laying date of ten egg-masses of *Lethocerus maximus*. (Cullen, 1969).

Batch	Number of eggs	Height above water level (cm)	Laying Date	
1	130	3	4 April	Dry season
2	194	6	25 April	Dry season
3	105	4.5	30 April	Dry season
4	119	5	12 June	Dry season
5	142	24	16 June	Wet season
6	257	33	3 July	Wet season
7	182	21.5	3 July	Wet season
8	146	30	5 July	Wet season
9	139	36.5	6 July	Wet season
10	86	31.5	10 July	Wet season

Under these conditions *L. maximus* bred on 10 different occasions over a time period of about 3 months (Table I) each egg batch being produced by a different female. Egg batch No.1 is shown in Fig.3. Oviposition took place at night and there was a clear difference in the positioning of the eggs in the dry and wet seasons (Table I). The higher position in the wet season is probably an adaptation to guard against drowning resulting from a sudden rising of the water level. Usually 100-200 eggs were laid in each batch.

The eggs of *L. maximus* hatched after 8 days during which time they increased in length from 5.6 to 7.4mm. Hatching always occurred at night. The first instar nymphs were fed tadpoles of *Hyla* and later graduated to small fish. To avoid cannibalism they were housed individually in wire cages which were suspended in large aquaria and through which fish could pass freely. The nymphs grew rapidly and reached the adult instar about 3 months after oviposition (Table II).

TABLE II

The average duration in days of the five nymphal instars of *L.maximus*, *H. columbiae* and *B. malkini*. (Cullen, 1969).

	L. maximus	*H. columbiae*	*B. malkini*
Egg	8	14	10
Instar I	6.5	7	7.9
Instar II	8.0	9	11.1
Instar III	13.7	13	16.0
Instar IV	18.5	24	21.6
Instar V	35.0	46	21.3
TOTAL (to nearest whole day)	90	113	88

Although *L. maximus* oviposited freely when kept in cages outdoors and although the nymphs grew well when housed outdoors, *L. maximus* never oviposited when kept indoors in the laboratory in Trinidad. This was the case even though they were given abundant food and oviposition sites *(Rhynospora)*.

The most obvious difference between the external conditions and laboratory conditions were meteorological, in particular differences in light intensity and humidity. For this reason, when the tropical room was built in Oxford, particular attention was paid to these two conditions. Light intensity was made comparable with that of tropical sunlight by having closely spaced fluorescent 200 watt tubular lighting completely covering the ceiling. An artificial dawn and twilight was provided by staggering the switching on and the switching off of the lights. High night time humidity was provided by a humidifier in the ventilation system. *L. maximus* was housed in large 75 gallon glass-fibre tanks planted with sedges and reeds and stocked with fish as potential prey.

Despite being provided with these apparently ideal conditions *L. maximus* has not successfully oviposited in the laboratory in Oxford. The reasons for this remain obscure. Female bugs which were dissected frequently contained apparently mature eggs in their oviducts and spermatozoa in their spermathecae and so seemingly had the potential to oviposit. It is possible that an essential constituent is absent from their laboratory diet although this seems unlikely. It also seems unlikely that a seasonal endocrine factor is involved because in their natural environment the breeding of *L. maximus* is not seasonal (Cullen 1969). It has to be concluded that we are still ignorant of an essential requirement for the breeding of *L. maximus* in the laboratory.

L. cordofanus

L. cordofanus is native to Central and East Africa from where the laboratory stock is obtained. The author has had no direct experience of breeding *L. cordofanus* and to his knowledge there is no published work on the breeding biology of this species. It is reasonable to suppose, however, that the breeding habits are similar to those of *L.maximus* and other species of *Lethocerus*, and that they oviposit out of the water on emergent vegetation.

The great majority of *L. cordofanus* which have been received in Oxford have been kept in individual cages in tanks containing tropical fish as prey. However, groups of mixed sex have been kept, in a similar way to *L. maximus*, in breeding chambers provided with sedges and other vegetation. These bugs have caught and digested fish and other bugs and have seemed healthy but have never laid eggs in the expected manner. However, female *L. cordofanus* frequently deposited bundles of eggs in the water unattached to the vegetation. None of these eggs hatched implying that they were infertile or that they drowned. When we noticed these bundles of eggs we often took them and attached them manually to a surface above the water level. Unfortunately they always collapsed and shrivelled whereas if they were viable, they would have maintained their rigidity and grown in length. Again this may have been due to their infertility or due to dessication possibly from the lack of a water-proofing component normally produced at oviposition. When female *L. cordofanus* were dissected they, like *L. maximus*, frequently contained apparently mature eggs. Why they have never been laid in the normal way remains unclear.

L. annulipes

L. annulipes, rarer in Trinidad than *L. maximus*, has never been kept in large numbers. In Trinidad *L. annulipes* was housed in large aquaria provided with growing vegetation. The aquaria were indoors and, perhaps because of this, oviposition did not take place. In Oxford they have been housed in the same way as *L. cordofanus* and like that species the females have frequently deposited egg batches in the water but these have never been hatched.

H. columbiae

Hydrocyrius, *Limnogeton* and *Belostoma* belong to that group of belostomatids in which the female bug lays the eggs on the back (hemelytra) of the male. The male bug then displays an altered behaviour pattern in lying horizontally at the water surface so that the eggs are provided with optimal incubation conditions. Eggs removed from a male giant water bug and placed in 10 cm. of non-aerated water fail to hatch. In normal conditions the eggs hatch after about 2 weeks and progression to the adult instar takes 12-13 weeks (Table II).

H. columbiae has been frequently bred in Oxford by placing equal numbers of male and female bugs (3-6 pairs) together in a well planted tank stocked with abundant food. (In the absence of alternative food, and sometimes in its presence, all belostomatids except *Limnogeton*, will actively prey upon others of the same species). Usually a male bug was found to be carrying eggs after a few days at which stage the other bugs were removed. The nymphal *H. columbiae* were fed on young fish and tadpoles, and were also found to prey upon the aquatic snail *Planorbis*.

Given sufficient prey, nymphal belostomatids will feed virtually continuously. This does not in fact imply a continuous ingestion of food because feeding, as in many true bugs (Hemiptera), is a two stage process. Saliva, containing a potent mixture of hydrolytic enzymes (Rees and Offord, 1969) is first injected into the prey and, after lysis of the tissues, the resulting 'soup' is sucked back into the bug. With large prey this process can take many hours. The time taken for complete metamorphosis is very variable and is probably directly related to prey availability. Whereas most bugs reared in Oxford completed metamorphosis in 3-4 months, Miller (1961) reported metamorphosis in *H. columbiae* taking three times this period.

Because *H. columbiae* bred fairly readily and second generation offspring could be produced, it was hoped that it might be used as a major source of laboratory reared insect muscle. Unfortunately this has not been possible because *H. columbiae* seems to possess fully mature flight muscle for only a short period of its life history (see Discussion). Consequently on nearly every occasion when a *H. columbiae* was dissected it was found to have either immature or degenerating flight muscles.

Limnogeton fieberi

Our experience of breeding *L. fieberi* has paralleled very closely that of breeding *H. columbiae*. However, because it is a smaller insect and only rarely received from our supplier no systematic effort was made to rear it.

L. fieberi is morphologically adapted to feeding on snails and is unable to capture fish or other active prey. In *Lethocerus* and *Hydrocyrius* the mouth parts are short and stout but in *Limnogeton* the rostral segments are long, slender and curved. This enables *L. fieberi* to reach a snail which has withdrawn deeply into its shell. The holding tanks in Oxford usually supported a large population of the snail *Planorbis* so *L. fieberi* could be reared without difficulty. However, as with *H. columbiae,* the adult flight muscles were almost invariably immature or degenerate.

Belostoma malkini

B. malkini is the smallest of the giant water bugs which have been bred in the laboratory. As with *H, columbiae* and *L. fieberi,* given plentiful prey, *B. malkini* can be bred without difficulty but also like those bugs the flight muscles are seldom found to be mature. The breeding biology of *B. malkini* was extensively studied in Trinidad and is described in an earlier paper (Cullen 1969).

DISCUSSION

The breeding of giant water bugs has reached the situation where those species with the most suitable muscles have not bred in the laboratory, while those species which do breed do not contain normal mature flight muscles.

In many insects there is a period after the final moult to the adult form during which the flight muscles enlarge and mature. This is followed by a period of intensive flight activity (dispersal phase) and this in turn is followed by a third period (reproductive phase) in which the flight muscles degenerate. This degeneration is a normal part of the life cycle of the insect. The relative importance and duration of these three phases is very variable from species to species.

The majority of the giant water bugs sent to Oxford, in particular *L. cordofanus,* have been caught during the second phase when they have been flying and have been attracted to light. Consequently when they are received they have normal mature flight muscles. The longer they are then kept in the laboratory, the greater is the chance of them entering the third phase. This tendency seems to be greater in *L. maximus* than in *L. cordofanus* and greater in the other belostomatid genera than in *Lethocerus*. *L. cordofanus* seems to be the best species for keeping for long periods without its flight muscles degenerating. If the two activities of flight and reproduction are mutually exclusive then this might offer some explanation as to why *L. cordofanus* is reluctant to breed in the laboratory. It is known that the flight activity of belostomatids is greatest during the wet season and around the full moon (Bowden, 1964; Cullen, 1969; Duviard, 1974). It is possible that *L. cordofanus* in its natural environment will fly on several consecutive nights until a pool or lake is found which signals to the insect that it is an appropriate place in which to breed. (It is noticeable that many bugs newly received to the laboratory are restless and attempt to leave their cages). It would be of great assistance in attempting to breed *L. cordofanus* if it was known what the signals for breeding are. These may be to do with the immediate environment (e.g. prey availability, oviposition site availability), they may be meteorological, or they may be endocrine in origin. It is quite likely that it is a combination of these. Clearly we need to know more about the breeding biology of *Lethocerus* before it can be routinely reproduced in the laboratory.

The other genera of belostomatids present a different problem. It is not known how long the period of flight muscle maturation lasts in the bugs in their natural environment, although those *Belostoma* examined in Trinidad were at least 15 days post-moult (Cullen 1969). In the laboratory the period of flight muscle maturation seems prolonged, and those insects collected in flight with mature muscles seem in undue haste to allow them to degenerate. It is unlikely that the increased provision of potential food will delay this event if it is part of the normal life cycle of the insect. It might be possible to interfere with this by endocrine surgery on the water bugs but this will not be possible until more is known about the normal physiology of the endocrine system in belostomatids.

REFERENCES

Bowden J. *J. ent. Soc. sth Afr* 16 (1964) 293.

Cullen M.J. *Proc.R.ent.Soc.Lond. (A)* 44 (1969) 123.

Duviard D. *Oecologia* 15 (1974) 321.

Miller P.L. *J. Insect Physiol* 6 (1961) 243.

Rees A.R. and Offord R.E. *Nature* 221 (1969) 675.

Fig.1. The six species of belostomatid bug obtained in Oxford.

Fig.2. *Lethocerus maximus* with wings extended.

Fig.3. Egg batch of *Lethocerus maximus*.

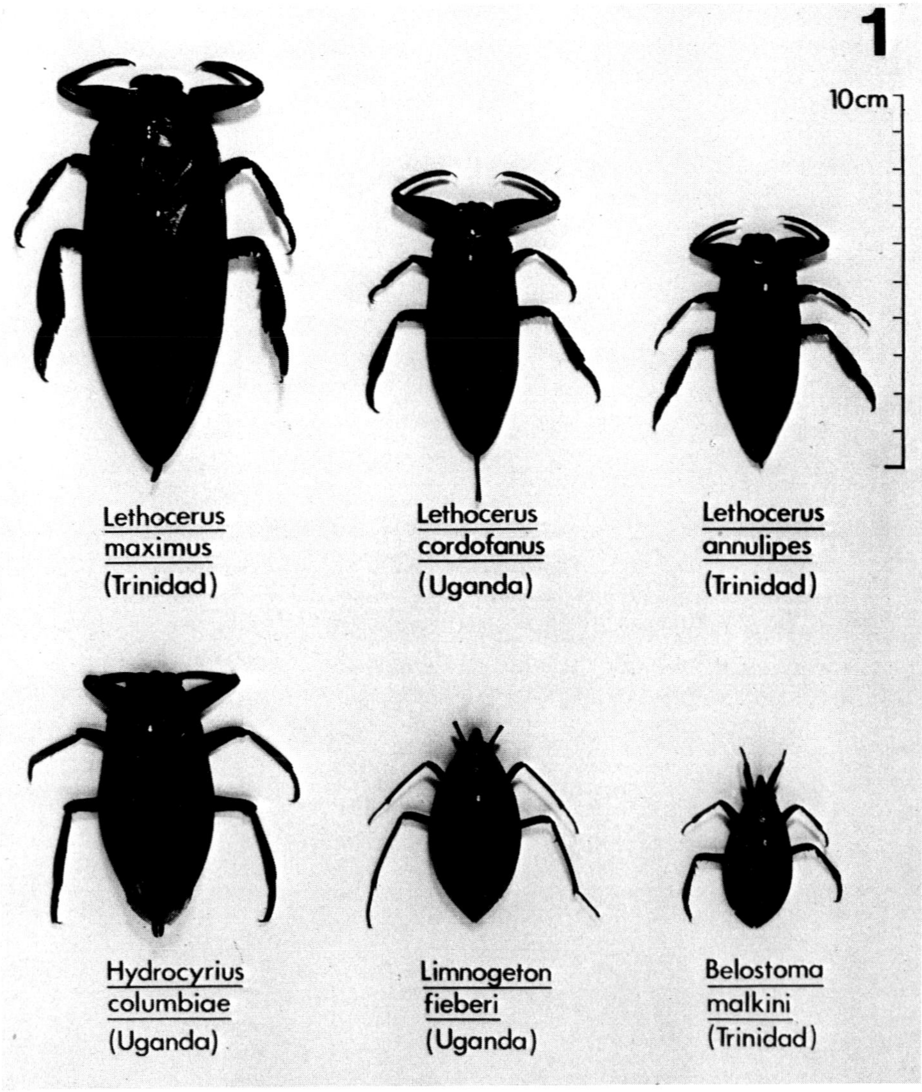
1
10cm
Lethocerus maximus (Trinidad)
Lethocerus cordofanus (Uganda)
Lethocerus annulipes (Trinidad)
Hydrocyrius columbiae (Uganda)
Limnogeton fieberi (Uganda)
Belostoma malkini (Trinidad)

2

Lethocerus maximus

3

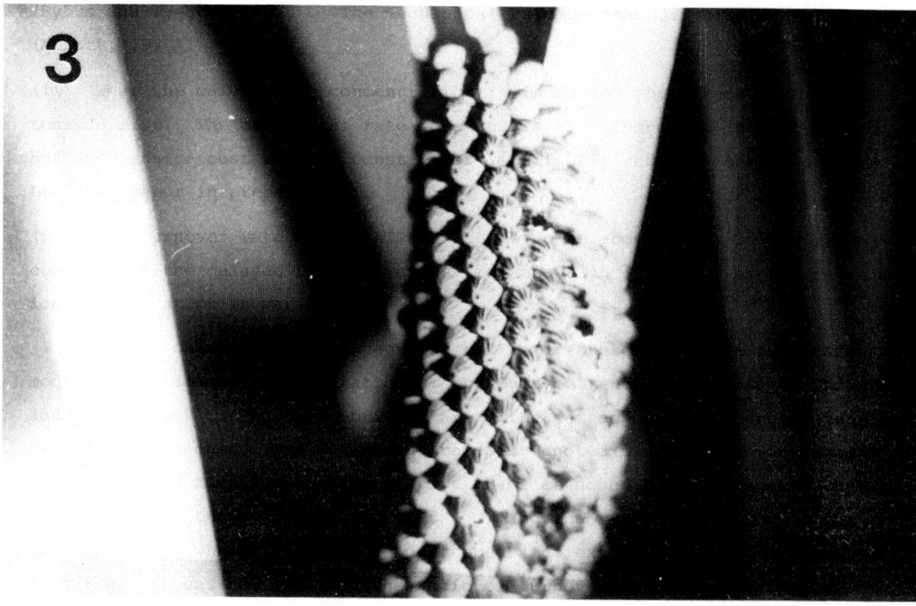

AUTHOR INDEX